Train Your Brain

Textbooks in Mathematics

Series editors:
Al Boggess, Kenneth H. Rosen

Nonlinear Optimization
Models and Applications
William P. Fox

Linear Algebra
James R. Kirkwood, Bessie H. Kirkwood

Real Analysis
With Proof Strategies
Daniel W. Cunningham

Train Your Brain
Challenging Yet Elementary Mathematics
Bogumił Kamiński, Paweł Prałat

Contemporary Abstract Algebra, Tenth Edition
Joseph A. Gallian

Geometry and Its Applications
Walter J. Meyer

Linear Algebra
What you Need to Know
Hugo J. Woerdeman

Introduction to Real Analysis, 3rd Edition
Manfred Stoll

Discovering Dynamical Systems Through Experiment and Inquiry
Thomas LoFaro, Jeff Ford

Functional Linear Algebra
Hannah Robbins

https://www.routledge.com/Textbooks-in-Mathematics/book-series/CANDHTEXBOOMTH

Train Your Brain —
Challenging Yet Elementary
Mathematics

By
Bogumił Kamiński
Paweł Prałat

CRC Press
Taylor & Francis Group
Boca Raton London New York

CRC Press is an imprint of the
Taylor & Francis Group, an **informa** business
A CHAPMAN & HALL BOOK

First edition published 2020
by CRC Press

6000 Broken Sound Parkway NW, Suite 300, Boca Raton, FL 33487-2742
and by CRC Press

2 Park Square, Milton Park, Abingdon, Oxon, OX14 4RN

© 2021 Taylor & Francis Group, LLC

CRC Press is an imprint of Taylor & Francis Group, LLC

ISBN: 978-0-367-56487-2 (pbk)
ISBN: 978-1-003-09798-3 (ebk)
ISBN: 978-0-367-67935-4 (hbk)

Typeset in Computer Modern font
by KnowledgeWorks Global Ltd.

Contents

8 Solutions 223

Contents

Introduction

The book contains carefully selected problems that are challenging, yet only require elementary mathematics. It is intended to prepare the readers for rigorous mathematics, but neither prior preparation nor any mathematical sophistication is required from them before reading this book. The book guides the readers to think and express themselves in a rigorous, mathematical way, to extract facts, analyze the problem, and identify main challenges. Moreover, it shows how to draw appropriate, true conclusions and helps to see a big picture. Despite the fact that this is not the main goal of this book, as a bi-product, the readers are provided with a firm foundation in a diverse range of topics that might be useful in their future work. Finally, we often use computer support to help us get a better intuition into discussed problems. This is a still rather unique approach in mathematics but is getting more and more popular in the current multidisciplinary and data driven world.

The presented material can be seen as a means to bridge the gap between introductory calculus/linear algebra courses and more advanced courses that are offered at universities. It improves the ability to read, write, and think in a rigorous, mature mathematical fashion. It provides a solid foundation of various topics that would be useful for more advanced courses. However, the book is not only intended for undergraduate students that would like to become professional research mathematicians. In almost any mathematically related work (such as computer programming, data science, machine learning, economics, engineering, etc.), precise reasoning, and understanding what logical steps need to be taken to transition from the assumptions to the desired conclusion, are crucial to be successful.

The content of this book is also suitable for high school students that are interested in competing in math competitions or simply for people of all ages and backgrounds who want to expand their knowledge and to challenge themselves with interesting questions. In fact, the problems are mostly selected from an extensive collection of problems from Polish Mathematical Olympics and a library of training problems from XIV High School of Stanislaw Staszic in Warsaw (Poland).

This book is clearly not the only one of its type. There are three main reasons for writing another book on this topic. First of all, we found that many interesting problems appear only in the Polish language and are not translated to other languages. Some of them are unique and might be interesting for a broader, English speaking, audience. We feel that they deserve to be popularized. More importantly, we grouped questions into six chapters representing various disciplines of mathematics. Each chapter consists of many sections devoted to a collection of related topics. Each of these sections starts with a problem that is followed by the necessary background (definitions and theorems used), careful and detailed solution, and discussion of possible generalizations. The sections finish with a number of additional related exercises that are solved at the end of the book. As a result, this book can be used as a textbook for a systematic and structured introduction to a fascinating world of high school math competitions, or as a book preparing university students for more advanced courses. Finally, with an increasing role of computational methods in mathematics, we decided to show a few examples when computer aid can be used to verify or guide the solutions to some problems. We present the related code for a few suitable problems, discuss the implementation details and, in the "Julia language companion" available on-line at www.ryerson.ca/train-your-brain/, we provide a thorough introduction to the Julia language that we use in this book, along with detailed explanations of the codes we present.

In order to help the reader to navigate in the text, for each problem, we clearly distinguish a few subsections, whose functions are listed below.

SOURCE

In this part, we provide the source of the problem.

PROBLEM

This part contains the statement of the problem.

THEORY

These parts are "sprinkled" across the whole book and appear "as needed"; that is, the first time a given definition or theorem is used, it is introduced and later on it is only referenced. Let us also mention that for some theorems presented in the book we provide proofs (especially if they are easy, insightful, or potentially useful for solving the problems) but often the proofs are omitted. Similarly, some definitions are accompanied by examples but some are not. As always, we try to select material that is helpful for the reader to prepare for future related questions rather than trying to be exhaustive.

SOLUTION

In this part, we provide a detailed solution. For some problems we provide more than one solution as the aim of the book is not to simply solve all problems, but rather to highlight the most important and common approaches and tricks needed to be successful in solving similar questions in the future.

REMARKS

Here we try to explain what is a typical thought process of the (successful) person trying to solve the question. It is often the case in mathematics that the solution of the question itself does not give us any insight on how it is actually found. After reading the solution, the reader is convinced that the claim holds but the reason for that and the process of "discovering" the solution may remain mysterious. Hence, this section is as important as the solution itself and should not be skipped.

EXERCISES

This part contains follow up exercises that use the same or similar concepts. They should serve as a good test whether the reader "digested" the content or needs more practice.

We tried to indicate the source for as many problems as possible. If the source is omitted, it means that the problem is either our own, is well-known, or we had it in our personal notes but were unable to recover the original source. We also tried to make sure that it is clear whether the solution is also taken from the source or is our own. We did our best to track back all the sources but please contact us if we missed anything. We would be more than happy to provide a more complete and accurate picture of the sources of all the problems we have presented in a later edition of the book. In particular, problems from Polish Mathematical Olympics and their solutions are marked under the acronym PLMO. We would like here to thank the organizers for granting us the right to use their translations in this book. In the chapter on geometry, we have extensively used an excellent collection of problems "Exercises in geometry" (in Polish) prepared by Waldemar Pompe who also kindly agreed to include their translations in this book.

If you find any errors or omissions in this book, then please kindly let us know and we will reflect it in the errata that will be available at www.ryerson.ca/train-your-brain/.

Finally, we would like to thank Calum MacRury for carefully reading the manuscript, and Igor Kamiński for helping with selecting topics and problems to include in this book.

Chapter 1

Inequalities

We begin the book with a chapter on inequalities. This is an exciting subject, as it very often requires reducing the problem to some other area of mathematics which may not initially seem to be related to the problem at hand. For example, it might turn out that one of the sides can be interpreted as the probability that some event holds, or that the side has some geometric interpretation.

Since this is the first chapter, let us start with introducing some basic definitions that will be used through the entire book.

THEORY

Let $\mathbf{R}$ denote the set of real numbers, let $\mathbf{N} = \{1, 2, \ldots\}$ denote the set of natural numbers, let $\mathbf{Z} := \{\ldots, -1, 0, 1, \ldots\}$ denote the set of integers, and let $\mathbf{Q} := \{a/b : a \in \mathbf{Z}, b \in \mathbf{N}\}$ denote the set of rational numbers. Let $[n]$ denote the set of the first n natural numbers; that is, $[n] := \{1, 2, \ldots, n\}$. We use subscript $+$ and $-$ to restrict the set to positive and negative numbers, respectively. For example, $\mathbf{R}_+$ denotes the set of positive real numbers. We will use $\ln(x)$ to denote the natural logarithm of x.

1

1.1   Convexity and Concavity

SOURCE

Problem: XLV PLMO – Phase 1 – Problem 3
Solution: our own

PROBLEM

Prove that if a, b, and c are the lengths of the sides of a triangle, then

$$\frac{1}{a} + \frac{1}{b} + \frac{1}{c} \leq \frac{1}{b+c-a} + \frac{1}{c+a-b} + \frac{1}{a+b-c} \,. \qquad (1.1)$$

THEORY

$\boxed{\text{Triangle Inequality}}$ If a, b, and c are the lengths of the sides of some triangle, then the **triangle inequality** states that

$$c \leq a + b \,.$$

Note that this statement permits the inclusion of degenerate triangles; that is, when $c = a + b$. However, usually this possibility is excluded, thus leaving out the possibility of equality.

$\boxed{\text{Cartesian Coordinate System}}$ A **cartesian coordinate system** is a coordinate system that specifies each point uniquely in a plane by a pair of numerical coordinates, which are the signed distances to the point from two fixed perpendicular directed lines, measured in the same unit of length. Each reference line is called a coordinate **axis** (plural **axes**) of the system, and the point where they meet is its **origin**, the ordered pair $(0,0)$.

$\boxed{\text{Convex and Concave Functions}}$ A function $f \colon \mathbf{R} \to \mathbf{R}$ is said to be **convex** on a connected set (interval) $D \subseteq \mathbf{R}$ if for all $x_1, x_2 \in D$ and $t \in [0,1]$, we have that

$$f\!\left(tx_1 + (1-t)x_2\right) \leq tf(x_1) + (1-t)f(x_2). \qquad (1.2)$$

Intuitively, a function is convex if for all $x_1, x_2 \in D$ its graph lies below a straight line connecting points $(x_1, f(x_1))$ and $(x_2, f(x_2))$. For example, function $f(x) = 1/x$ is convex on $D = \mathbf{R}_+$ and functions $g(x) = x^2$ and $h(x) = 2^x$ are convex on $D = \mathbf{R}$.

Similarly, a function $f \colon \mathbf{R} \to \mathbf{R}$ is said to be **concave** on a connected set (interval) $D \subseteq \mathbf{R}$ if for all $x_1, x_2 \in D$ and $t \in [0,1]$, we have that

$$f\!\left(tx_1 + (1-t)x_2\right) \geq tf(x_1) + (1-t)f(x_2). \qquad (1.3)$$

Examples of concave functions are: $f(x) = -2(x+7)^2 + 4$ on $D = \mathbf{R}$ or $f(x) = \ln n$ on $D = \mathbf{R}_+$.

Finally, let us mention that if a function is continuous, then it is enough to check that the condition (1.2) or (1.3) holds for $t = 1/2$ in order to establish that the corresponding function is convex or, respectively, concave. Using this fact, one can easily prove that function $f(x) = 2^x$ is convex. Indeed, notice that for all $x, y \in \mathbf{R}$, we have

$$0 \le \left(2^{x/2} - 2^{y/2}\right)^2 = 2^x - 2 \cdot 2^{(x+y)/2} + 2^y .$$

Rearranging this inequality, we obtain $\left(2^x + 2^y\right)/2 \ge 2^{(x+y)/2}$, but this is precisely the condition (1.2) with $t = 1/2$.

SOLUTION

It follows immediately from triangle inequality that all fractions on the right hand side of (1.1) are positive. In particular,

$$\frac{1}{b+c-a} + \frac{1}{c+a-b} = \frac{(c+a-b)+(b+c-a)}{(b+c-a)(c+a-b)} = \frac{2c}{c^2 - (a-b)^2} > 0 .$$

Since the numerator is positive (that is, $2c > 0$), it follows that the denominator is positive too (that is, $c^2 - (a-b)^2 > 0$). Moreover, clearly $(a-b)^2 \ge 0$, and so $c^2 - (a-b)^2 \le c^2$. Putting all of these observations together, we get that

$$\frac{1}{b+c-a} + \frac{1}{c+a-b} = \frac{2c}{c^2 - (a-b)^2} \ge \frac{2}{c} . \tag{1.4}$$

Moreover, the equality holds if and only if $a = b$. Similarly, we get

$$\frac{1}{c+a-b} + \frac{1}{a+b-c} \ge \frac{2}{a} , \quad \text{and} \tag{1.5}$$

$$\frac{1}{b+c-a} + \frac{1}{a+b-c} \ge \frac{2}{b} . \tag{1.6}$$

After summing the three inequalities (that is, (1.4), (1.5), and (1.6)) together and dividing by 2, we get the desired result. We additionally notice that equality holds if and only if $a = b = c$.

REMARKS

Note that if one starts from the left hand side of (1.1), it is not clear how to reach the right hand side of (1.1). (In particular, observe that $1/a + 1/b - 1/c$ may be greater than $1/(a+b-c)$; consider, for example, $a = 3$, $b = 5$, and $c = 4$.) However, when we look at the right hand side, we notice that sum of any two denominators is twice some denominator on the left hand side. This suggests that it might be easier to start from the right hand side and try to reach the left hand side. Using our observation, it makes sense to re-write the right hand side as

$$\left(\frac{1/2}{b+c-a}+\frac{1/2}{c+a-b}\right)+\left(\frac{1/2}{c+a-b}+\frac{1/2}{a+b-c}\right)+\left(\frac{1/2}{b+c-a}+\frac{1/2}{a+b-c}\right)$$

and continue from there.

Let us also observe that a more general inequality in fact holds. For any convex function $f\colon \mathbf{R} \to \mathbf{R}$ on a connected subset $D \subseteq \mathbf{R}$, it follows that

$$f(a) + f(b) + f(c) \leq f(b+c-a) + f(c+a-b) + f(a+b-c), \quad (1.7)$$

provided that $a, b, c \in D$. Our problem is a specific case when $f(x) = 1/x$, convex function on $D = \mathbf{R}_+$.

The proof of this more general inequality follows exactly the same argument as above. Point A has coordinates $(a + b - c, f(a + b - c))$ and point $B = (b+c-a, f(b+c-a))$. Now, point D is the midpoint between A and B; that is, D has the first coordinate equal to $\frac{1}{2}(a+b-c) + \frac{1}{2}(b+c-a) = c$, and the second coordinate equal to $\frac{1}{2}f(a+b-c) + \frac{1}{2}f(b+c-a)$. Finally, point C has the same first coordinate as D, but its second coordinate is equal to $f(c)$. Inequality (1.4) is a special case of the following observation illustrated in Figure 1.1: for any convex function f on $D = [x_1, x_2]$,

$$f\left(\frac{x_1}{2} + \frac{x_2}{2}\right) \leq \frac{f(x_1)}{2} + \frac{f(x_2)}{2}. \quad (1.8)$$

We apply this observation with $x_1 = b + c - a$ and $x_2 = c + a - b$ to obtain the desired inequality.

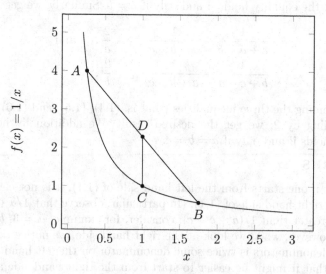

FIGURE 1.1: Illustration for proving (1.8).

THEORY

> **Jensen's Inequality** Let us point out that inequality (1.8) can be easily generalized to any number of numbers $x_1, \ldots, x_n$ (not only two), and to any weights (not only half). This generalization is known as **Jensen's inequality** and can be stated as follows: for any convex function $f(x) : D \to \mathbf{R}$, numbers $x_1, \ldots, x_n \in D$, and weights $a_1, \ldots, a_n \in \mathbf{R}_+$, we have that

$$f\left(\frac{\sum_{i=1}^{n} a_i x_i}{\sum_{i=1}^{n} a_i} \right) \leq \frac{\sum_{i=1}^{n} a_i f(x_i)}{\sum_{i=1}^{n} a_i}.$$

The inequality is reversed if f is concave. Rather,

$$f\left(\frac{\sum_{i=1}^{n} a_i x_i}{\sum_{i=1}^{n} a_i} \right) \geq \frac{\sum_{i=1}^{n} a_i f(x_i)}{\sum_{i=1}^{n} a_i}.$$

In both cases, equality holds if and only if $x_1 = \ldots = x_n$ or f is linear.

EXERCISES

1.1.1. Prove that for any $a, b, c \in \mathbf{R}$ such that $0 < a \leq b \leq c$,

$$\frac{1}{a} - \frac{1}{b} + \frac{1}{c} \geq \frac{1}{a + c - b}.$$

Illustrate the solution graphically. Does the same inequality hold for any function $f \colon \mathbf{R} \to \mathbf{R}$ that is convex on some connected subset of $\mathbf{R}$?

1.1.2. Prove that for any $n \in \mathbf{N}$ and any real number $s \geq 2$, the following inequality holds:

$$\frac{\sum_{k=1}^{n} k^s}{\sum_{k=1}^{n} k} \geq \left(\frac{2}{3}n + \frac{1}{3} \right)^{s-1}.$$

1.1.3. Prove that for any $x \in \mathbf{R}_+$,

$$\sqrt{x} + \sqrt{x + 2} < 2\sqrt{x + 1}.$$

1.2   Arithmetic-Geometric Inequality

SOURCE

Problem and idea for the solution: XXXI PLMO – Phase 2 – Problem 2

PROBLEM

Show that the following inequality holds for all $x_1, \ldots, x_n \in \mathbf{R}$:

$$\prod_{i=1}^{n} x_i \leq \frac{1}{2^n} + \sum_{i=1}^{n} \frac{x_i^{2^i}}{2^i}. \tag{1.9}$$

THEORY

| Geometric Sequence | A **geometric sequence** is a sequence of numbers where each term after the first is found by multiplying the previous one by a fixed, non-zero number called the **common ratio**. For example, the sequence $5, 10, 20, 40, \ldots$ is a geometric sequence with common ratio 2. Similarly $45, 15, 5, 5/3, \ldots$ is a geometric sequence with common ratio $1/3$. The common ratio of a geometric sequence may be negative, resulting in an alternating sequence; for example, $5, -10, 20, -40, \ldots$ is a geometric sequence with common ratio -2.

The general form of a geometric sequence is $a, ar, ar^2, \ldots$, where $r \neq 0$ is the common ratio and a is a **scale factor**, equal to the sequence's initial value. It follows immediately from the definition that a geometric sequence follows the following recursive relation: for every integer $i \geq 1$, $a_i = ra_{i-1}$. Hence, the i-th term is given by $a_i = ar^{i-1}$.

| Geometric Series | A **geometric series** is defined as $\sum_{i=1}^{n} a_i$. It is straightforward to see that

$$\sum_{i=1}^{n} a_i = \sum_{i=1}^{n} ar^{i-1} = \frac{a(1-r^n)}{1-r}, \tag{1.10}$$

provided $r \neq 1$. Indeed,

$$\begin{aligned}
\sum_{i=1}^{n} a_i &= \sum_{i=1}^{n} ar^{i-1} = \frac{1}{1-r} \cdot (1-r) \sum_{i=1}^{n} ar^{i-1} \\
&= \frac{1}{1-r} \cdot \left((a + ar + \ldots + ar^{n-1}) - (ar + ar^2 + \cdots + ar^n) \right) \\
&= \frac{a - ar^n}{1-r} = \frac{a(1-r^n)}{1-r}.
\end{aligned}$$

Arithmetic, Geometric, and Harmonic Means For any sequence of n numbers $x_1, \ldots, x_n \in \mathbf{R}$, the **arithmetic mean** is defined as

$$A(x_1, \ldots, x_n) := \frac{1}{n} \sum_{i=1}^{n} x_i .$$

For any sequence of n numbers $x_1, \ldots, x_n \in \mathbf{R}_+ \cup \{0\}$, the **geometric mean** is defined as

$$G(x_1, \ldots, x_n) := \left(\prod_{i=1}^{n} x_i \right)^{1/n} .$$

Finally, for any sequence of n numbers $x_1, \ldots, x_n \in \mathbf{R}_+$, the **harmonic mean** is defined as

$$H(x_1, \ldots, x_n) := \frac{n}{\sum_{i=1}^{n} 1/x_i} .$$

The following inequality relates the first two means and appears to be very useful. For any of n numbers $x_1, \ldots, x_n \in \mathbf{R}_+ \cup \{0\}$,

$$A(x_1, \ldots, x_n) = \frac{1}{n} \sum_{i=1}^{n} x_i \geq \left(\prod_{i=1}^{n} x_i \right)^{1/n} = G(x_1, \ldots, x_n) . \qquad (1.11)$$

The equality holds if and only if all x_i are equal.

In order to verify inequality (1.11), let us first note that the inequality trivially holds if $x_i = 0$ for some i. Hence, without loss of generality, we may assume that $x_i \in \mathbf{R}_+$ for all $i \in [n]$. Moreover, to prove inequality (1.11), it is enough to show that the following inequality holds:

$$\ln \left(\frac{1}{n} \sum_{i=1}^{n} x_i \right) \geq \ln \left(\left(\prod_{i=1}^{n} x_i \right)^{1/n} \right)$$

$$= \frac{1}{n} \cdot \ln \left(\prod_{i=1}^{n} x_i \right) = \frac{\sum_{i=1}^{n} \ln(x_i)}{n} . \qquad (1.12)$$

This is because $f(x) = \ln(x)$ is an increasing function on $D = \mathbf{R}_+$. But, inequality (1.12) follows immediately from Jensen's inequality applied to $f(x) = \ln(x)$, a concave function on $D = \mathbf{R}_+$, and $a_i = 1/n$ for all i. From this inequality, we also get that equality holds (in both inequalities (1.11) and (1.12)) if and only if all the x_i terms are equal.

Finally, let us consider the relationship between the harmonic and the geometric mean. We claim that

$$H(x_1, \ldots, x_n) = \frac{n}{\sum_{i=1}^{n} 1/x_i} \leq \left(\prod_{i=1}^{n} x_i \right)^{1/n} = G(x_1, \ldots, x_n).$$

Indeed, by substituting $y_i = 1/x_i$, we use inequality (1.11) to get

$$\frac{n}{\sum_{i=1}^{n} 1/x_i} = \left(\frac{1}{n}\sum_{i=1}^{n} y_i\right)^{-1} \leq \left(\prod_{i=1}^{n} y_i\right)^{-1/n} = \left(\prod_{i=1}^{n} x_i\right)^{1/n}.$$

As before, equality holds if and only if all the x_i terms are equal.

SOLUTION

First, let us note that, without loss of generality, we may assume that $x_i \in \mathbf{R}_+ \cup \{0\}$ for all $i \in [n]$. Indeed, if inequality (1.9) holds for all sequences $x_1, \ldots, x_n \in \mathbf{R}_+ \cup \{0\}$, then for any sequence $y_1, \ldots, y_n \in \mathbf{R}$, we get that

$$\prod_{i=1}^{n} y_i \leq \prod_{i=1}^{n} x_i \leq \frac{1}{2^n} + \sum_{i=1}^{n} \frac{x_i^{2^i}}{2^i} = \frac{1}{2^n} + \sum_{i=1}^{n} \frac{y_i^{2^i}}{2^i},$$

by setting $x_i = |y_i|$ for all $i \in [n]$.

We will start from the left hand side of inequality (1.9), and try to reach its right hand side. First, note that

$$\prod_{i=1}^{n} x_i = \left(1 \cdot \prod_{i=1}^{n} x_i^{2^n}\right)^{1/2^n}$$

$$= \left(1 \cdot \prod_{i=1}^{n} x_i^{2^{n-i} \cdot 2^i}\right)^{1/2^n} = \left(1 \cdot \prod_{i=1}^{n} \prod_{j=1}^{2^{n-i}} x_i^{2^i}\right)^{1/2^n}.$$

The product under the root has

$$1 + \sum_{i=1}^{n} 2^{n-i} = 1 + \sum_{j=0}^{n-1} 2^j = 1 + \frac{1-2^n}{1-2} = 1 + (2^n - 1) = 2^n$$

terms. (See equality (1.10) for the value of the geometric series.) Hence, we can apply the theorem relating the geometric and the arithmetic mean to get that

$$\left(1 \cdot \prod_{i=1}^{n} \prod_{j=1}^{2^{n-i}} x_i^{2^i}\right)^{1/2^n} \leq \frac{1}{2^n}\left(1 + \sum_{i=1}^{n} \sum_{j=1}^{2^{n-i}} x_i^{2^i}\right)$$

$$= \frac{1}{2^n} + \sum_{i=1}^{n} \frac{2^{n-i}}{2^n} x_i^{2^i} = \frac{1}{2^n} + \sum_{i=1}^{n} \frac{x_i^{2^i}}{2^i}.$$

This finishes the proof of the result.

REMARKS

As the left hand side of inequality (1.9) is a product that is smaller than the right hand side that is a sum of some kind, it is natural to try to apply the geometric-arithmetic inequality. The fact that the smallest term of the right hand side is $1/2^n$ suggests that we need 2^n terms, and exactly one of them is equal to 1. Then, we see that $x_i^{2^i}$ is divided by 2^i which means that we need 2^{n-i} such terms. Combining all these observations together, our goal is to transform our inequality so that the following properties hold: 1) there are 2^n terms in the sum, 2) x_i should be present in 2^{n-i} identical terms. So, starting from the right hand side of inequality (1.9), we get that

$$\frac{1}{2^n} + \sum_{i=1}^{n} \frac{x_i^{2^i}}{2^i} = \frac{1}{2^n} + \sum_{i=1}^{n} \sum_{j=1}^{2^{n-i}} \frac{1}{2^n} x_i^{2^i}.$$

Finally, the only other thing to notice is that there are exactly 2^n terms added together. This allows us to use the geometric-arithmetic inequality, thus completing the argument.

EXERCISES

1.2.1. Show that for any $a, b, c, d \in \mathbf{R}_+$, the following inequality holds:

$$(a + b + c + d) \left(\frac{1}{a} + \frac{1}{b} + \frac{4}{c} + \frac{16}{d} \right) \geq 64.$$

When does equality hold?
(Source of the problem: Student Circle – High School of Stanisław Staszic in Warsaw. Solution: our own.)

1.2.2. Show that for any n numbers $a_1, \ldots, a_n \in \mathbf{R}_+$, the following inequality holds:

$$\frac{a_1}{a_2 + 1} + \frac{a_2}{a_3 + 1} + \cdots + \frac{a_{n-1}}{a_n + 1} + \frac{a_n}{a_1 + 1} \geq \frac{n^2}{n + \alpha},$$

where $\alpha = \sum_{i=1}^{n} 1/a_i$.
(Source of the problem: Exam – Paweł Bechler – High School of Stanisław Staszic in Warsaw. Solution: our own.)

1.2.3. Prove that for any $a, b \in \mathbf{R}_+$, for which $ab = 1$, we have that

$$a^m + b^m \geq 2,$$

where $m \in \mathbf{R}_+$.

1.3   Mathematical Induction

SOURCE

Problem and solution: well-known problem

PROBLEM

Prove that for any integer $n \geq 2$ and any sequence of n real numbers $a_1, \ldots, a_n \in (1, \infty)$,

$$\sum_{i=1}^{n} \frac{a_i}{\ln(a_{i+1})} \geq \sum_{i=1}^{n} \frac{a_i}{\ln(a_i)}, \tag{1.13}$$

where $a_{n+1} = a_1$.

THEORY

$\boxed{\textbf{Mathematical Induction}}$ **Mathematical induction** is a powerful proof technique. It is typically used to prove that a property $P(n)$ holds for every integer $n \geq n_0$, where $n_0 \in \mathbf{Z}$. This method requires two things to be proven. First, one needs to check the **base case**; that is, to prove that the property holds for the smallest number n_0. Second, one needs to prove the **induction step**; that is, show that if the property holds for some $n \in \{n_0, n_0 + 1, \ldots\}$ (this assumption is often called **inductive hypothesis**), then it holds for $n + 1$. These two steps establish the property $P(n)$ for every integer $n \in \{n_0, n_0 + 1, \ldots\}$.

In order to illustrate the method, let us prove the following simple inequality (property $P(n)$): $2n + 1 \leq 2^n$ for all integers $n \geq 3$. The base case ($P(3)$) clearly holds:

$$7 = 2 \cdot 3 + 1 \leq 2^3 = 8.$$

Suppose that $P(n)$ holds: $2n + 1 \leq 2^n$ for some integer $n \geq 3$. We want to show that $P(n+1)$ holds: $2(n + 1) + 1 \leq 2^{n+1}$. This is true since

$$2(n+1) + 1 = (2n+1) + 2 \leq 2^n + 2 \leq 2^n + 2^n = 2^{n+1}.$$

(The first inequality holds by inductive hypothesis; the second one holds since $2 \leq 2^n$ for any $n \geq 3$.)

SOLUTION

We say that property $P(n)$ holds if inequality (1.13) holds for all sequences of n real numbers $a_1, \ldots, a_n \in (1, \infty)$. The following symmetry will turn out to be useful: inequality (1.13) applied to a sequence $a_1, \ldots, a_n \in (1, \infty)$ is equivalent to its application to the sequence $b_1, \ldots, b_n \in (1, \infty)$, where $b_i = a_{i+1}$. (Recall the convention that $a_{n+1} = a_1$.) In particular, it implies that, without loss of generality, we may assume that a_n is a smallest element in the sequence (as one can shift the initial sequence until smallest element is at the end).

We will prove by mathematical induction (on n) that $P(n)$ holds for all integers $n \geq 2$. First, let us check the base case ($n = 2$). We need to show that property $P(2)$ holds; that is, to prove that for any $a_1, a_2 \in (1, \infty)$, we have

$$\frac{a_1}{\ln(a_2)} + \frac{a_2}{\ln(a_1)} \geq \frac{a_1}{\ln(a_1)} + \frac{a_2}{\ln(a_2)}.$$

Clearly, this inequality holds if and only if

$$(a_1 - a_2)\left(\frac{1}{\ln(a_2)} - \frac{1}{\ln(a_1)}\right) \geq 0. \tag{1.14}$$

By our assumption that a_2 is a smallest element, we have that $a_1 - a_2 \geq 0$ and that $1/\ln(a_2) - 1/\ln(a_1) \geq 0$, so inequality (1.14) holds and the base case is finished.

For the induction step, assume that $P(n-1)$ holds for some integer $n \geq 3$; that is, for any $a_1, \ldots, a_{n-1} \in (1, \infty)$,

$$\sum_{i=1}^{n-1} \frac{a_i}{\ln(a_i)} \leq \sum_{i=1}^{n-1} \frac{a_i}{\ln(a_{i+1})} = \sum_{i=1}^{n-2} \frac{a_i}{\ln(a_{i+1})} + \frac{a_{n-1}}{\ln(a_1)}. \tag{1.15}$$

We want to show that $P(n)$ holds; that is, for any $a_1, \ldots, a_n \in (1, \infty)$,

$$\sum_{i=1}^{n} \frac{a_i}{\ln(a_i)} \leq \sum_{i=1}^{n} \frac{a_i}{\ln(a_{i+1})} = \sum_{i=1}^{n-2} \frac{a_i}{\ln(a_{i+1})} + \frac{a_{n-1}}{\ln(a_n)} + \frac{a_n}{\ln(a_1)}. \tag{1.16}$$

Fix any sequence $a_1, \ldots, a_n \in (1, \infty)$. Without loss of generality, we may assume that a_n is a smallest element. Starting from the left hand side of inequality (1.16) and using the inductive hypothesis (inequality (1.15)), we get that

$$\sum_{i=1}^{n} \frac{a_i}{\ln(a_i)} = \sum_{i=1}^{n-1} \frac{a_i}{\ln(a_i)} + \frac{a_n}{\ln(a_n)} \leq \sum_{i=1}^{n-2} \frac{a_i}{\ln(a_{i+1})} + \frac{a_{n-1}}{\ln(a_1)} + \frac{a_n}{\ln(a_n)}.$$

Hence, to get inequality (1.16) it is enough to show that

$$\frac{a_{n-1}}{\ln(a_1)} + \frac{a_n}{\ln(a_n)} \leq \frac{a_{n-1}}{\ln(a_n)} + \frac{a_n}{\ln(a_1)},$$

which is equivalent to

$$(a_{n-1} - a_n)\left(\frac{1}{\ln(a_n)} - \frac{1}{\ln(a_1)}\right) \geq 0.$$

Again, similarly to the argument used for inequality (1.14), it is straightforward to see that this inequality holds since it is assumed that a_n is a smallest element. The induction step is finished and so is the proof.

REMARKS

In fact, one can prove more general property. For any two increasing functions f and g on D the following is true: for any integer $n \geq 2$ and any sequence of n real numbers $a_1, \ldots, a_n \in D$,

$$\sum_{i=1}^{n} f(a_i)g(a_{i+1}) \leq \sum_{i=1}^{n} f(a_i)g(a_i).$$

Similarly, for any increasing function f and any decreasing function g on D,

$$\sum_{i=1}^{n} f(a_i)g(a_{i+1}) \geq \sum_{i=1}^{n} f(a_i)g(a_i).$$

Our question is a specific case of this general inequality when $f(x) = x$ and $g(x) = 1/\ln(x)$.

Finally, let us mention that the above two inequalities are a direct consequence of the following **rearrangement inequality**: for every two monotone sequences $x_1 \leq \ldots \leq x_n$ and $y_1 \leq \ldots \leq y_n$,

$$x_n y_1 + \ldots + x_1 y_n \leq x_{\sigma(1)} y_1 + \cdots + x_{\sigma(n)} y_n \leq x_1 y_1 + \ldots + x_n y_n,$$

where $\sigma : [n] \to [n]$ is any permutation of $[n]$. It is good to recall such general inequalities, and that they can be proven using mathematical induction. The idea for the proof of the initial problem then comes naturally.

EXERCISES

1.3.1. Prove that for any $a, b \in \mathbf{R}_+$,

$$a^b b^a \leq a^a b^b.$$

(Source of the problem: Lecture by Paweł Bechler – High School of Stanisław Staszic in Warsaw. Solution: our own.)

1.3.2. Prove that for any $a, b, c \in \mathbf{R}_+$,

$$\frac{ab}{c} + \frac{bc}{a} + \frac{ca}{b} \geq a + b + c.$$

(Source of the problem: Student Circle – High School of Stanisław Staszic in Warsaw. Solution: our own.)

1.3.3. Prove that for any $a, b, c \in \mathbf{R}_+$,

$$a^a b^b c^c \geq (abc)^{(a+b+c)/3}.$$

1.4 Bernoulli's Inequality

SOURCE

Problem and solution: our own

PROBLEM

Show that for any $0 < \alpha < \pi/2$,

$$\sqrt[\sin(\alpha)]{(2 - \cos^2(\alpha))} + \sqrt[\sin(\alpha)]{\cos^2(\alpha)} \geq 2.$$

THEORY

| Bernoulli's Inequality | If $x > -1$, then

$$(1 + x)^r \geq 1 + rx,$$

for $r \leq 0$ or $r \geq 1$, and

$$(1 + x)^r \leq 1 + rx,$$

for $0 \leq r \leq 1$. Strict inequalities hold, provided that $x \neq 0$ and $r \neq 0, 1$.

SOLUTION

Fix any $0 < \alpha < \pi/2$. Recall that $\sin^2(\alpha) + \cos^2(\alpha) = 1$. Substituting

$$r := 1/\sin(\alpha) \in (1, \infty)$$

and

$$x := \sin^2(\alpha) = 1 - \cos^2(\alpha) \in (0, 1),$$

we get that

$$\sqrt[\sin(\alpha)]{(2 - \cos^2(\alpha))} + \sqrt[\sin(\alpha)]{\cos^2(\alpha)} = (1 + x)^r + (1 - x)^r.$$

Now, since $x > -1$, $-x > -1$, and $r > 1$, we can apply Bernoulli's inequality to get

$$(1 + x)^r + (1 - x)^r \geq (1 + xr) + (1 - xr) = 2.$$

This finishes the proof.

REMARKS

Note that in this example we actually showed a slightly stronger inequality. Indeed, although x and r are related to each other (both are functions of α), the inequality is true even if they are not related. Such an approach of trying to prove a *stronger* result instead of the one we really care about is not uncommon in mathematics. It sometimes leads to a simpler proof of the result we care about.

A tricky part in this problem is to find a substitution $x = 1 - \cos^2(\alpha)$. In order to reach it, the first step is to check when the right hand side is equal to the left hand side, and we immediately see that this is the case when $\cos^2(\alpha) = 1$. It is then helpful to know that a typical trick in such cases is to consider a *deviation from the equality case*. From here, we obtain:

$$(1 + x)^{1/\sin(\alpha)} + (1 - x)^{1/\sin(\alpha)}.$$

Now, if one remembers Bernoulli's inequality, one immediately gets that it is at least 2 as long as $1/\sin(\alpha) > 1$. Fortunately, this is the case in our example. Alternatively, one can use Jensen's inequality as

$$\frac{(2 - \cos^2(\alpha)) + \cos^2(\alpha)}{2} = 1 \,,$$

and a^r is convex for $r > 1$.

EXERCISES

1.4.1. Prove that for any integer $n > 1$,

a) $$\left(\frac{n}{n+2}\right)^{n^2-n} < \frac{1}{2n-1} \,,$$

b) $$\left(\frac{n-1}{n}\right)^{n^2-1} < \frac{1}{n+2} \,.$$

(Source of the problem: Exam by Paweł Bechler – High School of Stanisław Staszic in Warsaw. Solution: our own.)

1.4.2. Prove that for any real number $x > -1$ and $n \in \mathbf{N}$,

$$\sqrt[n]{1+x} \leq 1 + \frac{x}{n} \,.$$

(Source of the problem: Lecture by Paweł Bechler – High School of Stanisław Staszic in Warsaw. Solution: our own.)

1.5 Euler's Number

SOURCE

Problem and solution: classic, well-known, problem

PROBLEM

Let m, n be any two natural numbers such that $m > n > 2$. Prove that

$$m^n < n^m.$$

THEORY

$\boxed{\text{Constant } e}$ The **constant** $e \approx 2.71828$ is a mathematical constant which appears in many different settings throughout mathematics. It can be defined as follows:

$$e = \lim_{n \to \infty} \left(1 + \frac{1}{n}\right)^n \qquad \text{or} \qquad e = \sum_{i=0}^{\infty} \frac{1}{i!}.$$

Moreover, the constant e is the unique real number such that

$$\left(1 + \frac{1}{x}\right)^x < e < \left(1 + \frac{1}{x}\right)^{x+1}$$

for all $x \in \mathbf{R}_+$.

Note that for $n \geq -a$, it follows from the arithmetic-geometric inequality that

$$\left(1 + \frac{a}{n}\right)^{n/(n+1)} = \left(1 \cdot \prod_{i=1}^{n}\left(1 + \frac{a}{n}\right)\right)^{1/(n+1)} < \frac{1 + n\left(1 + \frac{a}{n}\right)}{n+1} = 1 + \frac{a}{n+1}.$$

So, after raising both sides to the power of $n + 1$, we get that

$$\left(1 + \frac{a}{n}\right)^n < \left(1 + \frac{a}{n+1}\right)^{n+1}. \tag{1.17}$$

It follows that the sequence $x_n := \left(1 + \frac{a}{n}\right)^n$ is eventually increasing for any $a \neq 0$. Additionally, for $a > 0$,

$$\lim_{x \to \infty} \left(1 + \frac{a}{x}\right)^x = \lim_{x \to \infty} \left(\left(1 + \frac{1}{x/a}\right)^{x/a}\right)^a = \left(\lim_{x \to \infty} \left(1 + \frac{1}{x/a}\right)^{x/a}\right)^a = e^a,$$

and similarly for $a < 0$,

$$\lim_{x\to\infty} \left(1 + \frac{a}{x}\right)^x = \lim_{x\to\infty} \left(\frac{x}{a+x}\right)^{-x} = \lim_{x\to\infty} \left(1 + \frac{-a}{x+a}\right)^{-x}$$

$$= \left(\lim_{x\to\infty} \left(1 + \frac{-a}{x+a}\right)^{(x+a)-a}\right)^{-1} = (e^{-a})^{-1} = e^a.$$

There is one technical and subtle issue with the argument above. At some point, we switched from the limit of a sequence, $\lim_{n\to\infty} f(x_n)$, to the limit of a function, $\lim_{x\to\infty} f(x)$. There is a sequential characterization of limits of functions, namely, $\lim_{x\to a} f(x) = L$ (a could be infinity) if and only if $\lim_{n\to\infty} f(x_n) = L$ for *every* sequence $(x_n)_{n\geq 1}$ such that $\lim_{n\to\infty} x_n = a$. Note that it might be the case that $\lim_{x\to\infty} f(x)$ does not exist but $\lim_{n\to\infty} f(x_n)$ does; consider, for example, $f(x) = \sin(x)$ and $x_n = \pi n$.

In our situation, as constant e was defined by the limit of a sequence, we should have been slightly more careful and make sure we take a limit over integers. This is easy to verify after noting that

$$\left(1 + \frac{1}{\lceil n/a \rceil}\right)^{\lceil n/a \rceil - 1} \leq \left(1 + \frac{1}{n/a}\right)^{n/a} \leq \left(1 + \frac{1}{\lfloor n/a \rfloor}\right)^{\lfloor n/a \rfloor + 1}.$$

Since it is trivially true for $a = 0$, we can now safely claim that for any $a \in \mathbf{R}$,

$$\lim_{n\to\infty} \left(1 + \frac{a}{n}\right)^n = e^a. \tag{1.18}$$

There are many important and useful inequalities involving the constant e. We mention only a few here. For any $x \in \mathbf{R}$,

$$1 + x \leq e^x . \tag{1.19}$$

To see this we note that for natural $n > -x$ we can apply Bernoulli's inequality to get:

$$e^x \geq (1 + x/n)^n \geq 1 + n \cdot x/n = 1 + x .$$

On the other hand, for any $b \in \mathbf{R}_+$ and any $x \in [0, b]$, or for any $b \in \mathbf{R}_-$ and any $x \in [b, 0]$,

$$1 + \frac{e^b - 1}{b} \cdot x \geq e^x . \tag{1.20}$$

Indeed, one can use Jensen's inequality to show that

$$1 + \frac{e^b - 1}{b} \cdot x = \left(1 - \frac{x}{b}\right) \cdot e^0 + \frac{x}{b} \cdot e^b \geq e^{0(1-x/b)+b(x/b)} = e^x.$$

Both inequalities are illustrated on Figure 1.2. Once we introduce asymptotic notation, we will come back to these inequalities and prove inequalities (1.19) and (1.20) once more. However, as appropriate for the technique, we will

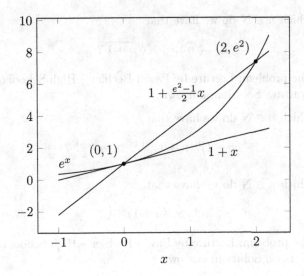

FIGURE 1.2: Illustration for inequalities (1.19) and (1.20).

concentrate on values of x close to zero and so the results obtained will be weaker.

SOLUTION

Inequality (1.19) is all we need to solve this problem:

$$m^n = n^n \left(\frac{m}{n}\right)^n = n^n \left(\frac{n+m-n}{n}\right)^n = n^n \left(1 + \frac{m-n}{n}\right)^n$$
$$\leq n^n \left(e^{\frac{m-n}{n}}\right)^n = n^n e^{m-n} \leq n^n n^{m-n} = n^m.$$

(The last inequality holds since $e < 3 \leq n$.)

REMARKS

In many problems that involve power function, we reach terms that can be expressed in the form $(1 + x/n)^n$. Then, it is often useful to remember that such terms (treated as sequences or functions of n with x fixed) are increasing, but bounded from above by e^x.

EXERCISES

1.5.1. Prove that for any integer $n > 2$,

$$\left(\frac{n}{n+2}\right)^{n^2-n} < \frac{1}{4^{n-1}}.$$

Can the constant 4 be improved for large n?

1.5.2. For which $n \in \mathbf{N}$ do we have that

$$\sqrt[n]{n} > \sqrt[n+1]{n+1} \ ?$$

(Source of the problem: Lecture by Paweł Bechler – High School of Stanisław Staszic in Warsaw. Solution: our own.)

1.5.3. For which $n \in \mathbf{N}$ do we have that

$$(n-1)^n (n+1)^{n+1} > n^{2n+1} \ ?$$

1.5.4. For which $n \in \mathbf{N}$ do we have that

$$n^n > (n+1)^{n-1} \ ?$$

(Source of the problem: Lecture by Paweł Bechler – High School of Stanisław Staszic in Warsaw. Solution: our own.)

1.5.5 Prove that for any $n \in \mathbf{N}$,

$$2 \leq \left(1 + \frac{1}{n}\right)^n \leq 3 \cdot \frac{n+1}{n+2} \ .$$

(Source of the problem: Lecture by Paweł Bechler – High School of Stanisław Staszic in Warsaw. Solution: our own.)

1.6 Asymptotics

SOURCE

Problem and solution: well-known problem

PROBLEM

Check for which $n \in \mathbf{N}$ the following statement holds: for all $x \in \mathbf{R}_+ \cup \{0\}$,

$$(1+x)^n \geq 1 + nx + \frac{(nx)^2}{2}.$$

THEORY

$\boxed{\textbf{Binomial Theorem}}$ The **binomial theorem** can be written as follows: for any $n \in \mathbf{N}$ and any $x, y \in \mathbf{R}$,

$$(x+y)^n = \sum_{i=0}^{n} \binom{n}{i} x^{n-i} y^i.$$

$\boxed{\textbf{Asymptotic Notation}}$ Let $f(x)$ and $g(x)$ be any two functions. In our applications, $f(x)$ is usually a complicated function whose behavior we would like to understand, and $g(x)$ has a simple form, and is positive for large enough x. We write:

- $f(x) = O(g(x))$ if there exists a positive constant C such that for all sufficiently large x we have that $|f(x)| \leq C|g(x)|$,

- $f(x) = \Omega(g(x))$ if there exists a positive constant c such that for all sufficiently large x we have that $|f(x)| \geq c|g(x)|$,

- $f(x) = \Theta(g(x))$ if $f(x) = O(g(x))$ and $f(x) = \Omega(g(x))$,

- $f(x) = o(g(x))$ if $\lim_{x \to \infty} f(x)/g(x) = 0$,

- $f(x) = \omega(g(x))$ if $\lim_{x \to \infty} |f(x)|/|g(x)| = \infty$,

- $f(x) \sim g(x)$ if $\lim_{x \to \infty} f(x)/g(x) = 1$.

As a simple example, consider $f(x) = 3x^{5/2} + 10x^2 + 10^{10}x \ln x$. For some moderate values of x, the last two terms are dominant but *eventually* the first one becomes much larger than both of them. Clearly, for any $x \geq 1$,

$$\begin{aligned} f(x) &= 3x^{5/2} + 10x^2 + 10^{10}x \ln x \\ &\leq 3x^{5/2} + 10x^{5/2} + 10^{10}x^{5/2} \leq 10^{11}x^{5/2}; \end{aligned}$$

it follows that $f(x) = O(x^{5/2})$. On the other hand, for any $x \geq 1$,

$$f(x) = 3x^{5/2} + 10x^2 + 10^{10}x \ln x \geq 3x^{5/2};$$

it follows that $f(x) = \Omega(x^{5/2})$ and so $f(x) = \Theta(x^{5/2})$. It is straightforward to see that, say, $f(x) = O(x^3)$ but $f(x) = \omega(x^2)$; that is, $f(x)$ is negligible compared to x^3 but grows faster than x^2. Finally, note that $f(x) \sim 3x^{5/2}$ as

$$\lim_{x\to\infty} \frac{f(x)}{3x^{5/2}} = \lim_{x\to\infty} \left(1 + \frac{10x^2}{3x^{5/2}} + \frac{10^{10}x \ln x}{3x^{5/2}}\right) = 1.$$

Alternatively, we could have observed that

$$\begin{aligned} f(x) &= 3x^{5/2} + O(x^2) + O(x \ln x) \\ &= 3x^{5/2} + O(x^2) = 3x^{5/2} + O(x^{5/2}) \sim 3x^{5/2}. \end{aligned}$$

Here are some useful properties.

- $O(f(x)) + O(g(x)) = O(|f(x)| + |g(x)|) = O(\max\{|f(x)|, |g(x)|\})$,
- $O(f(x)) \cdot O(g(x)) = O(f(x) \cdot g(x))$,
- $\Omega(f(x)) \cdot \Omega(g(x)) = \Omega(f(x) \cdot g(x))$,
- if $f : \mathbf{N} \to \mathbf{R}$ and $f(n) = O(1)$, then $f(n)$ is bounded by a constant.

One needs to be careful when working with asymptotic notation, as the notation presents many counterintuitive properties. For example, note that we cannot deduce that $\Omega(f(n)) + \Omega(g(n)) = \Omega(f(n))$. Indeed, if f and g are of the same order, it might not be true: $7n^2 + 3n = \Omega(n^2)$ and $-7n^2 + 10^{10} \ln n = \Omega(n^2)$ but $(7n^2 + 3n) + (-7n^2 + 10^{10} \ln n) = 3n + 10^{10} \ln n = \Omega(n)$.

The above definitions and examples assume that $x \to \infty$. However, sometimes we would like to understand the behavior of some function $f(x)$ when $x \to 0$. The notation introduced above can be easily adjusted to this situation. For example, we write $f(x) = O(g(x))$ if there exist positive constants C and M such that for all $x \in (0, M)$ we have $f(x) \leq Cg(x)$.

To illustrate this variant, let us come back to inequalities (1.19) and (1.20), and show their asymptotic counterparts. However, before we move to this task, let us show the following well-known fact: for any $x \in \mathbf{R}$,

$$e^x = \sum_{i=0}^{\infty} \frac{x^i}{i!} = 1 + x + \frac{x^2}{2!} + \frac{x^3}{3!} + \dots. \tag{1.21}$$

For $x = 0$ it trivially holds. Now fix any $x \in \mathbf{R} \setminus \{0\}$. Our goal is to show that $\lim_{n\to\infty} s_n = e^x$, where

$$s_n := \sum_{i=0}^{n} \frac{x^i}{i!}.$$

We will relate the sequence s_n to the sequence e_n defined as

$$e_n := \left(1 + \frac{x}{n}\right)^n,$$

for $n \in \mathbf{N}$. We also recall equality (1.18), whose importance will soon be seen:

$$e^x = \lim_{n \to \infty} \left(1 + \frac{x}{n}\right)^n = \lim_{n \to \infty} e_n.$$

For a fixed $n \in \mathbf{N}$, we use the binomial theorem to get that

$$e_n := \left(1 + \frac{x}{n}\right)^n = \sum_{i=0}^{n} \binom{n}{i} \frac{x^i}{n^i} = \sum_{i=0}^{n} \frac{x^i}{i!} \cdot \frac{n!}{(n-i)!n^i}.$$

We may assume that $n \geq 3|x|$ is large enough integer (but fixed) so that for any integer $i \geq n$ we have

$$0 \leq \frac{\frac{|x|^{i+1}}{(i+1)!}}{\frac{|x|^i}{i!}} = \frac{|x|}{i+1} \leq \frac{1}{3}.$$

This implies, that by equality (1.10), the value of the geometric series with the scale factor $1/3$ has

$$\left| \sum_{j=n+1}^{m} \frac{x^j}{j!} \cdot \frac{m!}{(m-j)!m^j} \right| \leq \sum_{j=n+1}^{m} \frac{|x|^j}{j!} \cdot \frac{m!}{(m-j)!m^j}$$

$$= \sum_{j=n+1}^{m} \frac{|x|^j}{j!} \cdot \frac{m}{m} \cdot \frac{m-1}{m} \cdot \ldots \cdot \frac{m-j+1}{m} \leq \sum_{j=n+1}^{m} \frac{|x|^j}{j!}$$

$$\leq \sum_{j=n+1}^{m} \frac{|x|^n}{n!} \cdot \frac{1}{3^{j-n}} \leq \frac{|x|^n}{2n!},$$

for any integer $m > n$. As a result, for any integer $m > n \geq 3|x|$,

$$\left| e_m - \sum_{i=0}^{n} \frac{x^i}{i!} \cdot \frac{m!}{(m-i)!m^i} \right| \leq \frac{|x|^n}{2n!}. \tag{1.22}$$

Next, observe that s_n can be made arbitrarily close to $\sum_{i=0}^{n} \frac{x^i}{i!} \frac{m!}{(m-i)!m^i}$ by making sure that m is large enough. To see this, note that there is a finite number of terms in the sum (namely, $n+1$), and that $\frac{m!}{(m-i)!m^i}$ tends to 1 as m tends to infinity. Therefore, as x and n are fixed, we can choose m large enough (that is, $m = f(n) \in \mathbf{N}$ for some function f) to ensure that

$$\left| s_n - \sum_{i=0}^{n} \frac{x^i}{i!} \frac{m!}{(m-i)!m^i} \right| \leq \frac{|x|^n}{2n!}. \tag{1.23}$$

(Let us comment that we choose $\frac{|x|^n}{2n!}$ as a convenient upper bound as it matches the bound in (1.22); however, here we could use any bound that tends to zero as $n \to \infty$.) Now, combining inequalities (1.22) and (1.23), we get that

$$|s_n - e_m| \leq \frac{|x|^n}{n!} .$$

Finally, as $e_m \to e^x$ as $m \to \infty$, for m large enough (that is, possibly after adjusting function f), we are guaranteed that

$$|e_m - e^x| \leq \frac{|x|^n}{n!} .$$

It follows that

$$|s_n - e^x| \leq \frac{2|x|^n}{n!},$$

which finishes the argument, as it implies that $s_n \to e^x$ as $n \to \infty$ (since, clearly $2|x|^n/n! \to 0$ as $n \to \infty$).

Now, to see that inequality (1.19) holds asymptotically, we use (1.21) and note that

$$e^x = 1 + x + \frac{x^2}{2} + O(x^3) \geq 1 + x + \frac{x^2}{4} \geq 1 + x,$$

provided that $x \in \mathbf{R}$ is sufficiently close to zero. (Of course, it holds for all $x \in \mathbf{R}$ but the aim here is to understand the behavior around zero.) Similarly, to see that the first part of inequality (1.20) holds asymptotically, note that for any $\epsilon > 0$,

$$e^x = 1 + x + O(x^2) \leq 1 + x + \epsilon x = 1 + (1 + \epsilon)x,$$

again, provided that $x \in \mathbf{R}_+$ is small enough.

SOLUTION

We will prove that the statement does *not* hold for any $n \in \mathbf{N}$; that is, for any $n \in \mathbf{N}$, there exists $x \geq 0$ such that

$$f(x) := 1 + nx + \frac{(nx)^2}{2} - (1 + x)^n > 0.$$

Clearly, for $n = 1$, we have

$$f(x) = 1 + x + \frac{x^2}{2} - (1 + x) = \frac{x^2}{2} > 0$$

for every $x > 0$. Similarly, for $n = 2$, we get that for any $x > 0$

$$f(x) = 1 + 2x + \frac{(2x)^2}{2} - (1 + x)^2 = x^2 > 0.$$

Now, let us fix any $n \geq 3$. This time we need a more sophisticated argument, as the statement clearly holds for large enough x but also for $x = 0$. However, it fails for x sufficiently close to zero (but not equal to zero). Using the binomial theorem, we get that

$$
\begin{aligned}
f(x) &= 1 + nx + \frac{(nx)^2}{2} - \sum_{i=0}^{n} \binom{n}{i} x^i \\
&= 1 + nx + \frac{n^2}{2} \cdot x^2 - \left(1 + nx + \frac{n(n-1)}{2} \cdot x^2 + \sum_{i=3}^{n} \binom{n}{i} x^i \right) \\
&= \frac{n}{2} \cdot x^2 - \sum_{i=3}^{n} \binom{n}{i} x^i .
\end{aligned}
\tag{1.24}
$$

Noting that for any $i \in \{3, \ldots, n\}$, $\binom{n}{i} \leq \binom{n}{\lfloor n/2 \rfloor} = \binom{n}{\lceil n/2 \rceil}$, we get that for any $x \in [0, 1/2]$,

$$
\begin{aligned}
f(x) &= \frac{n}{2} \cdot x^2 - \sum_{i=3}^{n} \binom{n}{i} x^i \geq \frac{n}{2} \cdot x^2 - \binom{n}{\lfloor n/2 \rfloor} \sum_{i=3}^{n} x^i \\
&\geq \frac{n}{2} \cdot x^2 - \binom{n}{\lfloor n/2 \rfloor} x^3 \sum_{i=0}^{n-3} x^i \geq \frac{n}{2} \cdot x^2 - \binom{n}{\lfloor n/2 \rfloor} x^3 \sum_{i=0}^{\infty} \left(\frac{1}{2} \right)^i \\
&= \frac{n}{2} \cdot x^2 - 2 \binom{n}{\lfloor n/2 \rfloor} x^3 = x^2 \left(\frac{n}{2} - 2 \binom{n}{\lfloor n/2 \rfloor} x \right) .
\end{aligned}
$$

Finally, we set $x_0 = n / (8 \binom{n}{\lfloor n/2 \rfloor}) \in (0, 1/2]$ to get the desired counter-example, that is, $f(x_0) = x_0^2 (n/2 - n/4) = x_0^2 \cdot n/4 > 0$.

REMARKS

Let us come back to our original question. Using the asymptotic notation (when $x \to 0$), one can continue the argument from equation (1.24) as follows, avoiding tedious calculations. Indeed, observe that

$$
\begin{aligned}
f(x) &= \frac{n}{2} \cdot x^2 - \sum_{i=3}^{n} \Theta(x^i) = \frac{n}{2} \cdot x^2 - \Theta(x^3) \\
&= \frac{n}{2} \cdot x^2 \left(1 + \Theta(x) \right) \sim \frac{n}{2} \cdot x^2 .
\end{aligned}
\tag{1.25}
$$

(Recall that n is perhaps large but fixed constant.) Hence, for sufficiently small x, we get $f(x) > 0$ and we are done. Actually, this asymptotic analysis suggests how to formalize the argument in the proof above, which only adds that we choose a specific (small) value for x to avoid asymptotic notation.

As mentioned in the theory part, we used above a non-standard notation when $x \to 0$. Of course, it is possible to avoid it and use a standard one with $y \to \infty$ by letting $x := 1/y \to 0$. Then, instead (1.25), we have

$$
\begin{aligned}
f(x) = f(1/y) &= \frac{n}{2} \cdot (1/y)^2 - \sum_{i=3}^{n} \Theta((1/y)^i) \\
&= \frac{n}{2} \cdot (1/y)^2 - \Theta((1/y)^3) \sim \frac{n}{2} \cdot (1/y)^2.
\end{aligned}
$$

The desired conclusion holds for sufficiently large y.

EXERCISES

1.6.1. Show that for any $n \in \mathbf{N}$, there exists a non-negative $x \in \mathbf{R}$ such that

$$
\prod_{i=1}^{n} (1+x)^i < 1 + \frac{n^2 + n + 1}{2} x .
$$

1.6.2. Prove that for any polynomial $W(x)$ and sufficiently large x we have that $(1 + x/n)^n > W(x)$, if $n \in \mathbf{N}$ is greater than the degree of W. What does it tell us about the function e^x?

1.7 Cauchy-Schwarz Inequality

SOURCE

Problem: LXIX PLMO – Phase 1 – Problem 10
Solution: our own

PROBLEM

Prove that for any integer $n \geq 3$ and any sequence of n numbers $x_1, \ldots, x_n \in \mathbf{R}_+$,

$$\frac{1 + x_1^2}{x_2 + x_3} + \frac{1 + x_2^2}{x_3 + x_4} + \ldots + \frac{1 + x_{n-2}^2}{x_{n-1} + x_n} + \frac{1 + x_{n-1}^2}{x_n + x_1} + \frac{1 + x_n^2}{x_1 + x_2} \geq n. \quad (1.26)$$

THEORY

$\boxed{\textbf{Cauchy-Schwarz Inequality}}$ The **Cauchy-Schwarz inequality** is an elementary inequality, and at the same time a powerful observation, which can be stated as follows. For any two sequences $a_1, \ldots, a_n \in \mathbf{R}$ and $b_1, \ldots, b_n \in \mathbf{R}$,

$$\left(\sum_{i=1}^{n} a_i^2 \right) \left(\sum_{i=1}^{n} b_i^2 \right) \geq \left(\sum_{i=1}^{n} a_i b_i \right)^2;$$

equality holds if and only if the two sequences are proportional; that is, there exists a constant $c \in \mathbf{R}$ such that $a_i = cb_i$ for all $i \in [n]$.

There are at least 12 different proofs of this inequality; here we present an elementary one. Note that

$$\begin{aligned}
0 &\leq \sum_{i=1}^{n} \sum_{j=1}^{n} (a_i b_j - a_j b_i)^2 = \sum_{i=1}^{n} \sum_{j=1}^{n} (a_i^2 b_j^2 - 2a_i a_j b_i b_j + a_j^2 b_i^2) \\
&= \sum_{i=1}^{n} a_i^2 \sum_{j=1}^{n} b_j^2 - 2 \sum_{i=1}^{n} a_i b_i \sum_{j=1}^{n} a_j b_j + \sum_{i=1}^{n} b_i^2 \sum_{j=1}^{n} a_j^2 \\
&= 2 \left(\sum_{i=1}^{n} a_i^2 \right) \left(\sum_{i=1}^{n} b_i^2 \right) - 2 \left(\sum_{i=1}^{n} a_i b_i \right)^2,
\end{aligned}$$

which immediately implies the desired inequality.

$\boxed{\textbf{Titu's Lemma}}$ The next inequality, known as **Titu's lemma**, is a direct consequence of Cauchy-Schwarz inequality. For any two sequences $x_1, \ldots, x_n \in \mathbf{R}$ and $y_1, \ldots, y_n \in \mathbf{R}$,

$$\sum_{i=1}^{n} \frac{x_i^2}{y_i} \geq \frac{\left(\sum_{i=1}^{n} x_i \right)^2}{\sum_{i=1}^{n} y_i}.$$

It is obtained by applying the substitutions $a_i = x_i/\sqrt{y_i}$ and $b_i = \sqrt{y_i}$ into the Cauchy-Schwarz inequality.

SOLUTION

First, let us note that the left hand side of (1.26) is equal to $A_1 + A_2$, where

$$A_1 \; := \; \frac{1^2}{x_2 + x_3} + \frac{1^2}{x_3 + x_4} + \cdots + \frac{1^2}{x_{n-1} + x_n} + \frac{1^2}{x_n + x_1} + \frac{1^2}{x_1 + x_2},$$

$$A_2 \; := \; \frac{x_1^2}{x_2 + x_3} + \frac{x_2^2}{x_3 + x_4} + \cdots + \frac{x_{n-2}^2}{x_{n-1} + x_n} + \frac{x_{n-1}^2}{x_n + x_1} + \frac{x_n^2}{x_1 + x_2}.$$

It follows from Titu's lemma (applied to two sequences, $1, 1, \ldots, 1$ and $x_2 + x_3, x_3 + x_4, \ldots, x_n + x_1, x_1 + x_2$) that

$$A_1 \; \geq \; \frac{(\sum_{i=1}^{n} 1)^2}{\sum_{i=1}^{n}(x_{i+1} + x_{i+2})} \; = \; \frac{n^2}{2\sum_{i=1}^{n} x_i};$$

as before, we used the convention that $x_{n+1} = x_1$ and $x_{n+2} = x_2$. Applying the lemma one more time (this time the first sequence is $x_1, x_2, \ldots, x_n$), we get that

$$A_2 \; \geq \; \frac{(\sum_{i=1}^{n} x_i)^2}{\sum_{i=1}^{n}(x_{i+1} + x_{i+2})} \; = \; \frac{(\sum_{i=1}^{n} x_i)^2}{2\sum_{i=1}^{n} x_i} \; = \; \frac{1}{2}\sum_{i=1}^{n} x_i.$$

Hence,

$$A_1 + A_2 \; \geq \; \frac{1}{2}\left(\frac{n^2}{\sum_{i=1}^{n} x_i} + \sum_{i=1}^{n} x_i\right) \; \geq \; \left(\frac{n^2}{\sum_{i=1}^{n} x_i} \cdot \sum_{i=1}^{n} x_i\right)^{1/2} \; = \; n,$$

where last inequality follows from the arithmetic-geometric mean inequality.

REMARKS

If one remembers Titu's lemma, then the solution comes to mind naturally after noticing that fractions on the left hand side of (1.26) contain squares in their numerators but there are no squares in their denominators. However, the question is what one needs to do without knowing the lemma (or without realizing that it can be applied to this problem).

Here is an elementary argument. Observe that

$$(2 - b)^2 + (2a - b)^2 \geq 0, \tag{1.27}$$

which is equivalent to

$$1 + a^2 \geq b + ab - b^2/2,$$

and if $b > 0$, we get that

$$\frac{1 + a^2}{b} \geq 1 + a - \frac{b}{2}. \tag{1.28}$$

Now, after substituting $a = x_i$ and $b = x_{i+1} + x_{i+2}$, we immediately get the desired inequality:

$$\sum_{i=1}^{n} \frac{1 + x_i^2}{x_{i+1} + x_{i+2}} \geq \sum_{i=1}^{n} \left(1 + x_i - \frac{x_{i+1} + x_{i+2}}{2} \right) = n.$$

The only question remaining is how to guess the starting point, that is, inequality (1.27)? One possible line of reasoning is as follows. We need to deal with the sum of n fractions, each of the form $\frac{1+a^2}{b}$, where $a = x_i > 0$ and $b = x_{i+1} + x_{i+2} > 0$. We observe that, although it is possible that some fractions are close to zero, the sum has to be large. Indeed, if one fraction is small, then its denominator must be large and so we expect the following fractions to be large. Our hope is that this will balance out and, on average, fractions have values at least 1. In fact, it is natural to conjecture that the left hand side of inequality (1.26) reaches its minimum for $x_1 = \ldots = x_n = 1$. But how do we turn it into a formal argument? Since the quadratic function is not the easiest to work with, the goal is to bound $f(a) = \frac{1+a^2}{b}$ from below by a simpler, linear, function $g(a) = ca + d$ with similar behavior, namely, if one term is small the other terms are forced to be large. It makes sense to make an approximation as tight as possible, so we want the line $g(a)$ to touch the parabola $f(a)$. However, what should be the touching point? The answer is relatively easy—as already mentioned, the original inequality (1.26) is tight when all x_i are equal (in fact, all are equal to 1). This implies that we want $a = b/2$ and so a touching point should be $(b/2, 1/b + b/4)$; see Figure 1.3 for an illustration. Hence, $g(a) = c(a - b/2) + 1/b + b/4$.

It remains to calculate the constant c. Since we want $f(a) \geq g(a)$, the function

$$f(a) - g(a) = \frac{1}{b} \cdot a^2 - c \cdot a + \frac{2cb - b}{4}$$

should be a quadratic function with its discriminant equal to zero. It follows that

$$c^2 - 4 \cdot \frac{1}{b} \cdot \frac{2cb - b}{4} = c^2 - 2c + 1 = 0,$$

and so $c = 1$. We get that $f(a) \geq g(a) = a - b/4 + 1/b$. Finally, since

$$\frac{b}{4} + \frac{1}{b} \geq 2\sqrt{\frac{b}{4} \cdot b} = 1,$$

we get that

$$\frac{1 + a^2}{b} \geq a - \frac{b}{4} + \frac{1}{b} = \frac{b}{4} + \frac{1}{b} + a - \frac{b}{2} \geq 1 + a - \frac{b}{2},$$

which is what we need to finish the proof—see (1.28).

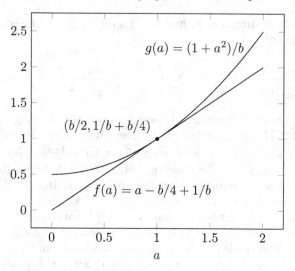

FIGURE 1.3: Illustration for tuning functions $f(a)$ and $g(a)$.

EXERCISES

1.7.1. Prove that for $a, b, c \in \mathbf{R}_+$,

$$(a+b+c)\left(\frac{1}{a} + \frac{1}{b} + \frac{1}{c}\right) \geq 9.$$

(Source of the problem: inspired by problem PLMO II – Phase 1 – Problem 6. Solution: our own.)

1.7.2. Prove that for any $a, b, c \in \mathbf{R}_+$ such that $a + b + c = 1$, we have that

$$\sqrt{2a+1} + \sqrt{2b+1} + \sqrt{2c+1} \leq \sqrt{15}.$$

(Source of the problem: Student Circle – High School of Stanisław Staszic in Warsaw. solution: our own.)

1.7.3. Prove that if $a, b, c \in \mathbf{R}$ are such that $a + b + c = 1$ and $\min\{a, b, c\} \geq -3/4$, then

$$\frac{a}{a^2 + 1} + \frac{b}{b^2 + 1} + \frac{c}{c^2 + 1} \leq \frac{9}{10}.$$

Does this inequality hold without the additional assumption that $\min\{a, b, c\} \geq -3/4$?

(Source of the problem and solution: XLVII OM – Phase 2 – Problem 3.)

1.8   Probability

SOURCE

Problem and solution: inspired by XLV PLMO – Phase 1 – Problem 10

PROBLEM

Prove that for any $x, y \in \mathbf{R}_+$ and any $m, n \in \mathbf{N}$,

$$((x+y)^m - x^m)^n + ((x+y)^n - y^n)^m \geq (x+y)^{nm}. \qquad (1.29)$$

In particular, when $x = y = 1$ and $n = m \in \mathbf{N}$, we get that

$$2(2^n - 1)^n \geq 2^{n^2}. \qquad (1.30)$$

THEORY

$\boxed{\textbf{Boole's Inequality}}$ The following elementary fact, known as **Boole's inequality** but also as the **union bound** is very useful. For any collection of events $A_1, \ldots, A_n$ in some probability space,

$$\mathbf{P}\left(\bigcup_{i=1}^{n} A_i\right) \leq \sum_{i=1}^{n} \mathbf{P}(A_i). \qquad (1.31)$$

We note that this inequality is sharp, since the equality holds for disjoint events.

SOLUTION

Proving the special case, inequality (1.30), is relatively easy. After dividing both sides by 2^{n^2+1}, we get an equivalent inequality

$$\left(1 - \frac{1}{2^n}\right)^n \geq \frac{1}{2}$$

that, in turn, after raising both sides to the power of $2^n/n$ is equivalent to

$$\left(1 - \frac{1}{2^n}\right)^{2^n} \geq \frac{1}{2^{2^n/n}}. \qquad (1.32)$$

Now, we notice that the left hand side of inequality (1.32) is an increasing function of n—see inequality (1.17). On the other hand, it is obvious that the right hand side of inequality (1.32) is a decreasing function of n. Hence the desired inequality holds if it holds for the smallest natural number, that is, for $n = 1$. For $n = 1$, both sides of inequality (1.32) are equal to $1/4$ and so we are done.

The proof of inequality (1.29) is more challenging. We start with dividing both sides by $(x+y)^{nm}$ and setting $p = x/(x+y) \in (0,1)$ to get an equivalent inequality: for any $p \in (0,1)$,

$$f(p) := (1 - p^m)^n + (1 - (1 - p)^n)^m \geq 1. \tag{1.33}$$

We are going to introduce a random process and two events, A and B, such that

$$\mathbf{P}(A) = (1 - p^m)^n, \tag{1.34}$$
$$\mathbf{P}(B) = (1 - (1 - p)^n)^m, \tag{1.35}$$

and argue that no matter what the outcome of the process is, at least one of the two events must hold. This will finish the proof, as then

$$1 = \mathbf{P}(A \cup B) = \mathbf{P}(A) + \mathbf{P}(B) - \mathbf{P}(A \cap B) \leq \mathbf{P}(A) + \mathbf{P}(B).$$

The last inequality is a specific case of Boole's inequality—see inequality (1.31).

Suppose that there are n students and m elective courses. Each student s is taking course c with probability p; all $n \cdot m$ events are independent of one another. Let A be the event that none of the students take all the courses. It is straightforward to see that equality (1.34) holds. Let B be the event that each course has at least one student taking it. Again, it is clear that equality (1.35) holds. The last piece missing is to show that $\mathbf{P}(A \cup B) = 1$. Suppose that B does *not* hold; that is, some course c is not taken by any student. This means that no student takes all the courses, and so A holds. Thus, at least one of the two events must hold, and so the proof is finished.

REMARKS

It is immediately apparent that the inequality can be reduced to one variable. Then, the question is what is most convenient way to do it. One natural approach is to make the right hand side of inequality (1.29) a constant. This directly gives

$$\left(1 - \left(\frac{x}{x+y}\right)^m\right)^n + \left(1 - \left(\frac{y}{x+y}\right)^n\right)^m \geq 1.$$

If $x/(x+y)$ is now the probability of some event, then $y/(x+y)$ is the probability of its complement. So we try to find a process involving $n \cdot m$ events, and two associated events.

Alternatively, one can solve this problem analytically. If $n = 1$ or $m = 1$, then the left hand side of inequality (1.29) is equal to its right hand side. Hence, we need to concentrate on the case $\min\{n, m\} > 1$. It will be more convenient to focus on inequality (1.33). The left hand side of inequality (1.33), function $f(p)$, has two terms: $(1 - p^m)^n$ and $(1 - (1 - p)^n)^m$. Clearly, $f(0) = f(1) = 1$.

The first term is a decreasing function of p, the second one is an increasing one, but it is not clear what the behavior of the sum of the two is. In order to show that $f(p) > 0$ for $p \in (0, 1)$, we will show that $f(p) > 0$ for *some* $p \in (0, 1)$ and that there is one extremum in the interval $(0, 1)$.

For the first property, note that for given n and m that are greater than 1, if p tends to zero, then

$$(1 - p^m)^n + (1 - (1 - p)^n)^m$$
$$= \left(1 - np^m + O(p^{2m})\right) + \left(1 - (1 - np + O(p^2))\right)^m$$
$$= \left(1 - np^m + O(p^{2m})\right) + n^m p^m \left(1 + O(p)\right)^m$$
$$= 1 + (n^m - n)p^m + O(p^{m+1})$$
$$= 1 + (n^m - n)p^m(1 + O(p)).$$

This implies that if p is greater than zero but sufficiently small, then $f(p) > 1$. The first property therefore holds.

For the second property, note that function $f(p)$ is differentiable on $[0, 1]$. Clearly,

$$f'(p) = mn\left((1 - (1 - p)^n)^{m-1}(1 - p)^{n-1} - p^{m-1}(1 - p^m)^{n-1}\right).$$

After setting $f'(p) = 0$ and rearranging terms, we get that

$$\left(\frac{1 - (1 - p)^n}{1 - (1 - p)}\right)^{m-1} = \left(\frac{1 - p^m}{1 - p}\right)^{n-1}.$$

Now, using the formula for geometric series (see (1.10)), we notice that it is equivalent to

$$\left(\sum_{i=0}^{n-1}(1 - p)^i\right)^{m-1} = \left(\sum_{i=0}^{m-1} p^i\right)^{n-1}.$$

Since the left hand side monotonically decreases from n^{m-1} (for $p = 0$) to 1 (when $p \to 1$) and the right hand side monotonically increases from 1 (for $p \to 0$) to m^{n-1} (for $p = 1$), there is exactly one point when the two sides are equal. It follows that $f'(p) = 0$ has only one solution in $(0, 1)$, which finishes the proof.

EXERCISES

1.8.1. Prove that for any $n \in \mathbf{N}$,

$$\frac{1}{2n + 1} \leq \frac{\binom{2n}{n}}{2^{2n}} \leq 1.$$

Can you improve these bounds for large n?

1.8.2. Prove that for $k, n \in \mathbf{N}$, such that $k \leq n$ and $p, q \in [0, 1]$, such that $p < q$ we have that

$$\sum_{i=k}^{n} \binom{n}{i}\left(q^i(1 - q)^{n-i} - p^i(1 - p)^{n-i}\right) \geq 0.$$

1.9   Geometry

SOURCE

Generalization of the problem from British Mathematical Olympiad 2006/7 – Phase 1 – Problem 5

PROBLEM

Prove that for any $a, b, c \in \mathbf{R}$,

$$(a^2 + b^2)^2 \geq (a + b + c)(a + b - c)(b + c - a)(c + a - b) . \qquad (1.36)$$

THEORY

$\boxed{\text{Heron's Formula}}$ **Heron's formula** states that the area of a triangle whose sides have lengths a, b, and c is

$$A = \sqrt{s(s - a)(s - b)(s - c)} ,$$

where s is the **semi-perimeter** of the triangle; that is,

$$s = \frac{a + b + c}{2} .$$

Heron's formula can also be written as

$$A = \frac{1}{4}\sqrt{(a + b + c)(-a + b + c)(a - b + c)(a + b - c)} .$$

SOLUTION

First, let us observe that there is a lot of symmetry in both the left hand side of inequality (1.36), the function $f(a, b) := (a^2 + b^2)^2$, and the right hand side, the function $g(a, b, c) := (a + b + c)(-a + b + c)(a - b + c)(a + b - c)$. In particular, both $f(a, b)$ and $g(a, b, c)$ are not affected by the sign of variables; for example,

$$\begin{aligned} g(-a, b, c) &= (-a + b + c)(-(-a) + b + c)(-a - b + c)(-a + b - c) \\ &= (-a + b + c)(a + b + c)(a + b - c)(a - b + c) = g(a, b, c). \end{aligned}$$

Hence, without loss of generality, it is enough to concentrate on non-negative values of a, b, c.

As in many earlier examples, it is convenient to translate the problem to another domain. We observe that we may assume that a, b, c are sides of some triangle, as otherwise $g(a, b, c) < 0$ and so the desired inequality trivially holds (since $f(a, b) \geq 0$ for any a, b). Hence, we can re-write the inequality as follows:

$$\frac{a^2 + b^2}{4} \geq \frac{1}{4}\sqrt{(a+b+c)(a+b-c)(b+c-a)(c+a-b)} \quad (1.37)$$
$$=: \quad A(a, b, c),$$

and use Heron's formula to notice that the right hand side of inequality (1.37), function $A(a, b, c)$, is the area of the considered triangle. Now, since a triangle with arms of lengths a and b has area less than or equal to $ab/2$ we get $A(a, b, c) \leq ab/2$. (The equality holds if and only if the angle between the two arms is $90°$.) By the geometric-arithmetic mean inequality, $ab/2 \leq (a^2 + b^2)/4$ and so the desired inequality holds. Finally, let us note that $ab/2 = (a^2 + b^2)/4$ if and only if $a = b$. So putting these observations together, we get that $f(a, b) = g(a, b, c)$ if and only if $a = b$ and $c = 2\sqrt{a}$.

REMARKS

Knowing Heron's formula turns out to be useful in this example and the way it can be applied comes to mind naturally. However, one can solve this problem without using it. After simplifying inequality (1.36), we get an equivalent inequality

$$h(a, b, c) := c^4 - 2(b^2 + a^2)c^2 + 2(b^4 + a^4) \geq 0,$$

and we note that $h(a, b, c)$ is a quadratic polynomial of c^2. Since the discriminant, Δ, satisfies

$$\Delta := (-2(b^2 + a^2))^2 - 4 \cdot 2(b^4 + a^4) = -4(b^2 - a^2) \leq 0,$$

the desired inequality holds. Moreover, as before, we get that the equality holds if and only if $|a| = |b| = |c|/\sqrt{2}$.

EXERCISES

1.9.1. Prove that for all sequences of n numbers $a_1, \ldots, a_n \in \mathbf{R}$ we have

$$\sqrt{n + \left(\sum_{i=1}^{n} a_i\right)^2} \leq \sum_{i=1}^{n} \sqrt{1 + a_i^2}.$$

When does equality hold?
(Source of the problem and solution: inspired by PLMO XXXVIII – Phase 1 – Problem 7.)

1.9.2. Prove that for any $x \in \mathbf{R}$ such that $1/4 \leq x \leq 1$ we have

$$\frac{x}{2}\sqrt{1 - x^2} + \frac{1}{16}\sqrt{16x^2 - 1} < \frac{4}{9}.$$

Chapter 2

Equalities and Sequences

As usual, we start the chapter with some basic definitions.

THEORY

Suppose that you are given an equation or a set of equations involving some number of variables. A natural question that is usually asked is to find a solution in some given domain. In general, there are the following three possible cases:

- no feasible solution exits,

- there exists a unique solution,

- there are multiple solutions (the number of them could be finite or infinite).

For example, consider a quadratic equation $ax^2 + bx + c = 0$, where $a \neq 0$. Our goal is to find all real solutions, that is, all values of $x \in \mathbf{R}$ that satisfy this equation. After multiplying both sides by $4a$, one can re-write it as follows:

$$(2ax - b)^2 = b^2 - 4ac.$$

It is now clear that if $b^2 - 4ac < 0$, then the equation has no real solutions, as the left hand side is non-negative. On the other hand, if $b^2 - 4ac = 0$, then the equation has exactly one solution, namely, $x = b/(2a)$. Finally, if $b^2 - 4ac > 0$, then we get two different solutions: $x_1 = (b + \sqrt{b^2 - 4ac})/(2a)$ and $x_2 = (b + \sqrt{b^2 + 4ac})/(2a)$. Let us mention that the value $b^2 - 4ac$, which allows deducing some properties of the roots without computing them, is called the **discriminant** and is often denoted as Δ.

Another important distinction is with respect to the number of equations. The problem we need to deal with could consist of

- one equation,

- more than one equation but finitely many—in this case we typically say that we have a **system of equations** (the word "system" indicates that the equations are to be considered collectively, rather than individually),

- infinitely many equations—this case is often represented as a recursive sequence of equations specifying the relationships between the involved variables.

Let us give a simple example of an infinite series of recursive equations, as this is not a typical situation. Let $x_0 = a$ for some non-zero real number a, and for $i \in \mathbf{N}$, let $x_{i+1} = 2x_i$. It is straightforward to see that $x_n = 2^n a$ is the only solution to this system of equations. Formally, one could prove it by induction on n.

Let us mention a specific family of equations which play a fundamental role in linear algebra, a subject which is used in most areas of modern mathematics. A **system of linear equations** is a collection of two or more linear equations involving the same set of variables. A system of non-linear equations can often be approximated by a linear system (such a process is called **linearization**), a helpful technique when designing a mathematical model or computer simulation of a complex system.

The equations of a linear system are **independent** if none of the equations can be derived algebraically from the others. In other words, when the equations are independent, each equation contains new information about the variables, and removing any of the equations increases the size of the solution set. Any system of n independent equations involving n variables has a unique solution.

The simplest method for solving a system of linear equations is to repeatedly eliminate variables. Consider, for example, the following system of linear equations:

$$\begin{cases} x + 2y + z = 3 \\ 2x + y + z = 3 \\ x - y = 0. \end{cases}$$

It is easy to see that the sum of the second and the third equation is equal to the first equation, so the system is *not* independent. After dropping the first equation we get the following, independent and equivalent, system:

$$\begin{cases} 2x + y + z = 3 \\ y = x. \end{cases}$$

We can now eliminate variable y from the first equation to get:

$$\begin{cases} z = 3 - 3x \\ y = x. \end{cases}$$

As we are not able to further reduce the system, we conclude that there are an infinite number of solutions: a triple $(x, y, z) = (t, t, 3 - 3t)$ satisfies the original system of equations for *any* $t \in \mathbf{R}$.

Finally, let us mention about the geometric interpretation. Let us start with a simple example, a linear system involving two variables, say x and y. Each linear equation determines a line on the xy-plane. Because a solution to a linear system must satisfy all of the equations, the solution set is the intersection of these lines, and is hence either a line (an infinite number of solutions), a single point (a unique solution), or the empty set (no solution). The three cases are illustrated on Figure 2.1. If there is only one equation $x + y = 2$, then any point $(x, y) = (t, 2 - t)$, $t \in \mathbf{R}$ satisfies this equation—see Figure 2.1(a). The following system

$$\begin{cases} x + y = 2 \\ x - y = 0. \end{cases}$$

has precisely one solution, the intersection of the corresponding two lines—see Figure 2.1(b). Finally, there is no solution to the following system

$$\begin{cases} x + y = 2 \\ x - y = 0 \\ y = 2, \end{cases}$$

as no point belongs to all of the corresponding three lines—see Figure 2.1(c). For three variables, each linear equation determines a plane in three-dimensional space, and the solution set is the intersection of these planes. In general, for n variables, each linear equation determines a hyperplane in n-dimensional space.

Situation is more complex for non-linear systems but one can still gain some intuition by representing each equation as a family of points that satisfy it. In the next section, we solve the following simple example algebraically:

$$\begin{cases} x^2 + y^2 = 2 \\ x + y = 2. \end{cases}$$

In Figure 2.2, we present the corresponding graphs that suggest the unique solution $(x, y) = (1, 1)$.

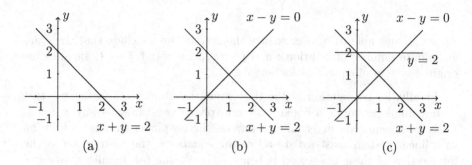

FIGURE 2.1: Geometric interpretation of linear systems.

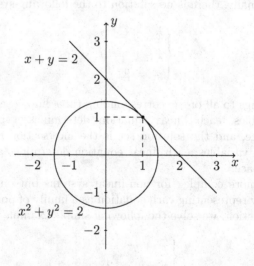

FIGURE 2.2: Graphs of $x^2 + y^2 = 2$ and $x + y = 2$.

2.1   Combining Equalities

SOURCE

Problem: PLMO LVI – Phase 1 – Problem 1
Solution: our own

PROBLEM

Solve the following system of equations, given that all variables involved are real numbers:

$$\begin{cases} x^2 = yz + 1 \\ y^2 = zx + 2 \\ z^2 = xy + 4 \,. \end{cases}$$

THEORY

A natural approach when solving systems of equations, that often turns out to be efficient, is to transform a given system into some other equivalent system that is easier to deal with.

Suppose that we are given n equations with unknown variables, represented by the vector $\mathbf{x}$, that is of the form $f_i(\mathbf{x}) = b_i$ for $i \in [n]$. For a given sequence of weights $c_i \in \mathbf{R}$, $i \in [n]$, if $c_j \neq 0$ for some $j \in [n]$, then one can take the original system of equations and replace the jth equation, $f_j(\mathbf{x}) = b_j$, with a linear combination of all equations, that is, with

$$\sum_{i=1}^{n} c_i f_i(\mathbf{x}) = \sum_{i=1}^{n} c_i b_i \,.$$

The resulting system of equations has identical solutions as the original system. Indeed, in order to see this, let us first assume that some $\mathbf{x}$ is a solution to the original system. It is clear that it is also a solution of the derived system. On the other hand, if some $\mathbf{x}$ is a solution of the derived system, then it must also satisfy $f_j(\mathbf{x}) = b_j$ as $c_j \neq 0$ and $\sum_{i \neq j} c_i f_i(\mathbf{x}) = \sum_{i \neq j} c_i b_i$.

To illustrate this technique, let us consider the following simple example of such a system, where $x, y \in \mathbf{R}$:

$$\begin{cases} x^2 + y^2 = 2 \\ x + y = 2 \,. \end{cases}$$

One can replace the first equation with the first equation minus two times the second equation to get the following equivalent system:

$$\begin{cases} x^2 - 2x + y^2 - 2y = 2 - 4 \\ x + y = 2 \,, \end{cases}$$

or equivalently,

$$\begin{cases} (x-1)^2 + (y-1)^2 = 0 \\ x+y = 2. \end{cases}$$

Since $(x-1)^2 \geq 0$ and $(y-1)^2 \geq 0$, the first equality holds only for $x = 1$ and $y = 1$. As a result, we may equivalently re-write the system as follows:

$$\begin{cases} x = 1 \\ y = 1 \\ x+y = 2. \end{cases}$$

It is obvious now that this system of equations is satisfied for $x = y = 1$. It follows that this is the only solution of the original system.

SOLUTION

After subtracting the first equation from the second one, we get that

$$(y-x)(x+y+z) = 1. \tag{2.1}$$

Similarly, from the second and third equation, we get that

$$(z-y)(x+y+z) = 2. \tag{2.2}$$

We observe that $x+y+z \neq 0$ and after subtracting (2.1) twice from (2.2), we get that

$$(z-3y+2x)(x+y+z) = 0.$$

It follows that $z = 3y - 2x$ and so we can reduce the system to two equations and two variables:

$$\begin{cases} x^2 = y(3y-2x) + 1 \\ y^2 = x(3y-2x) + 2, \end{cases}$$

or equivalently,

$$\begin{cases} (x+y)^2 = 4y^2 + 1 \\ 2(x+y)^2 = y^2 + 7yx + 2. \end{cases}$$

Substituting the first equation into the second one gives us the following sequence of equivalent equalities $2(4y^2 + 1) = y^2 + 7yx + 2$, $y^2 = yx$, and so $y(y-x) = 0$. It follows that $x = y$ or $y = 0$. Note that it is *not* the case that $x = y$, as then the left hand side of (2.1) is equal to 0 but the right hand side is equal to 1. If $y = 0$, then from $z = 3y - 2x$ we get that $z = -2x$ and from $(x+y)^2 = 4y^2 + 1$ we have $x^2 = 1$. This gives us two candidate solutions: $(x,y,z) = (-1,0,2)$ and $(x,y,z) = (1,0,-2)$. We can then directly check that both of them satisfy the original system of equations.

REMARKS

In order to gain some more experience, let us consider another problem to show how the technique we practice in this section can be applied. As an example, we use the problem from OM LVIII – Phase 1 – Problem 1. We are asked to solve the following system of equations, where variables involved are real numbers:

$$\begin{cases} x^2 + 2yz + 5x = 2 \\ y^2 + 2zx + 5y = 2 \\ z^2 + 2xy + 5z = 2. \end{cases}$$

Adding all equations together we get that $(x+y+z)(x+y+z+5) = 6$, so $x+y+z = 1$ or $x+y+z = -6$. We will independently consider the following two cases.

Case 1: all the numbers are equal. We get that $3x^2 + 5x - 2 = 0$ and so there are two possible values of x:

$$x_1 = \frac{-5+\sqrt{49}}{2 \cdot 3} = \frac{1}{3} \quad \text{and} \quad x_2 = \frac{-5-\sqrt{49}}{2 \cdot 3} = -2.$$

It follows that there are two solutions: $(x, y, z) = (1/3, 1/3, 1/3)$ and $(x, y, z) = (-2, -2, -2)$.

Case 2: not all the numbers are equal. Due to the symmetry, without loss of generality, we may assume that $x \neq y$—the other solutions will be obtained by permuting the solution vector (x, y, z). After comparing the left hand sides of the first and the second equation we get that $(x - y)(x + y - 2z + 5) = 0$. Since $x \neq y$, it follows that $x + y - 2z + 5 = 0$. We will now independently consider two sub-cases, depending on the value of $x + y + z$.

Sub-case 2a: $x + y + z = 1$. Since $x + y - 2z + 5 = 0$, $z = 2$ and $y = -1 - x$. After substituting this into the first equation, we get that

$$x^2 + 2(-1 - x) \cdot 2 + 5x = 2,$$

or equivalently that $0 = x^2 + x - 6 = (x - 2)(x + 3)$. We get two solutions: $(x, y, z) = (2, -3, 2)$ or $(x, y, z) = (-3, 2, 2)$.

Sub-case 2b: $x + y + z = -6$. This time, after combining this with $x + y - 2z + 5 = 0$, we get that $z = -1/3$ and $y = -17/3 - x$. After substituting this into the first equation, we get that

$$x^2 + 2(-17/3 - x) \cdot (-1/3) + 5x = 2,$$

or equivalently that

$$0 = x^2 + \frac{17}{3}x + \frac{16}{9} = \left(x + \frac{1}{3}\right)\left(x + \frac{16}{3}\right).$$

We get two more solutions: $(x, y, z) = (-1/3, -16/3, -1/3)$ and $(x, y, z) = (-16/3, -1/3, -1/3)$.

Combining all the cases together, we deduce that there are 8 candidate solutions:

$$(x, y, z) \in \left\{ (1/3, 1/3, 1/3), (-2, -2, -2), \right.$$
$$(-3, 2, 2), (2, -3, 2), (2, 2, -3),$$
$$(-16/3, -1/3, -1/3), (-1/3, -16/3, -1/3),$$
$$\left. (-1/3, -1/3, -16/3) \right\}.$$

We may then directly check that all of them meet the original system of equations.

EXERCISES

2.1.1. Solve the following system of equations, given that all variables involved are real numbers:
$$\begin{cases} a^3 + b = c \\ b^3 + c = d \\ c^3 + d = a \\ d^3 + a = b. \end{cases}$$
(Source of the problem and solution: PLMO LXIII – Phase 2 – Problem 1.)

2.1.2. Solve the following system of equations, given that all variables involved are real numbers:
$$\begin{cases} (x - y)(x^3 + y^3) = 7 \\ (x + y)(x^3 - y^3) = 3. \end{cases}$$
(Source of the problem and solution: PLMO LXII – Phase 2 – Problem 1.)

2.1.3. Solve the following system of equations, given that all variables involved are real numbers:
$$\begin{cases} x^2 - (y + z + yz)x + (y + z)yz = 0 \\ y^2 - (z + x + zx)y + (z + x)zx = 0 \\ z^2 - (x + y + xy)z + (x + y)xy = 0. \end{cases}$$
(Source of the problem and solution: PLMO LXI – Phase 2 – Problem 1.)

2.2 Extremal Values

SOURCE

Problem and method of solution: PLMO LVII – Phase 3 – Problem 1

PROBLEM

Solve the following system of equations, given that all variables involved are real numbers:

$$\begin{cases} a^2 = b^3 + c^3 \\ b^2 = c^3 + d^3 \\ c^2 = d^3 + e^3 \\ d^2 = e^3 + a^3 \\ e^2 = a^3 + b^3. \end{cases}$$

THEORY

Cyclic Systems of Equations A system of n equations with n variables $x_1, x_2, \ldots, x_n$ of the form

$$\begin{cases} f_1(x_1, x_2, \ldots, x_n) = 0 \\ f_2(x_1, x_2, \ldots, x_n) = 0 \\ \vdots \\ f_n(x_1, x_2, \ldots, x_n) = 0 \end{cases}$$

is called **cyclic** if the system does not change after replacing variable x_i with variable x_{i+1} for $i \in [n-1]$, and replacing variable x_n with variable x_1.

Let us note that any cyclic system with a finite number of variables $x_1, x_2, \ldots, x_n$ has the property that a circular permutation of any solution is also a solution. In other words, if $(x_1, x_2, \ldots, x_n) = (\bar{x}_1, \bar{x}_2, \ldots, \bar{x}_n)$ is a solution, then so is $(x_1, x_2, \ldots, x_n) = (\bar{x}_k, \bar{x}_{k+1}, \ldots, \bar{x}_n, \bar{x}_1, \bar{x}_2, \ldots, \bar{x}_{k-1})$, where $k \in [n]$. As a result, when solving cyclic systems of equations it is often useful to start with assuming that some variable attains the maximum or the minimum value among the whole set of all variables. Once the solution is found, one can simply recover the whole family of solutions that can be obtained by applying circular permutations to the particular solution.

In order to illustrate this technique in a simple setting, let us consider the following cyclic system of n equations and n variables $x_1, x_2, \ldots, x_n \in \mathbf{R}$, where $n \in \mathbf{N} \setminus \{1\}$. For $i \in [n]$, $x_i^3 + 2 = 3x_{i+1}$, assuming $x_{n+1} = x_1$. Using our observation, without loss of generality, we may assume that $x_1 = \max_{i \in [n]} x_i$. Since

$$x_1^3 + 2 = 3x_2 \leq 3x_1,$$

we get that

$$0 \geq x_1^3 + 2 - 3x_1 = (x_1^2 - 2x_1 + 1)(x_1 + 2) = (x_1 - 1)^2(x_1 + 2).$$

It follows that $x_1 = 1$ or $x_1 \leq -2$.

If $x_1 = 1$, then $x_2 = (x_1^3 + 2)/3 = (1 + 2)/3 = 1$ and, since the system is cyclic, we conclude that in fact $x_i = 1$ for each $i \in [n]$. Hence, $(x_1, x_2, \ldots, x_n) = (1, 1, \ldots, 1)$ is one solution to the system. Similarly, if $x_1 = -2$, then $x_2 = (x_1^3 + 2)/3 = (-8 + 2)/3 = -2$, and arguing as before we get that $(x_1, x_2, \ldots, x_n) = (-2, -2, \ldots, -2)$ is another solution to the system.

We will show that there are no more solutions. For a contradiction, suppose that there exists a solution for which $x_1 < -2$. In this case, we get that

$$x_2 - x_1 = \frac{x_1^3 + 2}{3} - x_1 = \frac{x_1^3 + 2 - 3x_1}{3} = \frac{(x_1 - 1)^2(x_1 + 2)}{3} < 0.$$

It follows that $x_2 < x_1 < -2$. Since the system is cyclic, we can keep repeating this argument to conclude that

$$-2 > x_1 > x_2 > \ldots > x_n > x_1,$$

which is clearly a contradiction.

SOLUTION

Without loss of generality, we may assume that b is the largest variable involved, that is, $b = \max\{a, b, c, d, e\}$. In particular, since function $f(x) = x^3$ is an increasing function, $b^3 \geq d^3$. By subtracting the second equation from the first one, we get that

$$a^2 - b^2 = (b^3 + c^3) - (c^3 + d^3) = b^3 - d^3 \geq 0, \qquad (2.3)$$

and so $a^2 \geq b^2$. Since $b \geq a$, we see that $a^2 \geq b^2$ holds only if $a \leq -|b|$ or $a = b$. We will independently consider both cases.

Case 1: $a \leq -|b|$. We get that $a^3 \leq (-|b|)^3 = -|b|^3$ and so

$$a^3 + b^3 \leq a^3 + |b|^3 \leq 0.$$

On the other hand, the fifth equation implies that $a^3 + b^3 = e^2 \geq 0$ and so $e = 0$. As a result, the fourth equation implies that $a^3 = d^2 \geq 0$ and so $a = 0$ and $b = 0$. Coming back to the original equations, we deduce that $d = 0$ and consequently that $c = 0$.

Case 2: $a = b$. It follows from (2.3) that $b^3 = d^3$ and so we also get that $b = d$. It follows from the third and the fourth equation that $c^2 = b^2$. If $c = b$ (that is, $a = b = c = d$), then we get from the second equation that $0 = 2b^3 - b^2 = b^2(2b - 1)$ and so $b = 0$ or $b = 1/2$ (and in both cases all other variables are equal). If $c = -b$ then we get from the first equation that $a = 0$, and again all variables are equal to 0.

In summary, there are only two solutions to our system of equations:

$$(a, b, c, d, e) \in \left\{ (0, 0, 0, 0, 0), (1/2, 1/2, 1/2, 1/2, 1/2) \right\}.$$

REMARKS

In all examples presented so far, all solutions had the property that all variables are equal, that is, when the minimum values is equal to the maximum value. Indeed, this is often the case but, of course, it does not have to be in general. Consider, for instance, the following cyclic system of equations where variables x_1, x_2, x_3 are real numbers:

$$\begin{cases} x_1(1 - x_2) = 1 \\ x_2(1 - x_3) = 1 \\ x_3(1 - x_1) = 1. \end{cases}$$

We get immediately that none of the variables x_1, x_2, x_3 is equal to 0 or 1; otherwise, the left hand side of one of the equations would be equal to 0. Hence, we can re-write the system as follows:

$$\begin{cases} x_2 = 1 - 1/x_1 \\ x_3 = 1 - 1/x_2 \\ x_1 = 1 - 1/x_3. \end{cases}$$

Note that if $x \notin \{0, 1\}$, then $f(x) := 1 - 1/x \notin \{0, 1\}$. Hence, after making a substitution, we get that

$$\begin{cases} x_2 = 1 - 1/x_1 \\ x_3 = 1 - 1/(1 - 1/x_1) = -1/(x_1 - 1) \\ x_1 = 1 - 1/(-1/(x_1 - 1)) = x_1. \end{cases}$$

It follows that any triple of a form $(x_1, x_2, x_3) = (t, 1 - 1/t, -1/(t - 1))$ is a solution to our system, provided that $t \in \mathbf{R}$ is *not* equal to 0 nor 1. In particular, $(x_1, x_2, x_3) = (1/2, -1, 2)$ is a solution so, indeed, there are non-constant solution vectors.

Finally, let us mention that the trick of assuming that one of the variables attains the maximum or the minimum does not only apply to cyclic systems— see, for example, Problem 2.2.3.

EXERCISES

2.2.1. Solve the following system of equations, given that all variables involved are real numbers:

$$\begin{cases} (x + y)^3 = 8z \\ (y + z)^3 = 8x \\ (z + x)^3 = 8y. \end{cases}$$

(Source of the problem: OM LXIII – Phase 1 – Problem 1. Solution: our own.)

2.2.2. Solve the following system of equations, given that all variables involved are real numbers:

$$\begin{cases} x^5 = 5y^3 - 4z \\ y^5 = 5z^3 - 4x \\ z^5 = 5x^3 - 4y. \end{cases}$$

(Source of the problem and solution: PLMO LIX – Phase 1 – Problem 1.)

2.2.3. Solve the following system of equations, given that all variables involved are *positive* real numbers:

$$\begin{cases} a^3 + b^3 + c^3 = 3d^3 \\ b^4 + c^4 + d^4 = 3a^4 \\ c^5 + d^5 + a^5 = 3b^5. \end{cases}$$

(Source of the problem and solution: PLMO LV – Phase 2 – Problem 1.)

2.3 Solving via Inequalities

SOURCE

Problem: PLMO LVI – Phase 3 – Problem 4 (slightly modified)
Solution: our own

PROBLEM

Suppose that $n \in \mathbf{N} \setminus \{1\}$ and $C \in (-2, 2)$. Find all solutions $(x_1, x_2, \ldots, x_n)$ of the following equation:

$$\sum_{i=1}^{n} \sqrt{x_i^2 + C x_i x_{i+1} + x_{i+1}^2} = \sqrt{C+2} \sum_{i=1}^{n} x_i, \qquad (2.4)$$

where $x_i \in \mathbf{R}$ for $i \in [n]$ and $x_{n+1} = x_1$.

THEORY

Consider an equation $f(\mathbf{x}) = g(\mathbf{x})$, where $\mathbf{x} = (x_1, x_2, \ldots, x_n)$ is a vector consisting of n unknown variables. Suppose that our goal is to find all vectors $\mathbf{x}$ that satisfy the equation. One possible approach that often turns out to be useful is to start with showing that $f(\mathbf{x}) \leq g(\mathbf{x})$ (or that $f(\mathbf{x}) \geq g(\mathbf{x})$). There are a lot of techniques for achieving this, many of them we already discussed in Chapter 1. More importantly, if one does this step carefully, it often turns out that the result is **sharp**, that is, best possible. In other words, there are vectors $\mathbf{x} = (x_1, x_2, \ldots, x_n)$ that satisfy $f(\mathbf{x}) = g(\mathbf{x})$. But this is exactly what we wanted! Hence, after careful investigation of the obtained inequality, we usually see all the "bottlenecks" that prevented us from improving the bound even further, and so all the necessary conditions for equality become apparent.

SOLUTION

Let us note that the right hand side of (2.4) can be rewritten as follows:

$$\sqrt{C+2} \sum_{i=1}^{n} x_i = \frac{\sqrt{C+2}}{2} \sum_{i=1}^{n} (x_i + x_{i+1}) = \frac{\sqrt{C+2}}{2} \sum_{i=1}^{n} |x_i + x_{i+1}| - \varepsilon$$

$$= \sum_{i=1}^{n} \sqrt{\frac{(C+2)(x_i + x_{i+1})^2}{4}} - \varepsilon,$$

where

$$\varepsilon := \frac{\sqrt{C+2}}{2} \left(\sum_{i=1}^{n} |x_i + x_{i+1}| - \sum_{i=1}^{n} (x_i + x_{i+1}) \right) \geq 0,$$

as $C > -2$. Our goal is to solve the following equation:

$$0 = \sum_{i=1}^{n}\left(\sqrt{x_i^2 + Cx_ix_{i+1} + x_{i+1}^2} - \sqrt{\frac{(C+2)(x_i+x_{i+1})^2}{4}}\right) + \varepsilon$$

$$= \sum_{i=1}^{n}\frac{x_i^2 + Cx_ix_{i+1} + x_{i+1}^2 - \frac{(C+2)(x_i+x_{i+1})^2}{4}}{\sqrt{x_i^2 + Cx_ix_{i+1} + x_{i+1}^2} + \sqrt{\frac{(C+2)(x_i+x_{i+1})^2}{4}}} + \varepsilon, \qquad (2.5)$$

where in the last step we used a standard method of removing square roots, that is, using the fact that

$$\sqrt{a} - \sqrt{b} = \left(\sqrt{a} - \sqrt{b}\right) \cdot \frac{\sqrt{a} + \sqrt{b}}{\sqrt{a} + \sqrt{b}} = \frac{a - b}{\sqrt{a} + \sqrt{b}}.$$

Starting from the right hand side of (2.5), we get that

$$\sum_{i=1}^{n}\frac{x_i^2 + Cx_ix_{i+1} + x_{i+1}^2 - \frac{(C+2)(x_i+x_{i+1})^2}{4}}{\sqrt{x_i^2 + Cx_ix_{i+1} + x_{i+1}^2} + \sqrt{\frac{(C+2)(x_i+x_{i+1})^2}{4}}} + \varepsilon$$

$$= \frac{1}{4}\sum_{i=1}^{n}\frac{4x_i^2 + 4Cx_ix_{i+1} + 4x_{i+1}^2 - (C+2)x_i^2 - 2(C+2)x_ix_{i+1} - (C+2)x_{i+1}^2}{\sqrt{x_i^2 + Cx_ix_{i+1} + x_{i+1}^2} + \sqrt{\frac{(C+2)(x_i+x_{i+1})^2}{4}}} + \varepsilon$$

$$= \frac{1}{4}\sum_{i=1}^{n}\frac{(2-C)x_i^2 - (4-2C)x_ix_{i+1} + (2-C)x_{i+1}^2}{\sqrt{x_i^2 + Cx_ix_{i+1} + x_{i+1}^2} + \sqrt{\frac{(C+2)(x_i+x_{i+1})^2}{4}}} + \varepsilon$$

$$= \frac{2-C}{4}\sum_{i=1}^{n}\frac{(x_i - x_{i+1})^2}{\sqrt{x_i^2 + Cx_ix_{i+1} + x_{i+1}^2} + \sqrt{\frac{(C+2)(x_i+x_{i+1})^2}{4}}} + \varepsilon \geq 0,$$

since $C \in (-2, 2)$. As $(2-C)/4 > 0$, it is now obvious that the equality holds if and only if all x_i are equal, that is, $(x_1, x_2, \ldots, x_n) = (t, t, \ldots, t)$ for some $t \in \mathbf{R}$, and

$$\varepsilon = \frac{\sqrt{C+2}}{2}\left(|2t| - 2t\right) = 0.$$

It follows that all solutions are of the form $(x_1, x_2, \ldots, x_n) = (t, t, \ldots, t)$ for some $t \in \mathbf{R}_+ \cup \{0\}$.

REMARKS

Since the right hand side of (2.4) has the term $\sqrt{C+2}$, we immediately see that the assumption that $C > -2$ is needed and natural. On the other hand, the assumption that $C < 2$ is not needed and so we deduce that it has to be an important condition that affects the solution. Indeed, if $C = 2$, then the left hand side of (2.4) is equal to $\sum_{i=1}^{n}|x_i+x_{i+1}|$ and so it is equal to $2\sum_{i=1}^{n}x_i$, the right hand side, provided that $x_i + x_{i+1} \geq 0$ for all i. In particular, any vector $(x_1, x_2, \ldots, x_n)$ with non-negative coefficients is a solution to the system.

This reasoning leads us to consider the following equality

$$\sum_{i=1}^{n} \sqrt{x_i^2 + Cx_i x_{i+1} + x_{i+1}^2} = \sum_{i=1}^{n} \sqrt{(x_i + x_{i+1})^2 - (2 - C)x_i x_{i+1}},$$

and to analyze the problem in terms of $x_i + x_{i+1}$. Now, one can notice that the right hand side is easy to transform to the following form:

$$\sqrt{C + 2} \sum_{i=1}^{n} x_i = \frac{\sqrt{C + 2}}{2} \sum_{i=1}^{n} (x_i + x_{i+1}).$$

Since the sum on the left hand side has each term in the form of the square root, it is natural to represent the right hand side the same way, obtaining

$$\sqrt{C + 2} \sum_{i=1}^{n} x_i \leq \sum_{i=1}^{n} \sqrt{\frac{(C + 2)(x_i + x_{i+1})^2}{4}},$$

as we did in the solution. To see this we used the fact that $z \leq |z|$ for all $z \in \mathbf{R}$. The remaining of the solution follows naturally by grouping the terms using x_i and x_{i+1}.

Finally, let us note that for $C \in (-2, 2)$ we have that $x_i^2 + Cx_i x_{i+1} + x_{i+1}^2$ is always non-negative, so the original problem is well defined for all $x_i \in \mathbf{R}$. In order to see this, note that

$$x_i^2 + Cx_i x_{i+1} + x_{i+1}^2 = \left(x_i + \frac{C}{2} x_{i+1}\right)^2 + \left(1 - \frac{C^2}{4}\right) x_{i+1}^2 \geq 0,$$

since $C \in (-2, 2)$.

EXERCISES

2.3.1. Solve the following equation

$$(x^4 + 3y^2)\sqrt{|x + 2| + |y|} = 4|xy^2|,$$

provided that $x, y \in \mathbf{R}$.
(Source of the problem: PLMO LXIV – Phase 2 – Problem 4. Solution: our own.)

2.3.2. Solve the following system of equations, given that all variables involved are real numbers:

$$\begin{cases} 3(x^2 + y^2 + z^2) = 1 \\ x^2 y^2 + y^2 z^2 + z^2 x^2 = xyz(x + y + z)^3. \end{cases}$$

(Source of the problem and solution approach: PLMO XLVIII – Phase 3 – Problem 2.)

2.3.3. Solve the following system of equations, given that all variables involved are real numbers:

$$\begin{cases} x^2y + 2 = x + 2yz \\ y^2z + 2 = y + 2zx \\ z^2x + 2 = z + 2xy. \end{cases}$$

(Source of the problem and solution: PLMO LXIX – Phase 1 – Problem 3.)

2.4   Trigonometric Identities

SOURCE

Problem and solution: PLMO LX – Phase 3 – Problem 6

PROBLEM

Let n be any natural number such that $n \geq 2$. Suppose that a sequence of non-negative numbers $(c_0, c_1, \ldots, c_n)$ satisfies the following condition:

$$c_p c_s + c_r c_t = c_{p+r} c_{r+s}$$

for all non-negative integers p, r, s, t such that $p + r + s + t = n$. Find c_2 under the assumption that $c_1 = 1$.

THEORY

If one encounters a cyclic system of equations, it is often useful to transform it into another system using variable substitution with trigonometric functions. The reason why such substitution has a chance to work is that trigonometric functions are periodic which can be used to exploit a cyclic nature of the system of equations at hand. This is especially useful when one has to deal with a system of equations for which the number of equations is a variable parameter.

In the problem we aim to solve in this section, the system is *not* cyclic but exhibits enough other similarities to justify trying such substitutions of trigonometric functions.

| Trigonometric Functions | The **trigonometric functions** are real functions which relate an angle of a right-angled triangle to ratios of two side lengths. They are among the simplest periodic functions, and as such are also widely used for studying periodic phenomena. The most widely used trigonometric functions are the **sine**, the **cosine**, and the **tangent**. Their reciprocals are respectively the **cosecant**, the **secant**, and the **cotangent**.

The oldest definitions of trigonometric functions, related to right-angle triangles, define them only for acute angles. Given an acute angle α of a right-angled triangle, the **hypotenuse** h is the side that connects the two acute angles—see Figure 2.3. The side b **adjacent** to α is the side of the triangle that connects α to the right angle. The third side a is said to be **opposite** to α.

If the angle α is given, then all sides of the right-angled triangle are well defined, up to a scaling factor. This means that the ratio of any two side lengths depends only on α. These six ratios define six functions of α, which

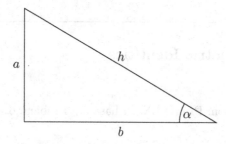

FIGURE 2.3: Classical definition of trigonometric functions.

are the trigonometric functions:

$$\sin(\alpha) \;=\; \frac{a}{h} \quad \text{(sine)}$$

$$\cos(\alpha) \;=\; \frac{b}{h} \quad \text{(cosine)}$$

$$\tan(\alpha) \;=\; \frac{a}{b} \quad \text{(tangent)}.$$

As a result, the reciprocal functions of cosine, sine, and tangent, are:

$$\csc(\alpha) \;=\; \frac{1}{\sin(\alpha)} = \frac{h}{a} \quad \text{(cosecant)}$$

$$\sec(\alpha) \;=\; \frac{1}{\cos(\alpha)} = \frac{h}{b} \quad \text{(secant)}$$

$$\cot(\alpha) \;=\; \frac{1}{\tan(\alpha)} = \frac{b}{a} \quad \text{(cotangent)}.$$

For extending these definitions to functions whose domain is the whole real line, one can use geometrical definitions using the standard unit circle (a circle with radius of 1)—see Figure 2.4.

In all the computations in this book, we always assume that the angle α is measured in **radians**, that is, the length of an arc of a unit circle defined by the angle and measured counter-clockwise. This is because radians are more "natural" and give more elegant formulation of a number of important results such as

$$\lim_{\alpha \to 0} \frac{\sin(\alpha)}{\alpha} = 1.$$

The trigonometric functions also have simple and elegant series expansions when radians are used. As a result, modern definitions express trigonometric functions as infinite series:

$$\sin(\alpha) \;=\; \alpha - \frac{\alpha^3}{3!} + \frac{\alpha^5}{5!} - \frac{\alpha^7}{7!} + \cdots$$

$$\cos(\alpha) \;=\; 1 - \frac{\alpha^2}{2!} + \frac{\alpha^4}{4!} - \frac{\alpha^6}{6!} + \cdots.$$

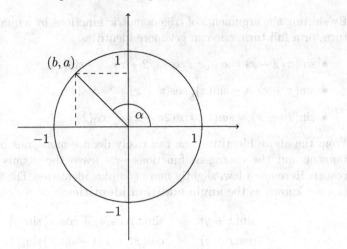

FIGURE 2.4: Definition of trigonometric functions in coordinate system. Note that $h = \sqrt{a^2 + b^2} = \sqrt{1} = 1$ and this time b is negative, so $\cos(\alpha) = b/h$ is also negative.

Trigonometric Identities The most useful trigonometric identities that can be exploited in such cases include the following basic relationship between the sine and the cosine that is called the **Pythagorean Identity**:

$$\sin^2(x) + \cos^2(x) = 1.$$

Dividing this identity by either $\sin^2(x)$ or $\cos^2(x)$ yields the other two Pythagorean identities:

$$1 + \tan^2(x) = \sec^2(x) \quad \text{and} \quad 1 + \cot^2(x) = \csc^2(x).$$

By examining the unit circle, we notice that reflections in the directions 0, $\pi/4$, $\pi/2$, and π generate equally looking results. As a result, the following properties of the trigonometric functions can be established:

- $\sin(-x) = -\sin(x),\ \cos(-x) = \cos(x)$;
- $\sin(\pi/2 - x) = \cos(x),\ \cos(\pi/2 - x) = \sin(x)$;
- $\sin(\pi - x) = \sin(x),\ \cos(\pi - x) = -\cos(x)$;
- $\sin(2\pi - x) = -\sin(x) = \sin(-x),\ \cos(2\pi - x) = \cos(x) = \cos(-x)$.

By shifting the arguments of trigonometric functions by a quarter turn, a half turn, or a full turn, one can get more identities:

- $\sin(\pi/2 + x) = \cos(x)$, $\cos(\pi/2 + x) = -\sin(x)$;
- $\sin(\pi + x) = -\sin(x)$, $\cos(\pi + x) = -\cos(x)$;
- $\sin(2\pi + x) = \sin(x)$, $\cos(2\pi + x) = \cos(x)$.

From the above identities, one can easily deduce analogous properties of the tangent and the cotangent functions—we leave the details details for the reader. Here are a few, slightly more complex identities. The first four identities are known as the **angle addition identities**:

$$\begin{aligned} \sin(x+y) &= \sin(x)\cos(y) + \cos(x)\sin(y), \\ \cos(x+y) &= \cos(x)\cos(y) - \sin(x)\sin(y); \end{aligned}$$

$$\begin{aligned} \tan(x+y) &= \frac{\tan(x) + \tan(y)}{1 - \tan(x)\tan(y)}, \\ \cot(x+y) &= \frac{\cot(x)\cot(y) - 1}{\cot(x) + \cot(y)}. \end{aligned}$$

The above formulas yield the **double-angle identities** by setting $x = y$. In particular,

$$\begin{aligned} \sin(2x) &= 2\sin(x)\cos(x) = \frac{2\tan(x)}{1 + \tan^2(x)}, \\ \cos(2x) &= \cos^2(x) - \sin^2(x) = 2\cos^2(x) - 1 = 1 - 2\sin^2(x). \end{aligned}$$

Let us finish with the **product-to-sum** and the **sum-to-product identities**:

$$\begin{aligned} \cos(x)\cos(y) &= \frac{\cos(x-y) + \cos(x+y)}{2}, \\ \sin(x)\sin(y) &= \frac{\cos(x-y) - \cos(x+y)}{2}, \\ \sin(x)\cos(y) &= \frac{\sin(x+y) + \sin(x-y)}{2}; \end{aligned}$$

$$\begin{aligned} \sin(x) + \sin(y) &= 2\sin\left(\frac{x+y}{2}\right)\cos\left(\frac{x-y}{2}\right), \\ \cos(x) + \cos(y) &= 2\cos\left(\frac{x+y}{2}\right)\cos\left(\frac{x-y}{2}\right), \\ \cos(x) - \cos(y) &= -2\sin\left(\frac{x+y}{2}\right)\sin\left(\frac{x-y}{2}\right). \end{aligned}$$

SOLUTION

By considering $p = n$ and $r = s = t = 0$, we get that $c_n c_0 + c_0^2 = c_n c_0$, which implies that $c_0 = 0$. Now, by considering $0 \leq p \leq n/2$, $s = p$, $t = 0$, and $r = n - 2p$, we get that $c_p^2 = c_{n-p}^2$. Since c_i are non-negative, we get that $c_p = c_{n-p}$ for each $0 \leq p \leq n/2$. In fact, by symmetry, $c_p = c_{n-p}$ for each $0 \leq p \leq n$. In particular, we have $c_n = c_0 = 0$ and $c_{n-1} = c_1 = 1$. Finally, consider $p = s = 1$, $r \in [n-2] \cup \{0\}$, and $t = n - r - 2$ to get that

$$c_{r+1}^2 = c_1^2 + c_r c_{n-r-2} = 1 + c_r c_{r+2}.$$

In particular, since all c_i are non-negative, it implies that $c_{r+1}^2 \geq 1$ and so $c_{r+1} \geq 1$. It follows that all c_i, except c_0 and c_n, are at least 1. We will not need this property but, in order to build an intuition, let us mention that with this stronger property at hand we get that for each $r \in [n-3]$ we have $c_{r+1}^2 \geq 2$ and so all c_i, except c_0, c_1, c_{n-1}, and c_n, are at least $\sqrt{2}$. One may continually recurse this argument to get a sequence of lower bounds.

From the identity above we get that for each $r \in [n-2]$, we have that

$$\frac{c_{r+2}}{c_{r+1}} = \frac{c_{r+1} - 1/c_{r+1}}{c_r} \tag{2.6}$$

(note that we excluded $r = 0$ as $c_0 = 0$). Let us summarize what we know so far: $c_0 = c_n = 0$, $c_1 = c_{n-1} = 1$, and (2.6) holds for each $r \in [n-2]$. Moreover, if there is a unique c_2 that satisfies the desired property, then in fact the whole sequence is defined uniquely. Indeed, by (2.6) we see that c_{r+2} is determined by c_{r+1} and c_r.

We will show that indeed the sequence is unique. For a contradiction, let us suppose that there are two sequences $(0, 1, c_2, c_3, \ldots, c_{n-2}, 1, 0)$ and $(0, 1, d_2, d_3, \ldots, d_{n-2}, 1, 0)$ that satisfy the desired property and $d_2 \neq c_2$. Without loss of generality, we may assume that $d_2 > c_2$. We will prove by (strong) induction that for $r \in [n-2]$ we have that $d_{r+1} > c_{r+1}$ and that $d_{r+1}/d_r > c_{r+1}/c_r$. This, in particular, will imply that $d_{n-1} > c_{n-1}$ which will give us the desired contradiction as $d_{n-1} = c_{n-1} = 1$.

The base case ($r = 1$) clearly holds; by our assumption, $d_2/d_1 = d_2 > c_2 = c_2/c_1$ and so both inequalities hold. For the inductive step, assume that both $d_{s+1} > c_{s+1}$ and $d_{s+1}/d_s > c_{s+1}/c_s$ hold for *all* $1 \leq s \leq r \leq n - 3$. In this case, we can use (2.6) to get that

$$\frac{d_{r+2}}{d_{r+1}} = \frac{d_{r+1} - 1/d_{r+1}}{d_r} = \frac{d_{r+1}}{d_r} - \frac{1}{d_{r+1}d_r}$$
$$> \frac{c_{r+1}}{c_r} - \frac{1}{c_{r+1}c_r} = \frac{c_{r+2}}{c_{r+1}}.$$

But this also immediately gives that $d_{r+2} > c_{r+2}$. This finishes the inductive proof and gives the desired contradiction. It follows that the sequence $(c_0, c_1, \ldots, c_n)$ is defined uniquely.

We will now show that the sequence $c_i = \sin(\pi i/n)/\sin(\pi/n)$, $i \in [n] \cup \{0\}$ satisfies the desired property, and so this is the only solution to our problem. Indeed, note that $c_0 = 0$, $c_1 = 1$, and for all non-negative integers p, r, s, t such that $p + r + s + t = n$, we have that

$$
\begin{aligned}
c_p c_s + c_r c_t &= \frac{\sin(\pi p/n)\sin(\pi s/n) + \sin(\pi r/n)\sin(\pi t/n)}{\sin^2(\pi/n)} \\[2mm]
&= \frac{\cos\left(\frac{\pi(p-s)}{n}\right) + \cos\left(\frac{\pi(p+s)}{n}\right) + \cos\left(\frac{\pi(r-t)}{n}\right) + \cos\left(\frac{\pi(r+t)}{n}\right)}{2\sin^2(\pi/n)} \\[2mm]
&= \frac{\cos\left(\frac{\pi(p-s)}{n}\right) + \cos\left(\frac{\pi(r-t)}{n}\right)}{2\sin^2(\pi/n)},
\end{aligned}
$$

since $\cos(\pi(r+t)/n) = \cos(\pi - (p+s)/n) = -\cos((p+s)/n)$. We continue in the following way, noting that $t = n - p - r - s$:

$$
\begin{aligned}
c_p c_s + c_r c_t &= \frac{\cos\left(\frac{\pi(p-s+r-(n-p-s-r))}{2n}\right)\cos\left(\frac{\pi(p-s-r+(n-p-s-r))}{2n}\right)}{\sin^2(\pi/n)} \\[2mm]
&= \frac{\cos\left(\frac{\pi(2p+2r-n)}{2n}\right)\cos\left(\frac{\pi(-2s-2r+n)}{2n}\right)}{\sin^2(\pi/n)} \\[2mm]
&= \frac{\sin\left(\frac{\pi(p+r)}{n}\right)\sin\left(\frac{\pi(s+r)}{n}\right)}{\sin^2(\pi/n)} = c_{p+r}c_{s+r}.
\end{aligned}
$$

This shows the desired equality, thus completing the proof.

REMARKS

Let us note that the solution of our problem is clearly divided into two separate steps: the proof of the uniqueness of the solution and then finding the actual sequence that satisfies the desired property. The main difficulty is the second part, and initially it is not clear how to guess a possible solution. From the first part, we have learnt that the sequence starts from $c_0 = 0$, then increases to $c_1 = 1$, and then $c_2 \geq \sqrt{2}$. However, at some point it must start decreasing as it is symmetric: $c_p = c_{n-p}$ for any $0 \leq p \leq n$. It is natural to guess that the sequence is first increasing, and then after reaching a "turning point" the monotonicity of the sequence changes. However, at this point it is only a conjecture. Now, being aware that trigonometric functions are possible substitutions, the natural guess is to use $c_i = \sin(\pi i/n)$ as it satisfies the requirements and is also non negative for each $i \in [n] \cup \{0\}$. However, since $c_1 = 1$, we have to normalize it by dividing by $\sin(\pi/n)$. All that is left now is to check that this guess satisfies the original equation, which we leave for the reader.

EXERCISES

2.4.1. Let $n \geq 2$ be any natural number. Find the number of sequences $(x_1, x_2, \ldots, x_n)$ of non-negative real variables that satisfy the following system of equations: for $i \in [n]$

$$x_{i+1} + x_i^2 = 4x_i,$$

where $x_{n+1} = x_1$.
(Source of the problem and solution: PLMO LI – Phase 3 – Problem 1.)

2.4.2. Let $n \in \mathbf{N}$. Find all solutions of the equation

$$|\tan(x)^n - \cot(x)^n| = 2n|\cot(2x)|.$$

(Source of the problem and solution: PLMO XLIX – Phase 1 – Problem 5, slightly modified.)

2.4.3. For a given $a \in \mathbf{R}$, let us recursively define the following sequence: $x_0 = \sqrt{3}$ and for all non-negative integers n,

$$x_{n+1} = \frac{1 + ax_n}{a - x_n}.$$

Find all values of a for which the sequence has a period equal to 8.
(Source of the problem and solution: PLMO LVI – Phase 1 – Problem 9.)

2.5   Number of Solutions

SOURCE

Problem and solution approach: PLMO LXII – Phase 3 – Problem 3

PROBLEM

Suppose that n is an odd natural number. Find the number of real solutions of the following system of equations:

$$\begin{cases} x_1(x_1 + 1) & = & x_2(x_2 - 1) \\ x_2(x_2 + 1) & = & x_3(x_3 - 1) \\ & \vdots & \\ x_{n-1}(x_{n-1} + 1) & = & x_n(x_n - 1) \\ x_n(x_n + 1) & = & x_1(x_1 - 1). \end{cases}$$

THEORY

There are many interesting problems in which one is asked to count the number of solutions to a given system of equations without actually finding any solution. As mentioned at the beginning of this chapter, one can distinguish four cases and each of them usually requires a different approach: there is no solution, there is exactly one solution, there are many solutions but a finite number of them, and there are infinitely many solutions. When more than one solution is expected, but finitely many, then typically the proof strategy requires the following three steps:

- deriving the rule for candidate solutions (that is, finding all solutions that meet the desired conditions of the problem);

- showing that any two candidate solutions are distinct;

- counting the candidate solutions.

For the last step, it is often useful to use the powerful technique called **double counting** that we discuss in detail in Section 4.3. The idea is to construct a bijection from the set of solutions to some other set that is easier to count.

In order to illustrate the double counting technique, let us consider the following problem. Suppose that k and n are natural numbers such that $k \leq n$. Our goal is to compute the number of solutions of the equation $\sum_{i=1}^{k} x_i = n$, provided that $x_i \in \mathbf{N}$ for each $i \in [k]$. We will show that there are $\binom{n-1}{k-1}$ distinct solutions. In order to see this, let us consider any solution $(x_1, x_2, \ldots, x_k)$ of the equation $\sum_{i=1}^{k} x_i = n$. For any $j \in [k]$, let us define $y_j := \sum_{i=1}^{j} x_i$. It is clear that $(y_1, y_2, \ldots, y_k)$ is an increasing sequence of natural numbers and $y_k = n$. More importantly, each increasing sequence $(y_1, y_2, \ldots, y_{k-1})$ of natural numbers, whose terms are bounded above by $n - 1$, uniquely yields one solution. In other words, there is a bijection from the set of solutions we want to count and the set of increasing sequences of natural numbers that are at most $n - 1$. The former one seems difficult to count but the latter one is easy to count as each increasing sequence is uniquely defined by an $(k-1)$-element subset of the set $[n - 1]$. Since there are $\binom{n-1}{k-1}$ subsets of size $k - 1$ selected from the set of $n - 1$ elements, the proof is finished.

Another standard and well-known counting problem is to count the number of ways to make change for a dollar. Our goal is to make a change of \$1 using ¢1, ¢5, ¢10, and ¢25 coins. There are many ways to solve this problem, including the one using generating functions which we discuss in Section 4.9. Here, we use a longer but elementary approach. Let x_n^1 be the number of ways one can make change of n cents using only ¢1 coins. Let x_n^2 be the number of ways one can make change of n cents using only ¢1 and ¢5 coins. Let x_n^3 be the number of ways one can make change of n cents using only ¢1, ¢5, and ¢10 coins. Finally, let x_n^4 be the number of ways one can make change of n cents using ¢1, ¢5, ¢10 and ¢25 coins. Our goal is to find x_{100}^4.

Clearly, $x_n^1 = 1$ for each non-negative integer n (including $x_0^1 = 1$ that is vacuously true—there is exactly one way to make a change of 0 cents, namely, do nothing). Moreover, we have $x_n^2 = \lfloor n/5 \rfloor + 1$, as at most $\lfloor n/5 \rfloor$ ¢5 coins can be dispensed and so there are $\lfloor n/5 \rfloor + 1$ ways to do it (including not using ¢5 coins at all). The remaining amount is uniquely delt with ¢1 coins.

We are now ready to compute x_{100}^4. We first observe that

$$x_{100}^4 = x_{100}^3 + x_{75}^3 + x_{50}^3 + x_{25}^3 + 1,$$

by independently considering 5 ways to use ¢25 coins. Similarly, by considering ¢10 coins, we get that

$$
\begin{aligned}
x_{25}^3 &= x_{25}^2 + x_{15}^2 + x_5^2 = 6 + 4 + 2 = 12 \\
x_{50}^3 &= x_{50}^2 + x_{40}^2 + x_{30}^2 + x_{20}^2 + x_{10}^2 + x_0^2 \\
&= 11 + 9 + 7 + 5 + 3 + 1 = 36 \\
x_{75}^3 &= x_{75}^2 + x_{65}^2 + x_{55}^2 + x_{45}^2 + x_{35}^2 + x_{25}^3 \\
&= 16 + 14 + 12 + 10 + 8 + 12 = 72 \\
x_{100}^3 &= x_{100}^2 + x_{90}^2 + x_{80}^2 + x_{70}^2 + x_{60}^2 + x_{50}^3 \\
&= 21 + 19 + 17 + 15 + 13 + 36 = 121.
\end{aligned}
$$

It follows that $x_{100}^4 = 121 + 72 + 36 + 12 + 1 = 242$.

Finally, let us mention that Larry King said in his *USA Today* column that there are 293 ways to make change for a dollar. Why did he get a different number? He included ¢50 coins! Though not commonly used today, half-dollar coins have a long history of heavy use alongside other denominations of coinage, but have faded out of general circulation for many reasons. With this additional coin available, the number of ways to make change for a dolar is

$$x_{100}^4 + x_{50}^4 + 1 \;=\; 242 + 49 + 1 \;=\; 292\,,$$

since

$$x_{50}^4 \;=\; x_{50}^3 + x_{25}^3 + 1 \;=\; 12 + 36 + 1 \;=\; 49\,.$$

We are still missing one way! The reason for that is Larry King included also a dollar coin which seems controversial. For example, Walter Wright said that a dollar coin cannot be considered change for a dollar bill, arguing after *Webster's New World Dictionary* that defines change as "a number of coins or bills whose total value equals a single larger coin or bill."

SOLUTION

Let n be any odd natural number. For convenience, let us use the convention that $x_{n+1} = x_1$. Then, for each $i \in [n]$ we can re-write the equation $x_i(x_i+1) = x_{i+1}(x_{i+1} - 1)$ as follows: $(x_i + x_{i+1})(x_{i+1} - x_i - 1) = 0$. It follows that $x_{i+1} = -x_i$ or $x_{i+1} = x_i + 1$. We may set $s_i = -1$ if $x_{i+1} = -x_i$ and set $s_i = 1$ if $x_{i+1} = x_i + 1$. As a result, the sequence $\mathbf{x} := (x_1, x_2, \ldots, x_n, x_{n+1})$ can be uniquely determined by the first term, x_1, and the sequence $\mathbf{s} := (s_1, s_2, \ldots, s_n)$. Indeed, since $x_{i+1} = s_i x_i + (s_i + 1)/2$ for each $i \in [n]$, we get that for each $\ell \in [n]$

$$x_{\ell+1} \;=\; x_1 \prod_{i=1}^{\ell} s_i + \sum_{i=1}^{\ell} \frac{s_i + 1}{2} \prod_{j=i+1}^{\ell} s_j\,.$$

There is clearly an infinite number of sequences $\mathbf{x}$ of this form but not all of them satisfy our system as we also require that $x_1 = x_{n+1}$, that is,

$$x_1 \;=\; x_{n+1} \;=\; x_1 \prod_{i=1}^{n} s_i + \sum_{i=1}^{n} \frac{s_i + 1}{2} \prod_{j=i+1}^{n} s_j\,. \tag{2.7}$$

Recall that n is odd. We will show that the number of terms in $\mathbf{s}$ that are equal to 1 is even. For a contradiction, suppose that it is *not* true, that is, the number of terms that are equal to -1 is even. In particular, $\prod_{i=1}^{n} s_i = 1$ and so (2.7) reduces to

$$0 \;=\; \sum_{i=1}^{n} \frac{s_i + 1}{2} \prod_{j=i+1}^{n} s_j\,.$$

However, since the number of terms for which $(s_i + 1)/2 = 1$ is odd and $(s_i + 1)/2 = 0$ otherwise, we deduce that the right hand side of the above equality is the sum of odd number of terms, each of them from $\{-1, 1\}$, and so it is *not* equal to zero. We get the desired contradiction and so, indeed, the number of terms in **s** that are equal to 1 is even. Since $\prod_{i=1}^{n} s_i = -1$, the required condition (2.7) implies that

$$x_1 = \sum_{i=1}^{n} \frac{s_i + 1}{4} \prod_{j=i+1}^{n} s_j \,,$$

and so x_1 is uniquely determined by the sequence **s** (and so is the whole sequence **x**). In fact, x_1 is the sum of even number of terms, each of which is equal to $1/2$ or $-1/2$, so x_1 is an integer (and so are all the entries of **x**).

Our next task is to show that each sequence **s** with an even number of 1's yields a unique sequence **x**. For a contradiction, suppose that $\mathbf{s} = (s_1, s_2, \ldots, s_n)$ and $\mathbf{s}' = (s_1', s_2', \ldots, s_n')$ yield the same sequence $\mathbf{x} = (x_1, x_2, \ldots, x_n)$. Since **s** and **s**' are different, let $i \in [n]$ be such that $s_i \neq s_i'$. Without loss of generality, we may assume that $s_i = 1$ and $s_i' = -1$. On the one hand, we get that $x_{i+1} = s_i x_i + (s_i + 1)/2 = x_i + 1$. On the other hand, $x_{i+1} = s_i' x_i + (s_i' + 1)/2 = -x_i$. It follows that $x_i + 1 = -x_i$ and so $x_i = -1/2 \notin \mathbf{Z}$. We get the desired contradiction (since $x_i \in \mathbf{Z}$ for all $i \in [n]$) and so there is a bijection between the sequences with even number of 1's and the set of solutions of our system.

It remains to count the number of such sequences **s**. We note that in order to generate a sequence of 1's and -1's with even number of 1's, one can take any sequence $(s_1, s_2, \ldots, s_{n-1})$ of 1's and -1's of length $n-1$ (there are clearly 2^{n-1} of them) and then the last term, s_n, is uniquely determined (in order to keep the number of 1's even). We conclude that there are 2^{n-1} solutions to the system of equations we deal with.

REMARKS

In the problem we deal with in this section, the key idea was to introduce the sequence s_i and write down an explicit solution to the recurrence relation for x_i. The sequence s_i is often called a **control sequence** of the sequence x_i.

The solution we have found is relatively complex. The idea is quite straightforward but there are many places and calculations where one can easily make a mistake. In such cases, to be on a safe side, it is strongly recommended to re-check some cases manually, or use computer to re-compute the number of solutions for some instances. Here is a program written in Julia that allows us to re-check the solution.

```
function find_sequence(s)
    x1 = sum((s[i]+1)/4*prod(s[i+1:end]) for i in 1:length(s))
    x = Int[x1]
    for si in s
        push!(x, si*x[end] + (si+1)/2)
    end
    println(x)
end
```

Now, we can test our solution for all scenarios, provided $n = 3$:

```julia
julia> find_sequence([1, 1, -1])
[-1, 0, 1, -1]

julia> find_sequence([1, -1, 1])
[0, 1, -1, 0]

julia> find_sequence([-1, 1, 1])
[1, -1, 0, 1]

julia> find_sequence([-1, -1, -1])
[0, 0, 0, 0]
```

And let us check just one more solution, this time for $n = 7$:

```julia
julia> find_sequence([1, 1, 1, 1, -1, 1, 1])
[-1, 0, 1, 2, 3, -3, -2, -1]
```

EXERCISES

2.5.1. For a given $a \in \mathbf{R}$, consider the following system of equations:

$$\begin{cases} x + y^2 + z^2 = a \\ x^2 + y + z^2 = a \\ x^2 + y^2 + z = a. \end{cases}$$

Find the number of real solutions (x, y, z) of this system as a function of a. (Source of the problem: PLMO XLVIII – Phase 2 – Problem 1. Solution: our own.)

2.5.2. Solve the following system of equations, given that all variables involved are positive real numbers:

$$(x^{2010} - 1)(y^{2009} - 1) = (x^{2009} - 1)(y^{2010} - 1).$$

(Source of the problem and solution: PLMO LXI – Phase 1 – Problem 1.)

2.5.3. Fix an integer $n \geq 2$, and consider the following system of n equations: for $i \in [n]$

$$x_{i+1}^2 + x_i^2 + 50 = 12x_{i+1} + 16x_i.$$

(As usual, we use the convention that $x_{n+1} = x_1$.) Find the number of solutions of this system, given that all the variables involved are integers. (Source of the problem and solution: PLMO L – Phase 3 – Problem 4.)

2.6 Sequence Invariants

SOURCE

Problem and solution approach: PLMO LVIII – Phase 3 – Problem 6

PROBLEM

Let $a_0 = -1$ and for each $n \in \mathbf{N}$ let

$$a_n = -\sum_{i=0}^{n-1} \frac{a_i}{n+1-i}.$$

Show that $a_n > 0$ for all $n \in \mathbf{N}$.

THEORY

An **invariant** of a sequence $\mathbf{s} = (s_i)_{i \geq 1}$ is a statement or predicate, denoted $p = p(s_i)$, such that $p(s_i)$ is true for all $i \in \mathbf{N}$. In our case we are asked to prove that the invariant is $a_i > 0$ for all $i \in \mathbf{N}$. The usual technique of proving invariants is via induction.

Perhaps surprisingly, it is quite common that proving a stronger invariant than the one originally suggested is simpler. Let us consider an example. The following sequence is defined recursively: $x_1 = 1$ and for each $n \in \mathbf{N}$ we have

$$x_{n+1} = x_n - \frac{1}{n(n+1)}.$$

Our goal is to prove that $x_n > 0$ for all n.

In this example, it is not enough to use the fact that $x_n > 0$ to prove that $x_{n+1} > 0$. Hence, this natural strategy will simply not work. On the other hand, a stronger property, namely that $x_n = 1/n$ for all n, is easy to prove by induction. Indeed, the base case clearly holds: $x_1 = 1 = 1/1$. For the inductive step, from the fact that $x_n = 1/n$ we easily get that $x_{n+1} = x_n - 1/(n(n+1)) = 1/n - 1/(n(n+1)) = 1/(n+1)$. The stronger property holds by induction; in particular, all the terms of the sequence are positive.

| **Squeeze Theorem** | Let us mention about one more useful tool that will be needed to solve one of the exercises. The **squeeze theorem**, also known as the **sandwich theorem**, is a theorem regarding the limit of a function or a sequence. It is used to confirm the limit of a function via comparison with two other functions whose limits are known or which can be easily computed.

Let I be a set having the point a as a limit point, that is, there exists a sequence of elements of I which converges to a. Let f, g, and h be functions defined on I, except possibly at a itself. Suppose that for every x in I not equal to a, we have

$$g(x) \leq f(x) \leq h(x)$$

and also suppose that

$$\lim_{x \to a} g(x) = \lim_{x \to a} h(x) = L.$$

In this case, $\lim_{x \to a} f(x) = L$. Let us note that a is *not* required to be an element from I. Indeed, if a is an endpoint of an open interval I, then the above limits are left-hand or right-hand limits. A similar statement holds for unbounded sets; for example, if $I = (0, \infty)$, then the conclusion holds taking the limits as $x \to \infty$.

As already mentioned, this theorem is also valid for sequences that corresponds to the case with $I = \mathbf{N}$ and $a = \infty$. Let $(x_n)_{n \in \mathbf{N}}$ and $(z_n)_{n \in \mathbf{N}}$ be two sequences converging to L. Suppose that $(y_n)_{n \in \mathbf{N}}$ is a sequence that satisfy the following property:

$$x_n \le y_n \le z_n$$

for all $n \ge N$, where $N \in \mathbf{N}$. Then, $(y_n)_{n \in \mathbf{N}}$ also converges to L.

Finally, let us mention that in many languages (for example, French, German, Italian and Russian), the squeeze theorem is also known as the two policemen and a drunk theorem. The story is that if two policemen are escorting a drunk prisoner between them, and both officers go to a cell, then (regardless of the path taken, and the fact that the prisoner may be wobbling about between the policemen) the prisoner must also end up in the cell.

SOLUTION

We will prove the desired property by induction on n. The base case is easy to verify: $a_1 = 1/2 > 0$. For the inductive step, suppose that $a_n > 0$ for some $n \in \mathbf{N}$. Our goal is to show that $a_{n+1} > 0$. Since $a_0 = -1$, we get that

$$a_{n+1} = -\sum_{i=0}^{n} \frac{a_i}{n+2-i} = \frac{1}{n+2} - \sum_{i=1}^{n-1} \frac{a_i}{n+2-i} - \frac{a_n}{2}.$$

Since

$$\frac{n+2}{n+1} \cdot \frac{1}{n+2-i} \le \frac{1}{n+1-i},$$

it follows that

$$a_{n+1} \ge \frac{1}{n+2} - \frac{n+1}{n+2} \sum_{i=1}^{n-1} \frac{a_i}{n+1-i} - \frac{a_n}{2}.$$

Since $a_0 = -1$, we get that

$$\sum_{i=1}^{n-1} \frac{a_i}{n+1-i} = \frac{1}{n+1} + \sum_{i=0}^{n-1} \frac{a_i}{n+1-i} = \frac{1}{n+1} - a_n,$$

and so

$$a_{n+1} \geq \frac{1}{n+2} - \frac{n+1}{n+2}\left(\frac{1}{n+1} - a_n\right) - \frac{a_n}{2} = a_n\left(\frac{n+1}{n+2} - \frac{1}{2}\right) > 0.$$

The proof by induction is finished.

REMARKS

The key idea in the proof was to try to replace the sum involving all a_i by the formula that involved only the previous term in the sequence. Such an approach is possible as the previous term was also defined in terms of the same (but one) terms a_i.

In our problem, in order to use this approach we had to transform $\frac{1}{n+2-i}$ into $\frac{1}{n+1-i}$ for $i > 1$ in a way that does not depend on i. To achieve this, we investigated the following ratio

$$\frac{1/(n+2-i)}{1/(n+1-i)} = \frac{n+1-i}{n+2-i} \leq \frac{n+1}{n+2},$$

and we checked that this universal bound is enough to derive the required claim.

EXERCISES

2.6.1. Let x_1 be any positive real number, and for each $n \in \mathbf{N}$ let

$$x_{n+1} = x_n + \frac{1}{x_n^2}.$$

Prove that $x_n/\sqrt[3]{n}$ has a limit and then find it.
(Source of the problem: PLMO L – Phase 1 – Problem 10. Solution: our own.)

2.6.2. Consider the following sequence defined recursively: $a_1 = 4$ and for each $n \in \mathbf{N}$, let $a_{n+1} = a_n(a_n - 1)$. Moreover, for each $n \in \mathbf{N}$, let $b_n = \log_2(a_n)$ and $c_n = n - \log_2(b_n)$. Prove that c_n is bounded.
(Source of the problem and solution: PLMO XLIX – Phase 1 – Problem 3.)

2.6.3. You are given two numbers $a, b \in \mathbf{R}$. Let $x_1 = a$, $x_2 = b$, and for each $n \in \mathbf{N}$ let $x_{n+2} = x_{n+1} + x_n$. Show that there exist $a, b \in \mathbf{R}$, $a \neq b$, for which there are at least 2,000 distinct pairs (k, ℓ), $k < \ell$, such that $x_k = x_\ell$. On the other hand, the number of such pairs is finite even if $a = b$, unless $a = b = 0$.
(Source of the problem: PLMO LII – Phase 3 – Problem 3, slightly modified. Solution: our own.)

2.7   Solving Sequences

SOURCE

Problem and solution idea: PLMO LXI – Phase 3 – Problem 6

PROBLEM

Let $C_0 > 1$ be a real number. Suppose that a sequence $(a_n)_{n \in \mathbf{N}}$ of positive real numbers satisfies the following properties: $a_1 = 1$, $a_2 = 2$, and for each $m, n \in \mathbf{N}$, we have that

$$a_{mn} = a_m a_n \quad \text{and} \quad a_{m+n} \leq C_0(a_m + a_n).$$

Prove that for each $n \in \mathbf{N}$, $a_n = n$.

THEORY

In this section, we concentrate on the following important types of sets, open sets and closed sets. In general, such sets can be very abstract but in practice, open sets are usually chosen to be similar to the open intervals of the real line. We will restrict ourselves to Euclidean space.

$\boxed{\text{Euclidean Space}}$ **Euclidean space** is the fundamental space of geometry. Originally, this was the three-dimensional space but in modern mathematics there are Euclidean spaces of any dimension that is a natural number, including the three-dimensional space, the two-dimensional Euclidean plane, and one-dimensional real line. One way to think of the Euclidean space is as a set of points satisfying certain relationships, expressible in terms of distance and angles. In Cartesian coordinates, if $\mathbf{p} = (p_1, p_2, \ldots, p_n) \in \mathbf{R}^n$ and $\mathbf{q} = (q_1, q_2, \ldots, q_n) \in \mathbf{R}^n$ are two points in Euclidean n-space, then the distance from p to q (or from q to p) is given by the Pythagorean formula:

$$d(\mathbf{p}, \mathbf{q}) = d(\mathbf{q}, \mathbf{p}) = \sqrt{\sum_{i=1}^{n}(q_i - p_i)^2}.$$

The following definitions will play an important role. An **(open)** n-**ball of radius** $r \in \mathbf{R}_+$ and centered at a point $p \in \mathbf{R}^n$, usually denoted by $B_r(p)$, is the set of all points at distance less than r from x. That is,

$$B_r(p) := \{x \in \mathbf{R}^n \mid d(x, p) < r\}.$$

A **closed** n-**ball of radius** $r \in \mathbf{R}_+ \cup \{0\}$, which is denoted by $B_r[p]$, is the set of all points at distance less than or equal to r away from p. In other words,

$$B_r[p] := \{x \in \mathbf{R}^n \mid d(x, p) \leq r\}.$$

Note that a ball (open or closed) always includes p itself. A subset of some space is **bounded** if it is contained in some ball.

| **Open and Closed Sets** | An open set is an abstract concept generalizing the idea of an open interval in the real line. One of the simplest examples are sets which contain a ball around each of their points but, as already mentioned above, an open set, in general, can be very abstract: any collection of sets can be called open, as long as the union of an arbitrary number of open sets is open, the intersection of a finite number of open sets is open, and the space itself is open. These conditions are very loose, and they allow enormous flexibility in the choice of open sets.

We restrict ourselves to Euclidean spaces. A subset S of the Euclidean n-space $\mathbf{R}^n$ is called **open** if for any point $x \in S$, there exists a real number $\varepsilon = \varepsilon(x) > 0$ such that $B_\varepsilon(x) \subseteq S$. One can show that the three desired properties are satisfied: a) the union of any number of open sets, or infinitely many open sets, is open, b) the intersection of a finite number of open sets is open, and c) $\mathbf{R}^n$ itself is open. On the other hand, note that infinite intersections of open sets need *not* be open. For example, the intersection of all intervals of the form $(-1/n, 1/n)$, where $n \in \mathbf{N}$, is the set $\{0\}$ which is *not* open in the real line $\mathbf{R}^1$.

The closure of a set S consists of all points in S together with all limit points of S. Formally, for a given set $S \subseteq \mathbf{R}^n$, x is a point of closure of S if every open ball centered at x contains a point of S (this point may be x itself), that is, for all $\varepsilon > 0$, $B_\varepsilon(x) \cap S \neq \emptyset$. The definition of a point of closure is closely related to the definition of a **limit point**. The difference between the two definitions is subtle but important, namely, in the definition of limit point, every neighborhood of the point x in question must contain a point of the set other than x itself. As a result, every limit point is a point of closure, but not every point of closure is a limit point. Finally, the **boundary** of a set S is the set of points which can be approached both from S and from the outside of S.

In general, a **closed set** is a set whose complement is an open set. However, there are other equivalent definitions that can be applied to Euclidean spaces. A set is closed if and only if it coincides with its closure. Equivalently, a set is closed if and only if it contains all of its limit points. Yet another equivalent definition is that a set is closed if and only if it contains all of its boundary points. Any intersection of closed sets is closed (including intersections of infinitely many closed sets). The union of finitely many closed sets is closed.

Note that both the empty set and the whole Euclidean space are both open and closed. Let us also mention that a set is **connected** if it *cannot* be represented as the union of two or more disjoint non-empty open sets.

Partially Ordered Sets In order to be able to introduce the definitions of infimum and supremum, we need to formalize the intuitive concept of an ordering of the elements of a given set. It will allow us to say that, for certain pairs of elements in the set, one of the elements precedes the other in the ordering. Note that the relation is called a "partial order" which immediately indicates that not every pair of elements needs to be comparable, that is, there may be pairs of elements for which neither element precedes the other in the partially ordered set. Partial orders thus generalize total orders, in which every pair is comparable.

Formally, let R be any **binary relation**, that is, a relation from some set S to the same set S. Such relations can be represented as a subset of $S \times S$. Indeed, element $x \in S$ is related to element $y \in S$ if and only if $(x, y) \in R$. A **partial order (poset)** is any binary relation $R \subseteq S \times S$ that satisfies the following three properties:

- R is **reflexive**: for any $x \in S$, x is related to x (each element is comparable to itself),

- R is **antisymmetric**: for any $x, y \in S$, if x is related to y and y is related to x, then $x = y$ (no two different elements precede each other),

- R is **transitive**: for any $x, y, z \in S$, if x is related to y and y is related to z, then x is related to z (the start of a chain of precedence relations must precede the end of the chain).

There are many interesting and important partial orders but here we only focus on relations on real numbers or vectors of real numbers. Let us first observe that the real numbers ordered by the standard less-than-or-equal relation $\leq$ is a partial order. (In fact, it is a totally ordered set as for any $x, y \in \mathbf{R}$, $x \leq y$ or $y \leq x$.) Indeed, $x \leq x$ for any $x \in R$, and so $\leq$ is reflexive. Since $x \leq y$ and $y \leq x$ implies that $x = y$, the relation is also antisymmetric. Finally, if $x \leq y$ and $y \leq z$, then $x \leq z$ and so $\leq$ is transitive.

For $\mathbf{x}, \mathbf{y} \in \mathbf{R}^n$, we usually say that

$$\mathbf{x} = (x_1, x_2, \ldots, x_n) \leq (y_1, y_2, \ldots, y_n) = \mathbf{y}$$

if and only if $x_i \leq y_i$ for all $i \in [n]$. We leave it for the reader to check that this relation is a partial order. However, note that it is *not* a total order! For example, for $\mathbf{x} = (1, 2) \in \mathbf{R}^2$ and $\mathbf{y} = (2, 1) \in \mathbf{R}^2$, neither $\mathbf{x} \leq \mathbf{y}$ nor $\mathbf{y} \leq \mathbf{x}$.

A set $S \in \mathbf{R}^n$ is said to be **bounded from below** if there exists $y \in \mathbf{R}^n$ such that $y \leq x$ for all $x \in S$. Similarly, a set $S \in \mathbf{R}^n$ is said to be **bounded from above** if there exists $y \in \mathbf{R}^n$ such that $x \leq y$ for all $x \in S$.

Infimum and Supremum The **infimum** of a set $S \subseteq \mathbf{R}^n$ is the greatest element in $\mathbf{R}^n$ that is less than or equal to all elements of S, if such an element exists. In other words, for a given $S \subseteq \mathbf{R}^n$, let $L_S \subseteq \mathbf{R}^n$ be defined as follows: $x \in L_S$ if and only if $x \leq y$ for all $y \in S$. If there exists $z \in L_S$ such that $x \leq z$

for all $x \in L_S$, then z is the infimum of S; otherwise, there is no infimum. Similarly, the **supremum** of a set $S \subseteq \mathbf{R}^n$ is least element in the subset of $\mathbf{R}^n$ that contains points that are greater than or equal to all elements of S, again, if such an element exists.

The two definitions are symmetric and, indeed, the infimum is in a precise sense dual to the concept of a supremum. As already mentioned, infima and suprema do not necessarily exist. However, if an infimum or supremum does exist, it is unique. Moreover, it follows immediately from the definition that the infimum of $S \subseteq \mathbf{R}^n$ exists if and only if it exists for S along every dimension of the $\mathbf{R}^n$ space. This, in particular, implies that S is bounded from below. By symmetry, similar observation holds for the supremum of S—set S has to be bounded from above for the supremum to exist. Finally, let us mention that the two definitions (infimum and supremum) naturally generalize to any partially ordered set, not just the relation $\leq$ on $\mathbf{R}^n$.

If an infimum (supremum) of a set S exists and it belongs to S, then we call it a **minimum** (respectively, **maximum**). The concepts of infimum and supremum are then closely related to minimum and maximum, but are more useful in analysis because they better characterize special sets which may have no minimum or maximum. For instance, the set of positive real numbers $\mathbf{R}_+$ does not have a minimum. Indeed, for any $x \in \mathbf{R}_+$, there exists $y \in \mathbf{R}_+$ such that $y < x$ (say, consider $y := x/2$). On the other hand, the infimum of $\mathbf{R}_+$ exists and is equal to 0. Indeed, it follows from the previous observation together with the fact that $0 \leq x$ for all $x \in \mathbf{R}_+$. Another simple examples are

$$\inf\{x \in \mathbf{R} \mid 0 < x < 1\} = 0$$
$$\inf\{x \in \mathbf{Q} \mid x^2 > 2\} = \sqrt{2}$$
$$\inf\left\{(-1)^n + \frac{1}{n} \in \mathbf{R} \mid n \in \mathbf{N}\right\} = -1.$$

SOLUTION

Let $(a_n)_{n \in \mathbf{N}}$ be any sequence of positive numbers that satisfies the desired properties. In particular, recall that there exists $C_0 > 1$ such that the sequence $(a_n)_{n \in \mathbf{N}}$ satisfies $a_{m+n} \leq C_0(a_m + a_n)$ for all $m, n \in \mathbf{N}$. Let S be the set of real numbers (not necessarily grater than 1) C for which $a_{m+n} \leq C(a_m + a_n)$ for all $m, n \in \mathbf{N}$. Observe that if $C \in S$, then $C' \in S$ provided that $C' > C$. It follows that there exists $C^* \in \mathbf{R}$ so that either $S = (C^*, \infty)$ (S is an open set) or $S = [C^*, \infty)$ (S is a closed set). We will show that the latter is true, that is, C^*, an infimum of S, belongs to S.

Let C^* be an infimum of S. For a contradiction, suppose that C^* does *not* belong to S. It follows that there exist $m, n \in \mathbf{N}$ such that $a_{m+n} > C^*(a_m + a_n)$. However, since the inequality is sharp, we get that there exists some $\varepsilon > 0$ such that $a_{m+n} > (C^* + \delta)(a_m + a_n)$ for all $0 \leq \delta \leq \varepsilon$. This contradicts the fact that C^* is an infimum. It follows that $C^* \in S$.

Let us note that $C^* \geq 1$, as $2 = a_2 \leq C^*(a_1 + a_1) = 2C^*$. We will show that, in fact, $C^* = 1$. Let us first observe that for any $n, m \in \mathbf{N}$,

$$
\begin{aligned}
a_{m+n}^2 \;=\; a_{(m+n)^2} \;&=\; a_{m^2+n^2+2mn} \;\leq\; C^*(a_{m^2} + a_{n^2+2mn}) \\
&\leq\; C^*(a_m^2 + C^*(a_{n^2} + a_{2mn})) \;=\; C^*(a_m^2 + C^*(a_n^2 + a_2 a_m a_n)) \\
&=\; C^*(a_m^2 + C^*(a_n^2 + 2a_m a_n)).
\end{aligned}
$$

Similar arguments give us the following system of inequalities:

$$
\begin{cases}
a_{m+n}^2 \;\leq\; C^*(a_m^2 + C^*(a_n^2 + 2a_m a_n)) \\
a_{m+n}^2 \;\leq\; C^*(a_n^2 + C^*(a_m^2 + 2a_m a_n)) \\
a_{m+n}^2 \;\leq\; C^*(2a_m a_n + C^*(a_m^2 + a_n^2)).
\end{cases}
$$

After combining the three inequalities together, we get that

$$
3a_{m+n}^2 \;\leq\; (C^* + 2(C^*)^2)(a_m + a_n)^2.
$$

Since all the terms are positive, after taking a square root of both sides, we get that

$$
a_{m+n} \;\leq\; \sqrt{\frac{2(C^*)^2 + C^*}{3}}(a_m + a_n).
$$

On the other hand, since the infimum C^* is contained in $\mathcal{S}$, for any $\varepsilon > 0$ there must exist $m, n \in \mathbf{N}$ for which $a_{m+n} > (C^* - \varepsilon)(a_m + a_n)$. Combining this with the above inequality (that holds for any $m, n \in \mathbf{N}$), we get that

$$
(C^* - \varepsilon)(a_m + a_n) < a_{m+n} \;\leq\; \sqrt{\frac{2(C^*)^2 + C^*}{3}}(a_m + a_n),
$$

or equivalently that

$$
C^* - \varepsilon \;<\; \sqrt{\frac{2(C^*)^2 + C^*}{3}}.
$$

Since this inequality holds for all $\varepsilon > 0$, we conclude that $\sqrt{\frac{2(C^*)^2+C^*}{3}} \geq C^*$, or equivalently that $2(C^*)^2 + C^* \geq 3(C^*)^2$. It follows that $0 \geq (C^*)^2 - C^* = C^*(C^* - 1)$ and so $C^* \leq 1$. Since $C^* \geq 1$, we get that $C^* = 1$ and so, in particular, $a_n \leq n$ for all $n \in \mathbf{N}$, which can be proved by induction on n.

Let us now note that for each $k \in \mathbf{N}$, $a_{2^k} = (a_2)^k = 2^k$. Using this and the bound we proved above, we get

$$
2^k = a_{2^k} \leq a_m + a_{2^k - m} \leq m + (2^k - m) = 2^k,
$$

and so $a_m + a_{2^k - m} = 2^k$ for all $1 \leq m \leq 2^k - 1$. But it means that for any $1 \leq m \leq 2^k - 1$, we get

$$
2^k \;=\; a_m + a_{2^k - m} \leq a_m + (2^k - m),
$$

and so $a_m \geq m$. The conclusion ($a_m = m$ for all $m \in \mathbf{N}$) follows as $a_m \leq m$.

REMARKS

The key idea that leads to the proof is to notice that there exists C^* that is a smallest value of C for which the desired bound for a_{m+n} is met. This follows from the fact that an intersection of closed sets is a closed set. What is this family of closed sets in our case? For any pair $m, n \in \mathbf{N}$ we want to have the bound $a_{m+n} \leq C(a_m + a_n)$ which clearly holds for any $C \geq a_{m+n}/(a_m + a_n)$. As a result, the set of values of C for which $a_{m+n} \leq C(a_m + a_n)$ for all $m, n \in \mathbf{N}$ is the set

$$S := \bigcap_{m,n \in \mathbf{N}} \left[\frac{a_{m+n}}{a_m + a_n}, \infty \right).$$

It follows that S is closed as it is an intersection of closed sets. In particular, $C^* \in S$.

With this observation, we know that for any $\varepsilon > 0$, there are natural numbers m and n such that

$$a_{m+n} > (C^* - \varepsilon)(a_m + a_n).$$

If we want to start using the fact that $a_{mn} = a_m a_n$, then it makes sense to try to square the above inequality and use the fact that $a_{(m+n)^2} = (a_{m+n})^2$. This then implies that

$$a_{(m+n)^2} = (a_{m+n})^2 > (C^* - \varepsilon)^2 (a_m + a_n)^2.$$

In order to use the fact that $a_{m+n} \leq C^*(a_m + a_n)$, consider three ways to distribute the terms in $(m+n)^2 = m^2 + n^2 + 2mn$, exactly like in the solution. If we do this, then this leads us to the following observation: for all $\varepsilon > 0$

$$\frac{2(C^*)^2 + C^*}{3} > (C^* - \varepsilon)^2.$$

Since this holds for any $\varepsilon > 0$, we have that in the limit, $2(C^*)^2 + C^* \geq 3(C^*)^2$, and so $C^* \geq (C^*)^2$. Therefore, we see that $C^* = 1$, as it cannot be less than 1.

EXERCISES

2.7.1. Find the number of infinite sequences $(a_i)_{i \in \mathbf{N}}$, such that $a_i \in \{-1, 1\}$ for all $i \in \mathbf{N}$, $a_{mn} = a_m a_n$ for all $m, n \in \mathbf{N}$, and each consecutive triple contains at least one 1 and one -1.
(Source of the problem and solution: PLMO LV – Phase 2 – Problem 3.)

2.7.2. Let us fix any real number a. We recursively define sequence $(a_n)_{n \in \mathbf{N}}$ as follows: let $a_1 = a$ and for each $n \in \mathbf{N}$, let $a_{n+1} = (a_n - 1/a_n)/2$ if $a_n \neq 0$ and $a_{n+1} = 0$ if $a_n = 0$. Prove that this sequence has infinitely many non-positive elements and infinitely many non-negative elements.
(Source of the problem and solution idea: PLMO XXIX – Phase 3 – Problem 5.)

2.7.3. Let n be any natural number such that $n \geq 3$. Find all sequences of real numbers $(x_1, x_2, \ldots, x_n)$ that satisfy the following conditions:

$$\sum_{i=1}^{n} x_i = n \quad \text{and} \quad \sum_{i=1}^{n} (x_{i-1} - x_i + x_{i+1})^2 = n,$$

where we set $x_0 = x_n$ and $x_{n+1} = x_1$.

(Source of the problem and solution idea: PLMO LX – Phase 2 – Problem 6.)

Chapter 3

Functions, Polynomials, and Functional Equations

As usual, we start the chapter with some basic definitions.

THEORY

Polynomials A **polynomial** is an expression consisting of variables and coefficients that involves only the operations of addition, subtraction, multiplication, and non-negative integer exponents of variables. An example of a polynomial of a single variable x is $x^2 - 7x + 3$. An example in three variables is $x^3 + xy^2z - xz + 7$. A polynomial in a single variable x can always be written (or rewritten) in the following form

$$P(x) = a_n x^n + a_{n-1} x^{n-1} + \ldots + a_1 x + a_0 = \sum_{i=0}^{n} a_i x^i,$$

where $a_0, a_1, \ldots, a_n$ are real constants and $a_n \neq 0$.

The **degree of a polynomial** is the highest degree of its **monomials** (individual terms) with non-zero coefficients. The degree of a term is the sum of the exponents of the variables that appear in it, and thus is a non-negative integer. For example, the polynomial $x^3 + xy^2z - xz + 7$ has 4 terms. The 4 terms have, correspondingly, degrees 3, 4, 2, and 0. As a result, the degree of this polynomial is 4. Polynomials of small degrees have names; in particular, degree 0 polynomial $P(x) = C$ is called **non-zero constant** (if $C \neq 0$) or **special case** (if $C = 0$), degree 1 polynomials are called **linear**, degree 2 ones are called **quadratic**, and degree 3 ones **cubic**.

In order to determine the degree of a polynomial that is not in standard form, one has to put it in standard form by expanding the products and combining the terms. For example,

$$(x+1)^2 - (x-1)^2 = 4x$$

if of degree 1 despite the fact that each summand has degree 2. This is not needed when the polynomial is expressed as a product of polynomials. One can easily see that the degree of a product is the sum of the corresponding degrees of all the factors. Similarly, the degree of the composition of two non-constant polynomials, say $P(x)$ and $Q(x)$, is the product of their degrees. For example, if $P(x) = x^2 - x$ and $Q(x) = x^3 + x$, then

$$\begin{aligned} P(x) \circ Q(x) &= P(Q(x)) = (x^3+x)^2 - (x^3+x) \\ &= x^6 + 2x^4 + x - x^3 - x = x^6 + 2x^4 - x^3 \end{aligned}$$

has degree $6 = 2 \cdot 3$.

$\boxed{\text{Complex Numbers}}$ In order to make our next observation about the number of roots of a given polynomial, it will be convenient to introduce a concept of complex numbers. However, we will not use them anymore in this book.

A **complex number** is a number that can be expressed in the form $a+bi$, where a and b are real numbers, and i is a solution of the equation $x^2 = -1$. Clearly, no real number satisfies this equation, and so i is called an **imaginary number**. For a given complex number $z = a + bi$, a is called the **real part**, and b is called the **imaginary part**.

Complex numbers allow solutions to certain equations that have no solutions in real numbers. For example, $x^2 - 4x + 8 = 0$ has no real solution. Indeed, it can be rewritten as follows $(x-2)^2 = -4$ and now it is clear that the square of a real number cannot be negative. On the other hand, since $i^2 = -1$, both $2+2i$ and $2-2i$ are solutions to this equation, as demonstrated below:

$$\begin{aligned} ((2+2i)-2)^2 &= (2i)^2 = 2^2 i^2 = 4(-1) = -4 \\ ((2-2i)-2)^2 &= (-2i)^2 = (-2)^2 i^2 = 4(-1) = -4. \end{aligned}$$

$\boxed{\text{Roots}}$ A **root** (or **zero**) of a real or complex function $f = f(x)$ is a number x from the domain of f such that $f(x) = 0$. In particular, a root of a polynomial $P(x)$ is a root of the corresponding polynomial function.

The **fundamental theorem of algebra** states that every non-constant single-variable polynomial with complex coefficients has at least one complex root. The theorem can be alternatively stated as follows: every non-zero, single-variable, degree n polynomial with complex coefficients has, counted with multiplicity, exactly n complex roots. The equivalence of the two statements can be proven through the use of successive polynomial division.

Clearly, the fundamental theorem of algebra can be applied to polynomials with real coefficients that we are concerned in this book, since every real number is a complex number with an imaginary part equal to zero. It follows that the number of complex roots of each such polynomial is exactly n and so the number of real roots is at most n. As observed earlier, it can be strictly less than n.

3.1　Vieta's Formulas

SOURCE

Problem and solution idea: PLMO LXVI – Phase 2 – Problem 4

PROBLEM

Suppose that real numbers x_1, x_2, x_3, and x_4 are roots of some polynomial $W(x)$ of degree 4 with all integer coefficients. Prove that if $x_3 + x_4$ is rational and $x_3 x_4$ is irrational, then $x_1 + x_2 = x_3 + x_4$.

THEORY

$\boxed{\text{Vieta's Formulas}}$ **Vieta's formulas** are formulas that relate the coefficients of a polynomial to sums and products of its roots. Consider a polynomial $P(x) = \sum_{i=0}^{n} a_i \cdot x^i$, where $a_1, a_2, \ldots, a_n \in \mathbf{R}$ and $a_n \neq 0$. Since $P(x)$ has degree n, by the fundamental theorem of algebra it has n (complex) roots $x_1, x_2, \ldots, x_n$ and can be written as $P(x) = a_n \prod_{i=1}^{n}(x - x_i)$. Vieta's formulas can be obtained by multiplying the factors in the second representation and then identifying the coefficients of each power of x in the two representations. It follows that for any $k \in [n]$, we have that

$$\sum_{1 \le i_1 < i_2 < \cdots < i_k \le n} \prod_{j=1}^{k} x_{i_j} = (-1)^k \frac{a_{n-k}}{a_n}. \tag{3.1}$$

Note that the indices i_k are in increasing order to ensure that each sub-product of roots is used exactly once. In particular, for $k = 1, k = 2$, and $k = n$ we get

$$\sum_{i=1}^{n} x_i = -\frac{a_{n-1}}{a_n}$$

$$\sum_{i=1}^{n} \sum_{j=i+1}^{n} x_i \cdot x_j = \frac{a_{n-2}}{a_n}$$

$$\prod_{i=1}^{n} x_i = (-1)^n \frac{a_0}{a_n}.$$

The following useful observation is an easy consequence of Vieta's formulas:

$$\sum_{i=1}^{n} x_i^2 = \frac{a_{n-1}^2 - 2 \cdot a_n \cdot a_{n-2}}{a_n^2}. \tag{3.2}$$

Indeed, note that

$$\left(\sum_{i=1}^{n} x_i \right)^2 = \sum_{i=1}^{n} x_i^2 + 2 \sum_{i=1}^{n} \sum_{j=i+1}^{n} x_i x_j$$

and so

$$\sum_{i=1}^{n} x_i^2 = \left(\sum_{i=1}^{n} x_i\right)^2 - 2\sum_{i=1}^{n}\sum_{j=i+1}^{n} x_i x_j = \left(-\frac{a_{n-1}}{a_n}\right)^2 - 2\frac{a_{n-2}}{a_n}.$$

SOLUTION

Suppose that $W(x) = \sum_{i=0}^{4} a_i x^i$ with $a_0, a_1, \ldots, a_4 \in \mathbf{Z}$ and $a_4 \neq 0$. We will first show that all roots of the considered polynomial are non-zero. Clearly, x_3 and x_4 are non-zero as their product is irrational. In order to derive a contradiction, suppose that $x_1 = 0$ or $x_2 = 0$. Without loss of generality, we may assume that $x_1 = 0$; x_2 may or may not be equal to 0. Using Vieta's formula (3.1) for $k = 1$, we get that $x_2 + x_3 + x_4 = -a_3/a_4$ is rational. As a result, since $x_3 + x_4$ is rational, x_2 is rational too. Similarly, Vieta's formula for $k = 3$ implies that $x_2 x_3 x_4 = -a_1/a_4$ is rational. Since $x_3 x_4$ is irrational, we get that $x_2 = 0$. But then the formula for $k = 2$ gives us that $x_3 x_4 = a_2/a_4$ is rational, which gives us the desired contradiction. Hence, all roots of the considered polynomial are indeed non-zero.

We will use Vieta's formulas again, in fact, a few times. Observe first that, since $x_3 + x_4$ is rational, $x_1 + x_2$ is also rational, as

$$(x_1 + x_2) + (x_3 + x_4) = \sum_{i=1}^{4} x_i = -a_3/a_4$$

is rational. Similarly, since $x_3 x_4$ is irrational, $x_1 x_2$ must also be irrational, as

$$(x_1 x_2)(x_3 x_4) = \prod_{i=1}^{4} x_i = a_0/a_4$$

is rational and $x_1 x_2 \neq 0$. Observe next that $x_1 x_2 + x_3 x_4$ is rational, as

$$(x_1 + x_2)(x_3 + x_4) + x_1 x_2 + x_3 x_4 = \sum_{i \neq j} x_i x_j = a_2/a_4$$

is rational. Since

$$(x_1 + x_2)x_3 x_4 + (x_3 + x_4)x_1 x_2 = \sum_{1 \leq i_1 < i_2 < i_3 \leq 4} x_{i_1} x_{i_2} x_{i_3} = -a_1/a_4$$

is rational,

$$(x_1+x_2)x_3 x_4+(x_3+x_4)x_1 x_2-(x_1+x_2)(x_1 x_2+x_3 x_4) = (x_3+x_4-x_1-x_2)x_1 x_2$$

is rational too. Finally, since $x_3 + x_4$ and $x_1 + x_2$ are both rational and $x_1 x_2$ is irrational, we conclude that $x_3 + x_4 - x_1 - x_2 = 0$ and so $x_1 + x_2 = x_3 + x_4$.

REMARKS

The key value of Viete's formulas is that one can use them to prove many useful properties of roots of polynomials without actually finding them. In the problem we consider in this section, the key observation is that if the coefficients of a given polynomial are integers, then all values following from Viete's formulas must be rational. In fact, one could relax the assumption and only assume that the coefficients are rational (not necessarily integers). Indeed, this clearly follows from the fact that multiplying any polynomial by a constant does not change its roots so any polynomial with rational coefficients can be transformed into a polynomial with integer coefficients.

Ones we observe that all values following from Viete's formulas are rational, the next step is to exhaustively write down all the facts about rationality or irrationality of different combinations of x_is. In practice, when solving problems of this flavour, one would write down many more facts that are possible to derive, most of them would turn out to be useless. This is hard to avoid as it typically difficult to predict which combination of them is crucial for the problem at hand.

EXERCISES

3.1.1. Find all sets of six real numbers $a_1, a_2, a_3, b_1, b_2, b_3$ with the property that for all $i \in [3]$, a_{i+1} and b_{i+1} are two different solutions of the equation $x^2 + a_i x + b_i = 0$ (here we let $a_4 = a_1$ and $b_4 = b_1$).
(Source of the problem and solution idea: PLMO LXX – Phase 1 – Problem 5.)

3.1.2. Let $n \geq 3$ be an integer. Prove that the polynomial

$$f(x) = x^n + \sum_{i=0}^{n-3} a_i x^i$$

has n real roots if and only if all a_i are equal to 0.
(Source of the problem and solution idea: PLMO LII – Phase 2 – Problem 3.)

3.1.3. Let x_1, x_2, x_3 be the roots of the equation $3x^3 + 6x^2 - 1 = 0$. Find the value of $\sum_{i=1}^{3} \frac{1}{x_i^4}$.
(Source of the problem and solution idea: Delta – ZM-1506.)

3.2 Functional Equations, Exploration

SOURCE

Problem and solution idea: PLMO LXIII – Phase 1 – Problem 8

PROBLEM

Find all functions $f: \mathbf{R} \to \mathbf{R}$, such that for all $x, y \in \mathbf{R}$, we have that

$$f(x + f(x + y)) = f(x - y) + f(x)^2. \tag{3.3}$$

THEORY

| Functional Equations | A **functional equation** is any equation in which the unknown represents a function. Typically, such equations relate the value of a function at some point with its values at other points. For example,

$$f(x + y) = f(x)f(y)$$

is satisfied by all exponential functions,

$$f(xy) = f(x) + f(y)$$

is satisfied by all logarithmic functions, and

$$f(xy) = f(x)f(y)$$

is satisfied by all power functions.

Solving functional equations can be very difficult. In order to warm up, let us consider the following very simple case. Suppose that our goal is to find all real-valued functions $f : \mathbf{R} \to \mathbf{R}$ that satisfy

$$f(x + y)^2 = f(x)^2 + f(y)^2$$

for all $x, y \in \mathbf{R}$. By considering the case $x = y = 0$, we get that $f(0)^2 = f(0)^2 + f(0)^2$ and so $f(0)^2 = 0$. From this it follows that $f(0) = 0$. Now, consider any $x \in \mathbf{R}$ and let $y = -x$. We get

$$f(x)^2 \le f(x)^2 + f(-x)^2 = f(x - x)^2 = f(0)^2 = 0$$

and so $f(x)^2 = 0$ and finally $f(x) = 0$.

| Continuous Functions | In order to solve our next example, we need to introduce continuous functions. A function $f : \mathbf{R} \to \mathbf{R}$ is **continuous** if sufficiently small changes of the argument x result in arbitrarily small changes of the value $f(x)$. Otherwise, a function is said to be **discontinuous**.

There are several different formal definitions of continuity of a function, all of them being equivalent. Below, we present a few which are commonly used.

Limits of functions: Function f is continuous at some point c of its domain if the limit of $f(x)$, as x approaches c through elements of the domain of f, exists and is equal to $f(c)$; that is,

$$\lim_{x \to c} f(x) \;=\; f(c).$$

Limits of sequences: One can instead require that for any sequence $(x_n)_{n \in \mathbf{N}}$ of points in the domain that converges to c, the corresponding sequence $(f(x_n))_{n \in \mathbf{N}}$ converges to $f(c)$; that is,

$$\lim_{n \to \infty} x_n \;=\; c \qquad \Rightarrow \qquad \lim_{n \to \infty} f(x_n) \;=\; f(c). \tag{3.4}$$

Epsilon–delta: For every $\epsilon > 0$ (arbitrarily small), there exists $\delta = \delta(\epsilon) > 0$ (which depends on ϵ) such that for all x in the domain of f with $c - \epsilon < x < c + \epsilon$, the function satisfies $f(c) - \delta < f(x) < f(c) + \delta$; that is,

$$|x - c| < \delta \qquad \Rightarrow \qquad |f(x) - f(c)| < \epsilon.$$

Finally, a function is continuous if it is continuous at all elements of its domain.

It is easy to see that the sum of two functions, continuous on some domain, is also continuous on this domain. The same holds for the product of continuous functions. Combining this with an easy fact that functions of the form $f(x) = Ax + B$ (for some constants $A, B \in \mathbf{R}$) are continuous, we get in particular that all polynomials are continuous functions.

| **Cauchy's Functional Equation and Additive Functions** | Let us consider one particular functional equation that is both interesting and important. **Cauchy's functional equation** is the functional equation

$$f(x + y) \;=\; f(x) + f(y). \tag{3.5}$$

Solutions to this equation are called **additive functions**.

It is clear that any linear functions $f : \mathbf{R} \to \mathbf{R}$ (that is, function of the form $f(x) = cx$ for some fixed constant $c \in \mathbf{R}$) is a solution. However, there are some other solutions that are extremely complicated. Having said that, if one is restricted to continuous functions, then linear functions are the only solutions to the Cauchy's functional equation. Hence, any other solution must be, in some sense, highly pathological.

Let $f : \mathbf{R} \to \mathbf{R}$ be any additive function that is also continuous, and let $c = f(1)$. We will first show that $f(q) = cq$ for any $q \in \mathbf{Q}$. It is convenient to independently consider the following three cases: $q = 0$, $q > 0$, and $q < 0$.

Case 1: $q = 0$. By setting $x = 0$ (or, in fact, any value) and $y = 0$ in (3.5) we get $f(0) = f(0) + f(0)$ and so $f(0) = 0$.

Case 2: $q > 0$. After applying (3.5) to $f(ax) = f(x + x + \cdots + x)$ repeatedly a times, we get that

$$f(ax) = af(x), \qquad \text{for any } a \in \mathbf{N}, x \in \mathbf{R}. \tag{3.6}$$

After substituting x with x/a in (3.6) and multiplying both sides by b/a, we get that

$$\frac{b}{a} f(x) = bf\left(\frac{x}{a}\right), \qquad \text{for any } a, b \in \mathbf{N}, x \in \mathbf{R}. \tag{3.7}$$

By combining (3.6) and (3.7), we get that

$$\frac{b}{a} f(x) = bf\left(\frac{x}{a}\right) = f\left(\frac{b}{a}x\right),$$

and finally, after setting $x = 1$, we have that for any $q = b/a \in \mathbf{Q}_+$,

$$f(q) = f\left(\frac{b}{a} \cdot 1\right) = \frac{b}{a} f(1) = \frac{b}{a} c = cq.$$

Case 3: $q < 0$. After applying (3.5) with $y = -x$, we get that

$$0 = f(0) = f(x - x) = f(x) + f(-x),$$

and so $f(-x) = -f(x)$. This allows us to reduce this case to the previous one. It follows that for any $q \in \mathbf{Q}_-$,

$$f(q) = f(-(-q)) = -f(-q) = -(c(-q)) = cq,$$

since $-q \in \mathbf{Q}_+$. The desired property follows, namely, $f(q) = cq$ for all $q \in \mathbf{Q}$.

It remains to show that, since f is continuous, $f(x) = cx$ for any $x \in \mathbf{R} \setminus \mathbf{Q}$. For that we will use the "limits of sequences" variant of the definition for a function to be continuous—see (3.4). Let $x \in \mathbf{R} \setminus \mathbf{Q}$. It is easy to construct a sequence $(x_n)_{n \in \mathbf{N}}$ of rational numbers that converges to x, for example, let $x_n = \lfloor xn \rfloor / n \in \mathbf{Q}$. It is clear that

$$0 \leq x - x_n = \frac{xn - \lfloor xn \rfloor}{n} < \frac{1}{n},$$

and so, indeed, $\lim_{n \to \infty} x_n = x$. From (3.4) it follows that

$$f(x) = \lim_{n \to \infty} f(x_n) = \lim_{n \to \infty} cx_n = c \lim_{n \to \infty} x_n = cx.$$

SOLUTION

After applying (3.3) for the specific case $y = f(0) - x$, we get that $f(x) = f(2x - f(0)) + f(x)^2$ which implies that

$$f(2x - f(0)) = f(x) - f(x)^2 = f(x)(1 - f(x)).$$

Since function $g(z) = z(1 - z)$ attains its maximum value, $1/4$, at $z = 1/2$, we get that $f(2x - f(0)) \leq \frac{1}{4}$. Since x is arbitrary and $f(0)$ fixed (once function is fixed; in fact, as argued below, it is always equal to 0 or -1), we conclude that function f has the following important property:

$$\text{for all } x \in \mathbf{R}, \qquad f(x) \leq 1/4. \qquad (3.8)$$

Applying (3.3) for another specific case, $x = f(0)$ and $y = -f(0)$, we get that $f(2f(0)) = f(2f(0)) + f(f(0))^2$ and so $f(f(0)) = 0$. On the other hand, for $x = y = 0$ we get $f(f(0)) = f(0) + f(0)^2$ so, combining the two observations, we get that there are only two possible values of $f(0)$: $f(0) = 0$ or $f(0) = -1$. We will consider these two cases independently.

Suppose first that $f(0) = -1$. In this case, $f(-1) = f(f(0)) = 0$ and (3.3) for $(x, y) = (0, 1)$ gives us $f(f(1)) = f(-1) + f(0)^2 = 0 + (-1)^2 = 1$. But this contradicts (3.8), and so no function f satisfies both the functional equation (3.3) and $f(0) = -1$.

Suppose then that $f(0) = 0$. Clearly, the constant function $f(x) = 0$ for $x \in \mathbf{R}$, satisfies (3.3); so this is certainly one solution. We will show that it is actually the only one. For a contradiction, suppose that for some $z \in \mathbf{R}$ we have $f(z) \neq 0$. Considering the case $x = y = z$ we get from (3.3) that $f(z + f(2z)) = f(z)^2$, which implies that $\delta := f(z + f(2z)) > 0$. We define recursively a sequence of numbers as follows:

$$x_1 = z + f(2z), \qquad x_{i+1} = x_i + f(x_i) \text{ for any } i \in \mathbf{N}.$$

In particular, $f(x_1) = \delta > 0$. By applying (3.3) with $x = x_{i+1}$ and $y = 0$, we get that

$$f(x_{i+1}) = f(x_i + f(x_i)) = f(x_i) + f(x_i)^2 \geq f(x_i).$$

It follows that for any $i \in \mathbf{N}$, $f(x_i) \geq f(x_{i-1}) \geq \ldots \geq f(x_1) = \delta > 0$. In particular, it means that both $(x_i)_{i \in \mathbf{N}}$ and $(f(x_i))_{i \in \mathbf{N}}$ are strictly increasing sequences. More importantly, it implies that

$$f(x_{i+1}) = f(x_i) + f(x_i)^2 \geq f(x_i) + \delta^2$$

and so for any $i \in \mathbf{N}$, $f(x_i) \geq f(x_1) + (i - 1)\delta^2$. As a result, the sequence $(f(x_i))_{i \in \mathbf{N}}$ is unbounded (that is, $\lim_{n \to \infty} f(x_n) = \infty$) which contradicts (3.8). Hence, indeed, $f(x) = 0$ ($x \in \mathbf{R}$) is the unique solution to our problem.

REMARKS

A standard approach used to solve functional equations is to prove some specific properties of the function involved. In particular, it is often useful to try to establish the value of the function for some carefully chosen, characteristic values. In our problem, this specific value was equal to 0. Another common approach is to try to reduce the functional equation involving, say, two independent variables (in our case, x and y) into a more general equation involving only one variable. In our case, we used this approach by setting $x = y$. It is important to remember that if we prove some property of a more general equation (in this case involving one variable), then we always have to check if the final solution meets the original equation.

EXERCISES

3.2.1. Find all functions $f: \mathbf{Z} \to \mathbf{Z}$ which satisfy the following condition:

$$f(a + b)^3 - f(a)^3 - f(b)^3 = 3f(a)f(b)f(a + b)$$

for all $a, b \in \mathbf{Z}$.
(Source and the problem and solution idea: PLMO LXV– Phase 1 – Problem 5.)

3.2.2. Find all pairs of functions $f: \mathbf{R} \to \mathbf{R}$ and $g: \mathbf{R} \to \mathbf{R}$ such that

$$g\big(f(x) - y\big) = f\big(g(y)\big) + x$$

for all $x, y \in \mathbf{R}$.
(Source of the problem and solution idea: PLMO LXIII – Phase 2 – Problem 4.)

3.2.3. Find all functions $f: \mathbf{R} \to \mathbf{R}$ such that for all $x, y \in \mathbf{R}$, we have that

$$f\big(f(x) - y\big) = f(x) + f\big(f(y) - f(-x)\big) + x.$$

(Source of the problem and solution idea: PLMO LIX – Phase 2 – Problem 3.)

3.3 Functional Equations, Necessary Conditions

SOURCE

Problem and solution idea: PLMO LV – Phase 1 – Problem 3 (modified)

PROBLEM

Find all bijections $f : \mathbf{R} \to \mathbf{R}$ that satisfy

$$f(x^2 + y) = xf(x) + f(y)$$

for all $x, y \in \mathbf{R}$.

THEORY

When solving functional equations, it is quite common that only some specific cases are considered. For example, one may fix some specific values in the given equation or relate one variable to some other variable or variables. Indeed, typically one starts from the original functional equation but then immediately transforms it into something more manageable and insightful. But, as a result, the resulting equations are *not* equivalent to the original ones. In other words, the obtained conditions are only necessary but often not sufficient. Therefore, it is always important to make sure that the final result actually satisfies the original functional equation.

In order to illustrate this issue, let us consider the following functional equation. Suppose that our goal is to find all functions $f : \mathbf{R}_+ \to \mathbf{R}_+$ that satisfy the following condition

$$\sqrt{f(x) + f(y)} = x + y,$$

for all $x, y \in \mathbf{R}$. First, after fixing $y = 0$ and squaring the equation, we get that

$$f(x) + f(0) = x^2 + 0 = x^2.$$

Now, we can set $x = 0$ to learn that $f(x) = 0$. We conclude that $f(x) = x^2$.

Let us stress it again that at this point we only proved that if the original equation has a solution, then this solution must be $f(x) = x^2$. However, in order to get this potential solution we have simplified and relaxed the original condition by setting $y = 0$ and squaring the equation. Therefore, $f(x) = x^2$ is only a candidate solution and now we have to check that it actually satisfies the original equation. Once we substitute the function into the original equation we get $\sqrt{x^2 + y^2} = x + y$. This equation clearly does not hold unless $x = 0$ or $y = 0$. As function $f(x) = x^2$ was the only potential solution, we conclude that there is no function f which satisfies the original functional equation.

SOLUTION

Let us start with rewriting the functional equation as follows: for all $x, y \in \mathbf{R}$,

$$f(y) = f(x^2 + y) - xf(x).$$

For $y = -x \in \mathbf{R}$, we get that

$$f(-x) = f(x^2 - x) - xf(x) = f(x^2 - x) - xf(1^2 + (x - 1)).$$

Since $x^2 - x = (x - 1)^2 + (x - 1)$, using the original equation we get that

$$f(x^2 - x) = f((x - 1)^2 + (x - 1)) = (x - 1)f(x - 1) + f(x - 1),$$

and so

$$
\begin{aligned}
f(-x) &= (x - 1)f(x - 1) + f(x - 1) - xf(1^2 + (x - 1)) \\
&= -(f(1^2 + (x - 1)) - f(x - 1))x.
\end{aligned}
$$

Going back to the original equation for the last time, we get that $f(-x) = -f(1)x$. It follows that the only functions that potentially satisfy the given functional equation have the form $f(x) = ax$, where a is a fixed real number.

We directly check that if $f(x) = ax$, then

$$f(x^2 + y) - xf(x) = ax^2 + ay - x \cdot ax = ay = f(y).$$

However, in our problem, our goal is to find bijections and so the case $a = 0$ has to be ruled out. We conclude that the only functions that meet all the conditions are of the form $f(x) = ax$ where $a \neq 0$.

REMARKS

In order to better understand how one can get ideas leading to the solution of similar problems, let us make some observations that we made during the process of solving the problem at hand. First, by fixing $x = 1$ we observed that for any $y \in \mathbf{R}$,

$$f(y + 1) = f(y) + f(1).$$

It follows that function f constantly increases by the same value, $f(1)$, when we increase its argument by 1.

Our second observation was that changing x to $-x$ does not affect the left hand side of the original equation and so $xf(x) = -xf(-x)$ for any $x \in \mathbf{R}$. Hence, for $x \neq 0$ we get that $f(x) = -f(-x)$. On the other hand, $f(0) = f(-1 + 1) = f(-1) + f(1) = -f(1) + f(1) = 0$, and so in fact $f(x) = -f(-x)$ for all $x \in \mathbf{R}$.

Our next idea was to try to find two different expressions of the form $x^2 + y$ that have identical value. One such expression was given in the solution above. Another identity that could have been used is $x^2 + x = (x + 1)^2 + (-1 - x)$. Applying it we would get that

$$
\begin{aligned}
f(x^2 + x) &= xf(x) + f(x) \\
f((x + 1)^2 + (-1 - x)) &= (x + 1)f(x + 1) + f(-1 - x).
\end{aligned}
$$

As a result,

$$
\begin{aligned}
f(x) + xf(x) &= f(-1-x) + (x+1)f(x+1) \\
&= -f(x+1) + (x+1)f(x+1) \\
&= xf(x+1) = xf(x) + xf(1),
\end{aligned}
$$

which yields $f(x) = f(1)x$, as required.

EXERCISES

3.3.1. Prove that if a function $f\colon \mathbf{R} \to \mathbf{R}$ satisfies the condition $f(x) = f(2x) = f(1-x)$ for all $x \in \mathbf{R}$, then it is periodic (that is, there exists $a \in \mathbf{R}_+$ such that $f(x+a) = f(x)$ for all $x \in \mathbf{R}$).
(Source of the problem and solution idea: PLMO LIII – Phase 2 – Problem 1.)

3.3.2. Given that function $f(x)$ satisfies $f\big(1/(1-x)\big) = xf(x) + 1$, find the value of $f(5)$.
(Source of the problem: question asked on Quora. Solution: our own.)

3.3.3. Suppose that a function $f(x,y,z)$ of three real arguments satisfies the following condition

$$
\sum_{i=1}^{5} f(x_i, x_{i+1}, x_{i+2}) = \sum_{i=1}^{5} x_i,
$$

where $x_{i+5} = x_i$. Prove that for all $n \geq 5$ we have

$$
\sum_{i=1}^{n} f(x_i, x_{i+1}, x_{i+2}) = \sum_{i=1}^{n} x_i,
$$

where $x_{i+n} = x_i$.
(Source of the problem and solution: PLMO LIX – Phase 3 – Problem 2.)

3.4   Polynomials with Integer Coefficients

SOURCE

Problem and solution idea: PLMO LVII – Phase 3 – Problem 6 (modified)

PROBLEM

Find all pairs of integers (a, b) with the property that there exists a polynomial $P(x)$ having integer coefficients such that

$$(x^2 + ax + b) \cdot P(x) = Q(x) = \sum_{i=0}^{n} c_i x^i, \qquad (3.9)$$

where $c_i \in \{1, -1\}$ for all $i \in [n] \cup \{0\}$.

THEORY

Rational Root Theorem Our next theorem, the **rational root theorem** (sometimes called the **rational root test**), states a constraint on rational solutions of a polynomial equation with integer coefficients. Consider any polynomial $P(x) = \sum_{i=0}^{n} a_i x^i$ with all integer coefficients, that is, $a_i \in \mathbf{Z}$ for all $i \in [n] \cup \{0\}$. Suppose that $P(x)$ has a rational root p/q, where $p \in \mathbf{Z}$ and $q \in \mathbf{Z}$ are co-prime (in other words, the fraction p/q is in its lowest terms). Then, $p \mid a_0$ and $q \mid a_n$.

Indeed, suppose that

$$0 = P(p/q) = \sum_{i=0}^{n} a_i (p/q)^i.$$

After multiplying this equation by q^n, we get that

$$p \cdot \left(\sum_{i=1}^{n} a_i p^{i-1} q^{n-i} \right) + a_0 q^n = 0.$$

Since p and q are co-prime, we get that $p \mid a_0$. Similarly, from the very same equation, we get that

$$a_n p^n + q \cdot \left(\sum_{i=0}^{n-1} a_i p^i q^{n-i-1} \right) = 0,$$

and we conclude that $q \mid a_n$.

An important application of this theorem is that it can be used to find all rational roots of a given polynomial. Indeed, it gives a finite number of possible fractions which can be checked to see if they are roots or not. If a rational root $x = r$ is found, a linear polynomial $(x - r)$ can be factored out of the polynomial using polynomial long division, resulting in a polynomial of a lower degree whose roots are also roots of the original polynomial.

Finally, let us mention that the rational root theorem is a special case (for a single linear factor) of **Gauss's Lemma** on the factorization of polynomials. For our purposes it is enough to use the following variant. If a polynomial $P(x)$ of degree greater than 1 with integer coefficients cannot be represented as a product of two non-constant polynomials with integer coefficients, then it also cannot be represented as a product of two non-constant polynomials with rational coefficients.

Now, if a polynomial $P(x)$ with integer coefficients has a rational root $x_0 = p/q$, where p and q are co-prime, then it is possible to represent it as $(x - p/q)R(x)$, where $R(x)$ has rational coefficients. If $R(x)$ is a rational constant r, then $P(x) = r(x-p/q) = a(qx-p)$, where $a = r/q$ is an integer, as by assumption r and rp/q are integers and p and q are co-prime. If $R(x)$ is not a constant, then the lemma of Gauss implies that $P(x) = P'(x)Q(x)$, where $P'(x)$ and $Q(x)$ are non constant and have integer coefficients and $P'(x)$ has a root p/q. If $P'(x)$ has degree 1, then clearly it can be written as $a(qx - p)$ where a is an integer and we are done. Otherwise, we have a polynomial $P'(x)$ of degree less than the degree of $P(x)$ with integer coefficients that has a root p/q. As $P(x)$ initially had a finite degree, by replacing $P(x)$ by $P'(x)$ and repeated application of this reasoning, we get that eventually we must reach the required factorization $P(x) = (qx - p)R'(x)$, where $R'(x)$ has integer coefficients. As a consequence, if $P(x) = \sum_{i=0}^{n} a_i x^i$, then $q \mid a_n$ and $p \mid a_0$, as stated above.

SOLUTION

Let $P(x)$ be a polynomial with integer coefficients that satisfies (3.9) for some pair of integers (a, b) and some set of coefficients $c_i \in \{-1, 1\}$. Let us first observe that if $Q(x) = 0$ for some $x \in \mathbf{R}$, then we have $-c_n x^n = \sum_{i=0}^{n-1} c_i x^i$ and so, in particular,

$$|x^n| = |x|^n = \left| \sum_{i=0}^{n-1} c_i x^i \right|.$$

We will now show that $|x| < 2$. For a contradiction, suppose that there exists $x \in \mathbf{R}$ such that $|x| \geq 2$ and the above equality holds. Since,

$$|x|^n = \left| \sum_{i=0}^{n-1} c_i x^i \right| \leq \sum_{i=0}^{n-1} |x|^i = \frac{|x|^n - 1}{|x| - 1} \leq |x|^n - 1,$$

we get that $0 \leq -1$ which gives us the desired contradiction. Since $Q(x)$ does *not* have roots which absolute values are greater than or equal to 2, the same property holds for polynomial $R(x) := x^2 + ax + b$. It follows that

$R(2) = 4 + 2a + b > 0$, $R(-2) = 4 - 2a + b > 0$, and so

$$-2 - \frac{b}{2} < a < 2 + \frac{b}{2}.$$

It follows immediately from (3.9), since $c_0 \in \{1, -1\}$ and $P(x)$ has integer coefficients, that b must be equal to 1 or -1. We will independently consider these two cases.

Case 1: $b = -1$. The possible values of a are -1, 0, and 1. If $a = -1$ or $a = 1$, then we can clearly fix $P(x) = 1$ to get the desired property. If $a = 0$, then one can take $P(x) = x + 1$ to get that

$$(x^2 + ax + b) \cdot P(x) = (x^2 - 1)(x + 1) = x^3 + x^2 - x - 1,$$

and the desired property holds.

Case 2: $b = 1$. This time there are more possible values of a to consider: -2, -1, 0, 1, and 2. If $a = -1$ or $a = 1$, then we again use $P(x) = 1$. If $a = 0$, then we use $P(x) = x + 1$ to get a desired property:

$$(x^2 + ax + b) \cdot P(x) = (x^2 + 1)(x + 1) = x^3 + x^2 + x + 1.$$

We are left with two cases, -2 and 2, for which we use $P(x) = x + 1$ and, respectively, $P(x) = x - 1$: $(x^2 - 2x + 1)(x + 1) = x^3 - x^2 - x + 1$ and $(x^2 + 2x + 1)(x - 1) = x^3 + x^2 - x - 1$.

Putting both cases together, we conclude that the set of solutions to our problem is

$$(a, b) \in \{(-2, 1), (-1, -1), (-1, 1), (0, -1), (0, 1), (1, -1), (1, 1), (2, 1)\}.$$

REMARKS

In order to solve the problem in this section, we used the geometry of the roots of a given polynomial, that is, the information about their localization (in the complex plane or on the real line, depending if we allow complex roots or restrict ourselves to real ones). It is perhaps surprising that one can actually deduce it from the degree and the coefficients of the polynomial. Some of these properties are important for many applications, such as upper bounds on the absolute values of the roots, which define a disk containing all roots, or lower bounds on the distance between two roots. Such bounds are widely used for root-finding algorithms for polynomials, either for limiting the regions where roots should be searched in, or for the computation of the computational complexity of these algorithms.

There are many upper bounds for the magnitudes of all complex roots. We will only mention two of them, Lagrange's and Cauchy's bounds. In our problem, we used the bound of Cauchy. Let $P(x) = \sum_{i=0}^{n} a_i x^i$ be a polynomial and let z be its root, that is, $P(z) = 0$. **Lagrange's Bound** is

$$|z| \leq \max\left\{1, \sum_{i=0}^{n-1} \left|\frac{a_i}{a_n}\right|\right\},$$

whereas **Cauchy's Bound** is

$$|z| \leq 1 + \max_{0 \leq i \leq n-1}\left\{\left|\frac{a_i}{a_n}\right|\right\}. \tag{3.10}$$

Lagrange's bound is smaller than Cauchy's one only when 1 is larger than the sum of all ratios $|a_i/a_n|$ but the largest which is relatively rare in practice. As a result, Cauchy's bound is more widely known and used than Lagrange's. We will prove the bound of Cauchy below.

Let z be any root of $P(x)$. If $|z| \leq 1$, then (3.10) is trivially satisfied so suppose that $|z| > 1$. Since $P(z) = 0$, we get that

$$-a_n z^n = \sum_{i=0}^{n-1} a_i z^i,$$

and so

$$|a_n||z|^n = \left|\sum_{i=0}^{n-1} a_i z^i\right| \leq \sum_{i=0}^{n-1} |a_i||z|^i \leq \left(\max_{0 \leq i \leq n-1} |a_i|\right) \cdot \sum_{i=0}^{n-1} |z|^i$$

$$= \max_{0 \leq i \leq n-1} |a_i| \frac{|z|^n - 1}{|z| - 1} \leq \max_{0 \leq i \leq n-1} |a_i| \frac{|z|^n}{|z| - 1},$$

as it is assumed that $|z| > 1$. It follows that

$$|a_n|(|z| - 1) \leq \max_{0 \leq i \leq n-1} |a_i|$$

which yields the desired bound (3.10).

EXERCISES

3.4.1. Let $f_1 = 0, f_2 = 1$, and $f_{n+2} = f_{n+1} + f_n$ for all $n \in \mathbf{N}$. Find all polynomials $P(x)$ having only integer coefficients with the property that for each $n \in \mathbf{N}$ there exists $k = k(n) \in \mathbf{Z}$ such that $P(k) = f_n$.
(Source of the problem: PLMO LX – Phase 1 – Problem 7. Solution: our own.)

3.4.2. Suppose that a polynomial $P(x)$ has all integer coefficients. Prove that if polynomials $P(P(P(x)))$ and $P(x)$ have a common real root, then $P(x)$ also has an integer root.
(Source of the problem and solution idea: PLMO LVIII – Phase 2 – Problem 1.)

3.4.3. Consider a polynomial $P(x) = x^2 + ax + b$ with $a, b \in \mathbf{Z}$. Suppose that for every prime number p, there exists $k \in \mathbf{Z}$ such that $P(k)$ and $P(k+1)$ are divisible by p. Prove that there exists $m \in \mathbf{Z}$ such that $P(m) = P(m+1) = 0$. (Source of the problem and solution idea: PLMO LVI – Phase 2 – Problem 4.)

3.5   Unique Representation of Polynomials

SOURCE

Problem and solution idea: PLMO LI – Phase 3 – Problem 6

PROBLEM

Find all polynomials $P(x)$ of odd degree that satisfy the following equation:

$$P(x^2 - 1) = (P(x))^2 - 1. \qquad (3.11)$$

THEORY

Lagrange Polynomials | Polynomials can be used to approximate complicated functions (for example, trigonometric functions) that are computationally difficult to deal with. Indeed, one can pick a few known data points, create a lookup table, and use polynomials to interpolate between those data points. This approach results in significantly faster computations. There are fast algorithms to compute numerically stable solutions, much faster than what is required by standard Gaussian elimination. Alternatively, one may write down the polynomial immediately in the form of Lagrange polynomials discussed next.

Suppose that one is given a set of n points (x_i, y_i), $i \in [n]$, with no two x_i values equal. The **Lagrange polynomial** is the polynomial of lowest degree that passes through all of these n points. It is easy to see that the interpolating polynomial of the least degree is unique and can be computed using the following formula:

$$P(x) = \sum_{j=1}^{n} y_j \prod_{\substack{1 \le k \le n \\ k \ne j}} \frac{x - x_k}{x_j - x_k}.$$

Indeed, it is easy to see that $P(x)$ goes through (x_i, y_i) as for $x = x_i$ all terms in the sum but the ith term vanish, and in the ith term all fractions in the product are equal to 1. In order to prove uniqueness, consider two polynomials $P(x)$ and $Q(x)$ of degree less than n that go through points (x_i, y_i). But this means that $P(x) - Q(x)$ is a polynomial of degree less than n and has n distinct roots, namely, x_i, $i \in [n]$. However, by the fundamental theorem of algebra, this is only possible if $P(x) - Q(x) = 0$ for all x and so $P(x) = Q(x)$ for all x.

In fact, the above proof of uniqueness naturally extends to the case of infinite number of points. We get the following useful fact. Suppose that there is an infinite set of points (x_i, y_i), $i \in \mathbf{N}$, with no two x_i values equal to each other. Note that it might be the case that there is no polynomial that goes through all of these points (consider, for example, the set

$\{(0,1), (1,0), (2,0), (3,0), \ldots\}$; there is no polynomial $P(x)$ that has an infinite number of roots, unless $P(x) = 0$ for all x, the special case). On the other hand, if there is a polynomial that passes through all of these points, then this polynomial is uniquely defined. In order to see this, consider any two polynomials $P(x)$ and $Q(x)$ that pass through these points. As for the finite case, by considering the polynomial $P(x) - Q(x)$, we get that $P = Q$, since both $P(x)$ and $Q(x)$ have a finite degree.

Finally, let us discuss one more useful fact. Suppose that a polynomial $P(x)$ has a root at point x_0 and x_0 is a local extremum (either local maximum or local minimum). We will show that the multiplicity of root x_0 is even. In order to see this, to derive a contradiction, suppose that the multiplicity of x_0 is some odd natural number k. It follows that one can represent polynomial $P(x)$ as $P(x) = (x-x_0)^k Q(x)$, where $Q(x)$ is some polynomial and $Q(x_0) \neq 0$. Since $Q(x)$ is continuous, there exists an open interval around x_0, $(x_0 - \epsilon, x_0 + \epsilon)$ for some $\epsilon > 0$, such that $Q(x)$ does *not* change sign on that interval. On the other hand, since k is odd, the polynomial $(x - x_0)^k$ *does* change the sign at x_0 and so $P(x)$ also changes it at that point. As a result, there is no extremum at x_0 and we get the desired contradiction. The statement holds and the proof is finished.

Even and Odd Functions Let us also add a remark on even and odd functions, as we will need the associated symmetry relations to solve our problem. A function $f \colon \mathbf{R} \to \mathbf{R}$ is **even** if $f(x) = f(-x)$ for all $x \in \mathbf{R}$. Geometrically speaking, the graph of an even function is symmetric with respect to the y-axis, meaning that its graph remains unchanged after reflecting it about the y-axis. Examples of even functions are $f(x) = |x|$, $f(x) = x^2$, and $f(x) = \cos x$. On the other hand, a function $f \colon \mathbf{R} \to \mathbf{R}$ is **odd** if $-f(x) = f(-x)$ for all $x \in \mathbf{R}$. Geometrically, the graph of an odd function has symmetry with respect to the origin, meaning that its graph remains unchanged after a rotation of 180 degrees about the origin. Examples of odd functions are $f(x) = x$, $f(x) = x^3$, $f(x) = \sin x$.

Let us mention about some basic but useful properties and their implications for polynomials. The sum of two even functions is even and the sum of two odd functions is odd. If $f(x)$ is even, then so is $-f(x)$; the same property holds for odd functions. As a result, the difference between two odd functions is odd and the difference between two even functions is even. The sum of an even and odd function is neither even nor odd, unless one of the functions is equal to zero over the whole domain.

From these observations we get that one can represent any polynomial $P(x) = \sum_{i=0}^{n} c_i x^i$ as a sum of two polynomials $Q(x)$ and $R(x)$ such that $Q(x)$ is odd and $R(x)$ is even. Indeed, one can partition all the terms of $P(x)$ into even and odd terms, that is,

$$Q(x) = \sum_{i=0}^{\lfloor (n-1)/2 \rfloor} c_{2i+1} x^{2i+1} \quad \text{and} \quad R(x) = \sum_{i=0}^{\lfloor n/2 \rfloor} c_{2i} x^{2i}.$$

It follows that $P(x)$ is even if and only if $Q(x) = 0$. Similarly, $P(x)$ is odd if and only if $R(x) = 0$. Moreover, since any polynomial has finite degree, in order to determine whether a polynomial is even or odd it is enough to check the corresponding conditions ($P(x) = P(-x)$ or, respectively, $P(x) = -P(-x)$) for infinitely many distinct points x (not necessarily for all $x \in \mathbf{R}$). Finally, note that if a function is even and odd, it must be equal to 0 everywhere. As a result, the only polynomial that passes both tests for infinitely many distinct points is the polynomial $P(x) = 0$, $x \in \mathbf{R}$.

SOLUTION

Let us first observe that

$$(P(x))^2 \;=\; P(x^2 - 1) + 1 \;=\; P((-x)^2 - 1) + 1 \;=\; (P(-x))^2.$$

It follows that either $P(x) = P(-x)$ holds for an infinite number of values of x or the equation $P(x) = -P(-x)$ does. Note that both of these properties cannot hold simultaneously, unless $P(x) = 0$ everywhere, which we directly check that is impossible. Finally, as noted above, satisfying any of these properties for infinitely many points implies that it actually holds for the whole real line $\mathbf{R}$.

Since $P(x)$ has odd degree, it follows that $P(x) = -P(-x)$ for all $x \in \mathbf{R}$, that is, $P(x)$ is symmetric around the origin. In particular, we get that $P(0) = 0$. Using the original equation with $x = 0$, we get that $P(-1) = P(0^2 - 1) = (P(0))^2 - 1 = -1$ and so $P(1) = 1$.

The next property that we will prove is that $P(y) \geq -1$ for all $y \geq -1$. (However, we will only use it for $y \geq 1$.) Let $y \geq -1$ be any real number. Since $x = x(y) := \sqrt{y + 1}$ satisfies $y = x^2 - 1$, we get that

$$P(y) \;=\; P(x^2 - 1) \;=\; (P(x))^2 - 1 \;\geq\; -1.$$

Our next task is to show that $P(x) = x$ holds for an infinitely many values of x and so the only solution is the polynomial $P(x) = x$. In order to see this, let us recursively define the following sequence of numbers: $x_1 = 1$ and for each $n \in \mathbf{N} \setminus \{1\}$, we define $x_n = \sqrt{x_{n-1} + 1}$. It is straightforward to see that this sequence is increasing. In fact, it is tending to $(\sqrt{5} + 1)/2 \approx 1.618$, the unique solution to the equation $x = \sqrt{x + 1}$. However, we will not need this property. We will show by induction on n that for all $n \in \mathbf{N}$, we have $P(x_n) = x_n$, which will finish the proof.

The base case is trivial: note that $P(x_1) = P(1) = 1 = x_1$. For the inductive step, suppose that $P(x_{n-1}) = x_{n-1}$ for some $n \in \mathbf{N} \setminus \{1\}$. It follows from the original equation (3.11) that

$$(P(x_n))^2 \;=\; P(x_n^2 - 1) + 1 \;=\; P(x_{n-1}) + 1.$$

From the inductive hypothesis, we get that

$$(P(x_n))^2 \;=\; P(x_{n-1}) + 1 \;=\; x_{n-1} + 1 \;=\; x_n^2,$$

and so $P(x_n) = x_n$ or $P(x_n) = -x_n$. Finally, since $x_n > x_1 = 1$ and $P(x_n) \geq -1$, we get that $P(x_n) = x_n$ and so the proof is finished.

REMARKS

In our problem, we were restricted to polynomials of odd degree. Let us now relax this assumption and consider polynomials of even degree. Constant polynomials (that is, polynomials of degree 0) are easy to investigate. If $P(x) = c$ for some $c \in \mathbf{R}$, then c must satisfy the equation $c = c^2 - 1$ and so there are only two constant polynomials that satisfy the equation (3.11): $P(x) = (1 + \sqrt{5})/2$ and $P(x) = (1 - \sqrt{5})/2$.

Let us now consider any non-constant polynomial $P(x)$ of even degree that satisfies the equation (3.11). Arguing as before, we get that $(P(x))^2 = (P(-x))^2$ for all $x \in \mathbf{R}$. This time, it follows that $P(x) = P(-x)$ for all $x \in \mathbf{R}$ (that is, $P(x)$ is even), and so $P(x)$ does *not* have any non-zero odd terms. As a result, there exists a polynomial $Q(x)$ such that $P(x) = Q(x^2)$. It is convenient to introduce the polynomial $R(x) := Q(x + 1)$, $x \in \mathbf{R}$, so that $P(x) = Q(x^2) = R(x^2 - 1)$. From this observation, using additionally the fact that $P(x)$ satisfies the equation (3.11), we get that for all $x \in \mathbf{R}$,

$$R((x^2 - 1)^2 - 1) = P(x^2 - 1) = (P(x))^2 - 1 = (R(x^2 - 1))^2 - 1.$$

Substituting $y = x^2 - 1$ we get that $R(y^2 - 1) = (R(y))^2 - 1$ and, since the function $f : [0, \infty) \to [-1, \infty)$ defined as $f(x) := x^2 - 1$ is a bijection, we get that the polynomial $R(x)$ satisfies (3.11) for all $x \in [-1, \infty)$. Arguing as before, we get that $(R(x))^2 = (R(-x))^2$ for all $x \in [-1, 1]$ and so either $R(x) = R(-x)$ holds for an infinite number of values of $x \in [-1, 1]$, or $R(x) = -R(-x)$ does. It follows that $R(x)$ is either even or odd on the whole real line $\mathbf{R}$. It follows that $R(x)$ in fact satisfies (3.11) for all $x \in \mathbf{R}$, not only for those $x \in [-1, \infty)$.

Let us note that the degree of $R(x)$ is the same as the degree of $Q(x)$ and so it is less than the one of $P(x)$. As a result, repeating this reduction process we will eventually reach the case when $R(x)$ has an odd degree, as it is impossible that $R(x)$ is constant if $P(x)$ were not constant. Hence, if this happens, then $R(x) = x$ since we showed earlier that this is the only solution for the odd case. Therefore, all non-constant solutions of even degree have the form $T^{(n)}(x)$ for some $n \in \mathbf{N}$, and $T(x) = x^2 - 1$. Note that $T^{(n)}(x) = T \circ \ldots \circ T(x)$ is the composition of the function $f(x)$ performed n times—see Section 3.7 for more details.

EXERCISES

3.5.1. Find all polynomials $P(x)$ with real coefficients that satisfy the following property: if $x + y$ is rational, then $P(x) + P(y)$ is also rational.
(Source of the problem: PLMO LIV – Phase 1 – Problem 9. Solution: our own.)

3.5.2. Let $P(x)$ be a polynomial with real coefficients. Prove that if there exists an integer k such that $P(k)$ is not an integer, then there are infinitely many such integers.

(Source of the problem and solution idea: PLMO LXVI – Phase 3 – Problem 2.)

3.5.3. Let $F(x)$, $G(x)$, and $H(x)$ be some polynomials of degree at most $2n+1$ with real coefficients. Moreover, suppose that the following properties hold:

(1) for all $x \in \mathbf{R}$, $F(x) \le G(x) \le H(x)$,

(2) there exist n different numbers $x_i \in \mathbf{R}$, $i \in [n]$, such that $F(x_i) = H(x_i)$ for all $i \in [n]$,

(3) there exists $x_0 \in \mathbf{R}$, different than x_i for $i \in [n]$, such that $F(x_0) + H(x_0) = 2G(x_0)$.

Prove that for all $x \in \mathbf{R}$, $F(x) + H(x) = 2G(x)$.

(Source of the problem and solution: XVIII Mathematical Olympics of Baltic Countries – Problem 3.)

3.6 Polynomial Factorization

SOURCE

Problem and solution: PLMO LXII – Phase 2 – Problem 6

PROBLEM

Suppose that $P_1(x)$ and $Q_1(x)$ are two different polynomials with real coefficients that satisfy the following condition: $P_1(Q_1(x)) = Q_1(P_1(x))$ for all $x \in \mathbf{R}$. For $n \in \mathbf{N} \setminus \{1\}$, let $P_n(x) := P_1(P_{n-1}(x))$ and $Q_n(x) = Q_1(Q_{n-1}(x))$. Prove that $P_1(x) - Q_1(x)$ divides $P_n(x) - Q_n(x)$ for all $n \in \mathbf{N}$.

THEORY

Consider any non-zero polynomial $P(x) = \sum_{i=0}^{n} p_i x^i$ of degree n and any point $x_0 \in \mathbf{R}$. We can then re-write $P(x)$ as follows: $P(x) = (x - x_0)Q(x) + r$, where $Q(x) = \sum_{i=0}^{n-1} q_i x^i$ is a polynomial of degree $n - 1$ and $r \in \mathbf{R}$ is a constant. In order to see this, observe that we can find the coefficients q_i and the constant r explicitly by comparing the corresponding coefficients of the two polynomials. We get that $q_{n-1} = p_n$, $q_{i-1} = x_0 q_i + p_i$ for $i \in [n-1]$, and $r = p_0 + q_0 x_0$. Hence, we can successively calculate q_i (staring from q_{n-1} and finishing with q_0), and at the end then compute r. Let us also note that $r = P(x_0)$. As a result, the above result could also be obtained in a different way. Consider the polynomial $R(x) := P(x) - P(x_0)$. Since $R(x_0) = 0$, we get that $R(x) = (x - x_0)Q(x)$ for some polynomial $Q(x)$, and so $P(x) = (x - x_0)Q(x) + P(x_0)$.

In particular, if $x_0 \in \mathbf{R}$ is a root of some non-zero polynomial $P(x)$, then the constant r has to be equal to zero and we get that $P(x) = (x - x_0)Q(x)$ for some polynomial of degree $n - 1$.

SOLUTION

Let us first observe that for any two different polynomials $G(x)$ and $H(x)$, and any polynomial $F(x) = \sum_{i=0}^{n} c_i x^i$, we have that $G(x) - H(x)$ divides $F(G(x)) - F(H(x))$. Indeed, note that

$$
\begin{aligned}
F(G(x)) - F(H(x)) &= \sum_{i=0}^{n} c_i (G(x))^i - \sum_{i=0}^{n} c_i (H(x))^i \\
&= \sum_{i=0}^{n} c_i \left((G(x))^i - (H(x))^i \right) \\
&= (G(x) - H(x)) \sum_{i=0}^{n} c_i \sum_{j=0}^{i-1} (G(x))^j (H(x))^{i-1-j} .
\end{aligned}
$$

Our second observation is that $P_n(Q_1(x)) = Q_1(P_n(x))$ for each $n \in \mathbf{N}$. This can be easily proved by induction on n. Indeed, the base case $(n = 1)$ follows immediately from our assumption that $P_1(Q_1(x)) = Q_1(P_1(x))$. For the inductive step, suppose that $P_{n-1}(Q_1(x)) = Q_1(P_{n-1}(x))$ for some $n \in \mathbf{N} \setminus \{1\}$. Using the inductive hypothesis and our assumption that $P_1(Q_1(x)) = Q_1(P_1(x))$, we get that

$$
\begin{aligned}
P_n(Q_1(x)) &= P_1(P_{n-1}(Q_1(x))) = P_1(Q_1(P_{n-1}(x))) \\
&= Q_1(P_1(P_{n-1}(x))) = Q_1(P_n(x)),
\end{aligned}
$$

and the second claim holds.

We are now ready to come back to our main task of showing that $P_1(x) - Q_1(x)$ divides $P_n(x) - Q_n(x)$ for all $n \in \mathbf{N}$. We will prove it by induction on n. The base case $(n = 1)$ is trivial. For the inductive step, suppose that the claim holds for some $n \in \mathbf{N}$, that is, $P_1(x) - Q_1(x)$ divides $P_n(x) - Q_n(x)$. Without loss of generality, we may assume that $P_1(x)$ is not constant (note that, since $P_1(x)$ and $Q_1(x)$ are different, they also cannot be both constant). Note that

$$
\begin{aligned}
P_{n+1}(x) - Q_{n+1}(x) &= P_1(P_n(x)) - Q_1(Q_n(x)) \\
&= P_n(P_1(x)) - Q_n(Q_1(x)) \\
&= (P_n(P_1(x)) - P_n(Q_1(x))) \\
&\quad + (P_n(Q_1(x)) - Q_n(Q_1(x))) ,
\end{aligned}
$$

where the second equality holds because the composition of functions is associative—see Section 3.7 for more details. We will independently show that both terms are divisible by $P_1(x) - Q_1(x)$. The first observation implies that the first term, $P_n(P_1(x)) - P_n(Q_1(x))$, is divisible by $P_1(x) - Q_1(x)$ since $P_1(x)$ and $Q_1(x)$ are different and P_1 is not constant. Using the second observation, we may re-write the second term as follows: $P_n(Q_1(x)) - Q_n(Q_1(x)) = Q_1(P_n(x)) - Q_1(Q_n(x))$. If $P_n(x)$ and $Q_n(x)$ are identical, then this term vanishes. Otherwise, we apply the first observation one more time to get that this term is divisible by $P_n(x) - Q_n(x)$, and so also by $P_1(x) - Q_1(x)$, by the inductive hypothesis.

REMARKS

The problem we considered in this section belongs to a large and important family of problems where one assumes that some property (or a set of properties) holds and the goal is to show that some other property also holds. However, it is important to keep in mind that formally the statement we aim to prove is an example of the conditional statement $P \to Q$. Moreover, the statement $P \to Q$ is true when P is false, regardless whether Q is false or true. In such examples, we say that the conditional statement is **vacuously true** or **true by default**, which may lead to situations not necessarily intended by the author. As an example, consider the following two statements: 1) *All* *the banks we robbed are in Canada*, and 2) *All the banks we robbed are outside of Canada*. Since we actually did not rob any bank, regardless whether in Canada or not, both statements are vacuously true.

Therefore, in practice it is important to make sure that there are objects that satisfy the assumed properties of the theorem. The problem in this section did not ask us to verify this so let us now make sure that we did not prove a statement that is vacuously true. Indeed, in our problem one can clearly take $Q_1(x) = x$ and any $P_1(x)$ different than $Q_1(x)$ to satisfy the assumptions of our problem. Less trivial example is the pair $Q_1(x) = x^3 - 3x$ and $P_1(x) = x^2 - 2$. Indeed, for this choice we get that

$$\begin{aligned} Q_1(P_1(x)) &= (x^2 - 2)^3 - 3(x^2 - 2) \\ &= x^6 - 6x^4 + 9x^2 - 2 \\ &= (x^3 - 3x)^2 - 2 = P_1(Q_1(x)). \end{aligned}$$

On the other hand, it might be the case that there are actually *no* objects that satisfy the assumed properties of the theorem. In such situations, conditional statements can often help us to formally prove it. Indeed, one can show that the statement $P \to Q$ is true and then that Q is false. The conclusion is that P has to be false since that is the only possibility for the conditional statement $P \to Q$ to be true. Such reasoning is called **proof by contradiction** and we often use it in this book.

In order to illustrate this technique, let us consider the following example related to the problem from this section. We will first prove the following conditional statement: *if* a polynomial $P : \mathbf{R} \to \mathbf{R}$ of odd degree satisfies the equation $P(x) = Q(x^2)$ for some polynomial $Q(x)$, *then* $P(x)$ is even. Indeed, it is clear that for each $x \in \mathbf{R}$ we have $P(x) = Q(x^2) = Q((-x)^2) = P(-x)$ so $P(x)$ is even. The conditional statement is true. However, any polynomial $P(x)$ of odd degree has the property that $\lim_{x \to \infty} P(x) = \infty$ and $\lim_{x \to -\infty} P(x) = -\infty$, or vice versa, $\lim_{x \to \infty} P(x) = -\infty$ and $\lim_{x \to -\infty} P(x) = \infty$. As a result, there exists $x \in \mathbf{R}$ such that $P(x) \neq P(-x)$, and so $P(x)$ is not even. In other words, we showed that no polynomial of odd degree is even. The conclusion is that no polynomial $P(x)$ of odd degree satisfies $P(x) = Q(x^2)$ for some polynomial $Q(x)$.

EXERCISES

3.6.1. Find all real numbers m for which the polynomial $f(x) = 2x^4 - 7x^3 + mx^2 + 22x - 8$ has two real roots whose product is equal to 2.
(Source of the problem and solution: PLMO XXX – Phase 1 – Problem 5.)

3.6.2. Given the polynomial $P(x) = x^4 - 3x^3 + 5x^2 - 9x$, $x \in \mathbf{R}$, find all pairs of integers a and b such that $a \neq b$ and $P(a) = P(b)$.
(Source of the problem and solution idea: PLMO LIV – Phase 2 – Problem 3.)

3.6.3. Find all polynomials $P(x)$ with real coefficients that satisfy the following property: for all $x \in \mathbf{R}$, $P(x^2) \cdot P(x^3) = (P(x))^5$.
(Source of the problem and solution idea: PLMO LIX – Phase 1 – Problem 6.)

3.7   Polynomials and Number Theory

SOURCE

Problem: PLMO LXX – Phase 2 – Problem 3
Solution: our own

PROBLEM

Let $f(t) = t^3 + t$ for $t \in \mathbf{R}$. Consider the family of iterated functions defined as follows: $f^{(0)}(t) = t$, $t \in \mathbf{R}$, and for each $i \in \mathbf{N}$ we define $f^{(i)}(t) = f(f^{(i-1)}(t))$, $t \in \mathbf{R}$. Decide if there exist rational numbers x and y and natural numbers m and n such that $xy = 3$ and $f^{(m)}(x) = f^{(n)}(y)$.

THEORY

$\boxed{\text{Function Composition}}$ Problem in this section requires considering repeated application of the same function to itself. In general, consider any two functions $f \colon X \to Y$ and $g \colon Y \to Z$. Then, the function g can be applied to the result of applying the function f to x. Formally, the **composition** of these two functions is the function $g \circ f \colon X \to Z$ defined as follows: $(g \circ f)(x) := g(f(x))$ for all $x \in X$. This process is called **function composition**.

If $f \colon X \to Y$ and $Y \subseteq X$, then one may compose function f with itself. The result is often denoted by $f^{(2)}$, that is, $f^{(2)} = f \circ f$. More generally, for any $n \in \mathbf{N} \setminus \{1\}$, the nth **functional power** is defined inductively by $f^{(n)} := f \circ f^{(n-1)}$. Repeated composition of such a function with itself is called **iterated function**.

Function composition has several useful properties. First of all, the composition of functions is always associative; that is, if f, g, and h are any three functions with suitably chosen domains and codomains, then $h \circ (g \circ f) = (h \circ g) \circ f$. An implication for iterated functions is that for any $k, \ell \in \mathbf{N}$, we have

$$f^{(k+\ell)} = f^{(k)} \circ f^{(\ell)} = f^{(\ell)} \circ f^{(k)}.$$

Moreover, it is easy to show that if f and g are one-to-one, then also $g \circ f$ is one-to-one. Similarly, the composition of two onto functions is always onto. As a result, if f and g are bijections, then $g \circ f$ is a bijection. The inverse function of a composition (assumed it is invertible) has the property that

$$(g \circ f)^{-1} = g^{-1} \circ f^{-1}.$$

Finally, let us mention that in order to solve the problem, we will use the concept of an **invariant**, which is discussed in more detail in Section 4.2.

SOLUTION

Let us first observe that $f^{(1)}(t) = f(f^{(0)}(t)) = f(t)$ so, indeed, we deal with iterated functions. Recall also that the composition of functions is associative, that is, for each $i \in \mathbf{N} \setminus \{1\}$, we have $f^{(i)} = f^{(1)} \circ f^{(i-1)} = f^{(i-1)} \circ f^{(1)}$.

Let us now define a function $s : \mathbb{Q} \to \{0,1\}$ in the following way. Let $r = a/b$ be a rational number expressed in lowest terms; that is, $a \in \mathbf{Z}, b \in \mathbf{N}$, and $\gcd(a,b) = 1$. Then, $s(r) = 0$ if $3 \mid a$; otherwise, $s(r) = 1$.

We will show that the value of s does *not* change under transformation f, which is the key observation that will allow us to solve the problem. Indeed, consider any rational number $r = a/b$ expressed in lowest terms. Then,

$$s(f(r)) = s(r^3 + r) = s\left(\frac{a^3}{b^3} + \frac{a}{b}\right) = s\left(\frac{a(a^2 + b^2)}{b^3}\right).$$

Suppose first that $s(r) = 0$, that is, $3 \mid a$. Since $\gcd(a,b) = 1$, we get that b is not divisible by 3. It follows that b^3 is not divisible by 3 whereas $3 \mid a(a^2+b^2)$, and so we get that $s(f(r)) = 0$. On the other hand, if $s(r) = 1$, then 3 does not divide a. It follows that also 3 does not divide $a(a^2+b^2)$, as $a^2 + b^2$ would be divisible by 3 if and only if both a and b were divisible by 3 which is not the case, and so $s(f(r)) = 1$.

Our final observation is that if $xy = 3$ for some rational numbers x, y, then $s(x) \neq s(y)$. Indeed, suppose that $x = a/b$ is a rational number expressed in lowest terms. If $s(x) = 0$ (that is, 3 divides a but 3 does not divide b), then $a = 3k$ for some $k \in \mathbf{Z}$ and so $y = 3/x = 3b/a = b/k$. Since 3 does not divide b we get that $s(y) = 1$. On the other hand, if $s(x) = 1$ (that is, 3 does not divide a), then in the fraction $y = 3b/a$ the numerator must be divisible by 3 and so $s(y) = 0$. However, this means that there is no solution to the equation defined in the problem, as no matter which n and m we select, function s evaluated at $f^{(m)}(x)$ is different than the one at $f^{(n)}(y)$ as long as $xy = 3$. In particular, it implies that $f^{(m)}(x) \neq f^{(n)}(y)$.

REMARKS

Let us point out that the assumption in our problem that x and y are rational numbers is crucial. Indeed, if x and y are allowed to be any real numbers, then for any $n, m \in \mathbf{N}$, we can easily find $x, y \in \mathbf{R}$ such that $xy = 3$ and $f^{(m)}(x) = f^{(n)}(y)$.

In order to see this note that for any $n \in \mathbf{N}$, the function $f^{(n)}(x)$ satisfies the following properties: i) $f^{(n)}(0) = 0$, ii) $f^{(n)}(x)$ is increasing on the interval $[0, \infty)$, iii) $\lim_{x \to \infty} f^{(n)}(x) = \infty$, and iv) $f^{(n)}(x)$ is a polynomial and so it is a continuous function. Now, fix any $n, m \in \mathbf{N}$ and consider function

$$g(x) := f^{(n)}(x) - f^{(m)}(3/x).$$

Using the properties of $f^{(n)}(x)$ we get that $g(x)$ is continuous on $(0, \infty)$, $\lim_{x \to 0+} g(x) = -\infty$, and $\lim_{x \to \infty} g(x) = \infty$. It follows that there exists

$x_0 \in (0, \infty)$ such that $g(x_0) = 0$. Hence, there exists a pair $x = x_0 \in \mathbf{R}_+$ and $y = 3/x_0 \in \mathbf{R}_+$ such that $xy = 3$ and $f^{(m)}(x) = f^{(n)}(y)$.

EXERCISES

3.7.1. Prove that there are no polynomials $P_1(x), P_2(x), P_3(x), P_4(x)$ with rational coefficients that satisfy

$$\sum_{i=1}^{4}(P_i(x))^2 = x^2 + 7 \qquad \text{for all } x \in \mathbf{R}.$$

(Source of the problem and solution idea: PLMO LXII – Phase 3 – Problem 6.)

3.7.2. Consider a polynomial $f(x) := x^2 + bx + c$, where $b, c \in \mathbf{Z}$. Prove that if $n \in \mathbf{N}$ divides $f(p)$, $f(q)$, and $f(r)$ for some $p, q, r \in \mathbf{Z}$, then

$$n \mid (p - q)(q - r)(r - p).$$

(Source of the problem and solution idea: PLMO LXIV – Phase 2 – Problem 1.)

3.7.3. Consider a polynomial $P(x)$ with integer coefficients that satisfies the following property: if $a, b \in \mathbf{Q}$ and $a \neq b$, then $P(a) \neq P(b)$. Does it mean that $P(a) \neq P(b)$ for all $a, b \in \mathbf{R}$, $a \neq b$?
(Source of the problem and solution idea: PLMO LXIV – Phase 2 – Problem 5.)

Chapter 4

Combinatorics

As usual, we start the chapter with some basic definitions.

THEORY

Graphs Some of our examples will be from graph theory and so here we introduce a few basic definitions. A (simple) **graph** $G = (V, E)$ is a pair consisting of a **vertex set** $V = V(G)$ and an **edge set** $E = E(G)$ consisting of pairs of vertices; that is,

$$E(G) \subseteq \{\{u, v\} : u, v \in V(G), u \neq v\}.$$

We write uv if u and v form an edge, and say that u and v are **adjacent** or **joined**. We refer to u and v as **endpoints** of the edge uv. The **order** of a graph is $n := |V(G)|$, and its **size** is $m := |E(G)|$.

If u and v are the **endpoints** of an edge, then we say that they are **neighbors**. The **neighborhood** of a vertex v, denoted $N(v)$, is the set of all neighbors of v. The **degree** of a vertex v, written $\deg(v)$, is the number of neighbors of v; that is, $\deg(v) := |N(v)|$. The numbers

$$\delta = \delta(G) := \min_{v \in V(G)} \deg(v)$$

$$\Delta = \Delta(G) := \max_{v \in V(G)} \deg(v)$$

are the **minimum degree** and, respectively, the **maximum degree** of G. A graph is called k**-regular**, provided each of its vertices has degree k.

A **clique** (sometimes called a **complete graph**) is a set of pairwise-adjacent vertices. The clique of order n is denoted by K_n. An **independent** set (sometimes called an **empty graph**) is a set of pairwise-nonadjacent vertices. The **path** on n vertices, denoted by P_n, consists of n vertices, $v_1, \ldots, v_n$, and $n-1$ edges, $v_i v_{i+1}$ for $i \in [n-1]$. The **cycle** on n vertices, denoted by C_n, consists of n vertices, $v_1, \ldots, v_n$, and n edges, $v_n v_1$ and $v_i v_{i+1}$ for $i \in [n-1]$.

A graph $G' = (V', E')$ is a **subgraph** of $G = (V, E)$ if $V' \subseteq V$ and $E' \subseteq E$. If $V' \subseteq V$, then

$$G[V'] = \left(V', \{uv \in E : u, v \in V'\} \right)$$

is the subgraph of G **induced** by V'.

| **Bipartite Graphs** | A graph G is **bipartite** if the vertex set can be partitioned into two sets, X and Y (that is, $V(G) = X \cup Y$, where $X \cap Y = \emptyset$), and every edge is of the form xy, where $x \in X$ and $y \in Y$. Here X and Y are called **partite sets**. This definition can be easily generalized to r-**partite** graphs. This time, $V(G) = X_1 \cup X_2 \cup \ldots \cup X_r$ for some $r \geq 2$ and there is no edge of the graph with both endpoints in X_i, for any $i \in [r]$. The **complete** r-**partite graph** $K_{n_1, n_2, \ldots, n_r}$ is the graph with partite sets $X_1, \ldots, X_n$ with $n_i = |X_i|$ ($i \in [r]$) and edges between every pair of vertices from different partite sets.

| **Matchings** | A **matching** in a graph G is a collection of disjoint edges. The vertices of G incident to the edges of a matching M are called **saturated** or **matched** by M; the other edges are **unsaturated** or **unmatched**. A matching is **maximal** if it cannot be extended by adding an edge. A matching is **maximum** if it contains the largest possible amount of edges. In particular, a **perfect matching** in a graph G is a (maximum) matching in G that saturates all vertices of G.

4.1 Enumeration

SOURCE

Problem and solution idea: PLOM XVIII – Phase 3 – Problem 3

PROBLEM

There are 100 students at the party. Each student knows at least 67 other students. Prove that there are at least four students that all know each other. We assume that this relationship is symmetric; that is, if student A knows student B, then student B knows student A.

THEORY

| Constructive Argument | A **constructive argument** is a method of proving a statement that demonstrates the existence of a mathematical object by creating (or providing a method for creating) the object. For example, a common way of showing that the set of prime numbers is infinite is a famous, constructive argument due to Euclid. For a contradiction, suppose that the set of prime numbers is finite, in which case there is the largest prime number that we denote by n. But then, since $n! + 1 > n$, $n! + 1$ is not prime. On the other hand, clearly, all of its prime factors are greater than n which gives us the desired contradiction.

| Non-constructive Argument | The approach mentioned above is in contrast to a **non-constructive argument** (also known as an **existence proof**) which proves the existence of a particular kind of object without providing an explicit example. To illustrate this method, we will show that there exist two irrational numbers, x and y, such that x^y is rational. (Recall that x is **rational** if $x = a/b$ for some $a \in \mathbf{Z}$ and $b \in \mathbf{N}$; otherwise, x is **irrational**. We will also use the fact that $\sqrt{2}$ is irrational.)

Indeed, if $\sqrt{2}^{\sqrt{2}}$ is rational, then we are immediately done: $x = y = \sqrt{2}$ is the pair that has the desired property. Otherwise, $x = \sqrt{2}^{\sqrt{2}}$ and $y = \sqrt{2}$ have the desired property as

$$x^y = \left(\sqrt{2}^{\sqrt{2}} \right)^{\sqrt{2}} = \sqrt{2}^{\sqrt{2} \cdot \sqrt{2}} = \sqrt{2}^2 = 2 \, .$$

Let us stress the fact that based on the above argument, we do not know which of the two pairs of x and y satisfies the desired property but we do know that precisely one of them does. In fact, it turns out that $\sqrt{2}^{\sqrt{2}}$ is irrational but this fact is not needed to claim the correctness of the statement.

| Greedy Algorithm | Let us come back to constructive arguments. The easiest approach one can try is to construct the desired object by making locally optimal choices at each stage with the intent of finding a global optimum. Such approach is often called **greedy strategy** (or **greedy algorithm**). Let us stress the fact that a greedy strategy does not usually produce an optimal solution but it may yield one or at least a good approximation of it.

SOLUTION

We will perform a greedy search for four students that mutually know each other. Start with *any* student A from the set of all students. Now, select *any* student B, different than A, that knows A. Clearly, it is possible since there are at least 67 students that know A.

We will now show that there exists a student C that knows both A and B. At most $100 - 67 - 1 = 32$ students do not know A and, similarly, at most 32 students do not know B. Since there are 98 students different than A and B and $32 + 32 < 98$, there must exist a student that knows both A and B, as claimed. We select *any* such student and call it C.

We continue this greedy selection process to find student D that knows students A, B, and C. Arguing as before, there are 97 students to chose from but only at most $3 \cdot 32 = 96$ of them do not know at least one of A, B, or C. Hence, at least one such student exists and the process is finished.

REMARKS

The property stated in the problem is best possible in the following sense. Suppose that there are still 100 students but this time each of them knows at least 66 other students, instead of 67. With this slightly weaker assumption, it is possible that there are no four students who know one another. Indeed, let us partition students into three sets of sizes 33, 33, and 34, respectively, and assume that a student from one set knows only students from the other two sets.

This property can be reformulated in the language of graph theory as follows: there exists a graph on $n = 100$ vertices, minimum degree 67, and without K_4, the complete graph on 4 vertices, as a subgraph. Moreover, this example is, in fact, the well-known Turán graph $T(100, 3)$, related to an important problem in extremal combinatorics. We will come back to such problems in Section 4.4. In general, the **Turán graph** $T(n, r)$ is a graph formed by partitioning a set of n vertices into r subsets, with sizes as equal as possible, and connecting two vertices by an edge if and only if they belong to different subsets. The number of edges in this graph is at most $(1 - 1/r)n^2/2$ with equality holding if and only if n is divisible by r; that is, all sets have equal sizes.

It is well-known (and, for completeness, we will prove it now) that the Turán graph has the maximum possible number of edges among all graphs on $n \geq r$ vertices with the property that no $r + 1$ vertices induce K_{r+1}, the complete graph on $r + 1$ vertices. Moreover, the Turán graph is the unique graph on n vertices that satisfies this property, while having this maximum number of edges. In order to prove this uniqueness property, suppose that $G = (V, E)$ is such extremal graph. We will start with proving the following property.

Observation: There are no $a, b, c, \in V$ such that $ab \in E$, $ac \notin E$, and $bc \notin E$.
Proof: For a contradiction, suppose that the opposite is true; that is, that there exist $a, b, c, \in V$ such that $ab \in E$, $ac \notin E$, and $bc \notin E$. Without loss of generality, we may assume that $\deg(a) \geq \deg(b)$. Let us independently consider the following two cases.

Case 1: $\deg(c) < \deg(a)$. Construct G' from G by replacing vertex c with vertex a', a copy of vertex a; that is, a' is adjacent to $v \in V \setminus \{c\}$ if and only if a is. In particular, a and a' are *not* adjacent. Note that G' has more edges than G. Moreover, G' does not contain K_{r+1}. Indeed, no $S \subseteq V(G') = (V \setminus \{c\}) \cup \{a'\}$, $|S| = r+1$, induces the complete graph in G' if $a' \notin S$ (otherwise, S would induce the complete graph in G). The same holds if both a and a' are in S (since a and a' are not adjacent in G'). Finally, we argue that this is true if $a' \in S$ but $a \notin S$ (otherwise, $(S \setminus \{a'\}) \cup \{a\}$ would induce the complete graph in G). This contradicts the fact that G has the maximum number of edges.

Case 2: $\deg(c) \geq \deg(a)$. This time we construct G' from G by removing vertices a and b and adding vertices c' and c'', two copies of vertex c; that is, c' and c'' are adjacent to $v \in V \setminus \{a, b\}$ if and only if c is, and are not adjacent to each other. In particular, $\{c, c', c''\}$ induce *no* edge. Note that G' has more edges than G. Indeed, note that we removed $\deg(a) + \deg(b) - 1 \leq 2\deg(a) - 1$ edges (since a and b were adjacent in G), less than the number of edges added, namely, $2\deg(c) \geq 2\deg(a)$. Moreover, arguing as in the previous case, G' does not contain K_{r+1} and so we get the desired contradiction.

It follows from the observation that for any three vertices $a, b, c \in V$, if $ab \notin E$ and $bc \notin E$, then $ac \notin E$. Hence, one can partition the vertex set V into k disjoint subsets $V_1, V_2, \ldots, V_k$ such that vertices from V_i are adjacent to all vertices in $V \setminus V_i$ but to no vertex in V_i. Since no $r + 1$ vertices induce K_{r+1}, we know that $k \leq r$; otherwise, one could pick one vertex from each set $V_1, V_2, \ldots V_{r+1}$ to form K_{r+1}.

Using the notation $n_i = |V_i|$, we clearly have $n = \sum_{i=1}^{k} n_i$. The number of edges of G is

$$\frac{1}{2} \sum_{i=1}^{k} n_i (n - n_i) = \frac{1}{2} \left(n \sum_{i=1}^{k} n_i - \sum_{i=1}^{k} n_i^2 \right) = \frac{1}{2} \left(n^2 - \sum_{i=1}^{k} n_i^2 \right).$$

By Jensen's inequality (see Section 1.1), the sum $\sum_{i=1}^{k} n_i^2$ is minimized for $n_i = n/k$, so the number of edges of G is less than or equal to

$$\frac{1}{2} \left(n^2 - k \left(\frac{n}{k} \right)^2 \right) = \left(1 - \frac{1}{k} \right) \frac{n^2}{2},$$

which is maximized for $k = r$. Note that it might happen that n/k is not an integer and so the above construction cannot be achieved. Hence, in fact, the unique graph that maximizes the number of edges has the actual sizes of V_i's selected in such a way that they differ by at most 1. This finishes the proof.

EXERCISES

4.1.1. There are $2n$ members of a chess club; each member knows at least n other members (knowing a person is a reciprocal relationship). Prove that it is possible to assign members of the club into n pairs in such a way that in each pair both members know each other.
(Source of the problem and solution idea: PLMO XLV – Phase 1 – Problem 9, modified.)

4.1.2. There are 17 players in the tournament in which each pair of two players compete against each other. Every game can last 1, 2, or 3 rounds. Prove that there exist three players who have played exactly the same number of rounds with one another.
(Source of the problem and solution: well-known, classic problem related to Ramsey numbers.)

4.1.3. Consider a group of people with the following property. Some of them know each other, in which case the corresponding pair of people mutually like each other or dislike each other. Moreover, there is a person who knows at least six other people. Interestingly, for each person the number of people he or she likes is equal to the number of people he or she dislikes. Prove that it is possible to remove some, but not all, like/dislike links such that it is still the case that each person has the same number of liked and disliked acquaintances.
(Source of the problem: PLMO LXIX – Phase 1 – Problem 7, modified. Solution: our own.)

4.2 Tilings

SOURCE

Problem: PLMO LXX – Phase 1 – Problem 3 (slight modification)
Solution: our own

PROBLEM

You are given a square grid of size 128×128.

a) Prove that it is possible to tile it with 3,276 blocks of size 5×1 and one block of size 2×2.

b) Prove that it is impossible to do it when the block of size 2×2 touches the border of the square grid.

THEORY

| Invariant | An **invariant** is a property held by a class of mathematical objects, which remains unchanged when transformations of a certain type are applied to the objects. Invariants are used in diverse areas of mathematics such as geometry, topology, algebra, and discrete mathematics.

In order to formally define an invariant we have to define an object, its property, and a transformation under which this property is invariant. Here are some classical examples, where in each of them we highlight the object, the property and, the transformation:

1. the distance (property) between two points on a number line (object) is not changed by adding the same quantity to both numbers (transformation);

2. the area (property) of a figure (object) is invariant with respect to translation (transformation);

3. the degree (property) of a polynomial (object) is invariant subject to multiplication by non-zero number (transformation);

4. the measure of angle (property) based on a given circle arc (object) is invariant with respect to the choice of location of the vertex on this arc (transformation).

In solving tilings problems we often rely on finding an insightful invariant of some mathematical property. In the problem we deal with in this section, the invariant will ensure that, after we appropriately assign numbers to all cells, no matter how a 5×1 block is placed it covers cells with the same sum of numbers. On the other hand, the corresponding sum for the 2×2 block will not have this property. This difference will turn out to be a key observation to get the proof.

SOLUTION

Part a) is rather straightforward. Observe that one can easily cover the 120×128 rectangular grid with 3,072 tiles of size 5×1 (since 120 is divisible by 5). So we will be done if we can cover the remaining 8×128 rectangular grid. In fact, since covering the 8×120 grid with 192 tiles of size 5×1 is equally easy, we may reduce the problem of covering the 128×128 grid to the one of covering the 8×8 grid (this time with 12 blocks of size 5×1 and one block of size 2×2). The tiling presented in Figure 4.1 uses the allowed blocks which finishes part a).

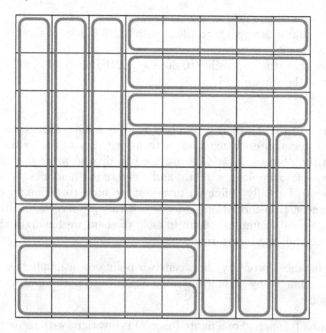

FIGURE 4.1: Illustration for Problem 4.2, part a).

Part b) is more interesting. Label the grid so that the bottom left cell has label $(1, 1)$ and the top right one has label $(128, 128)$. Starting from the bottom left corner, assign numbers from the set $\{0, 1, 2\}$ to all cells of the 128×128 square grid using the pattern presented in Figure 4.2. (For example, cells with labels (x, y) for x and y that both give the remainder of 2 when divided by 5 will get number 2 assigned.) Observe that rows $126, 127, 128$ and columns $126, 127, 128$ use only part of the pattern, namely, the part restricted to the first three rows and, respectively, the first three columns.

Let us first calculate the sum of the numbers assigned to the whole grid. It contains $25 \cdot 25$ complete copies of our 5×5 pattern, 25 copies of a part of our pattern consisting of its first three bottom rows, 25 copies of a part of the pattern consisting of its three leftmost columns, and one piece containing the

0	0	0	1	1
0	0	0	1	1
1	0	1	0	0
0	2	0	0	0
1	0	1	0	0

FIGURE 4.2: Illustration for Problem 4.2, part b).

first three bottom rows and three leftmost columns. Counting (independently) the sums of numbers in the four respective strips, we get that the total sum is equal to $25^2 \cdot 10 + 25 \cdot 6 + 25 \cdot 6 + 6 = 6{,}556$.

Let us now notice that no matter how we place 5×1 block it will always cover numbers that sum up to 2. Hence, regardless how we place 3,276 such blocks, they cover numbers that sum up to $2 \cdot 3{,}276 = 6{,}552$. This is the desired invariant that leads us to the solution of this problem. It follows that the remaining 2×2 block must cover numbers that sum up to 4. However, given the pattern in Figure 4.2 that we used, it is only possible if it lies in the top right corner of the pattern. Given the way we used the pattern to cover the 128×128 square grid, all of these positions do not lie on the border of the square grid. This proves part b) of the problem.

In fact, we not only solved part b) of the problem but proved something stronger. Namely, there are only 25^2 possible places where the 2×2 block can be placed. Moreover, by adjusting the process described in part a) of this problem, we can easily see that each of those 25^2 locations are possible.

REMARKS

The solution presented above is nice and easy to follow. However, it is not clear how to attack similar problems in the future. Hence, a natural question is how to guess the pattern presented in Figure 4.2. There are several possible methods of deriving such patterns but all of them aim to propose a setup that is repeating in terms of one of the blocks; in our problem, it is 5×1 block. A natural starting point is the straightforward pattern presented in Figure 4.3 and repeating it as described in the solution. Clearly, in this pattern 5×1 block covers exactly one 1 and four 0's.

Repeating the reasoning presented in the previous solution, we get that the number of 1's in the whole square grid is $25^2 \cdot 5 + 25 \cdot 3 + 25 \cdot 3 + 3 = 3{,}278$. On the other hand, 5×1 blocks cover 3,276 squares with 1's, which means

0	0	0	0	1
0	0	0	1	0
0	0	1	0	0
0	1	0	0	0
1	0	0	0	0

FIGURE 4.3: Illustration for Problem 4.2, part b)—starting pattern.

that the 2×2 block must cover exactly two 1's. It follows that if this block lies on the border, then there exists some $i \in \{0, \ldots, 25\}$ such that it lies:

1. on the two bottom rows and columns $1 + 5i$ and $2 + 5i$; or

2. on the two leftmost columns and rows $1 + 5i$ and $2 + 5i$; or

3. on the two top rows and columns $2 + 5i$ and $3 + 5i$; or

4. on the two rightmost columns and rows $2 + 5i$ and $3 + 5i$.

We observe now that one can repeat the argument when 1's form the diagonal from the top left cell to the bottom right one (instead of from the bottom left to the top right). In particular, the conclusion is that if the 2×2 block lies on the two bottom rows, it must lie on column $2 + 5i$ and $3 + 5i$ for some $i \in \{0, \ldots, 25\}$. Hence, there is no solution with the 2×2 block touching the bottom border of the square grid. The solutions for the other three borders can be eliminated the same way.

The solution presented earlier merges the two arguments by simply introducing the pattern obtained from the two diagonals.

EXERCISES

4.2.1. Consider a square grid of size 25×25 that has a smaller square grid of size 5×5 cut out from its bottom left corner. Can you cover the remaining cells with 100 blocks of size 1×6 or 2×3?
(Source of the problem: Polish Junior Mathematical Olympics X – Phase 1 – Problem 6. Solution: our own.)

4.2.2. Prove that it is impossible to cover a square grid of size 9×9 with tiles of size 1×5 or 1×6.
(Source of the problem: Letters of Polish Junior Mathematical Olympics, September 2014. Solution: our own.)

4.2.3. Can you cover a square grid of size 10×10 with 25 "T-shaped" blocks consisting of 4 small squares?

(Source of the problem and solution idea: Letters of Polish Junior Mathematical Olympics, September 2014.)

4.3   Counting

SOURCE

Problem and solution: PLMO LXIX – Phase 2 – Problem 5

PROBLEM

There are various clubs in a class consisting of 23 students. Each club has exactly 5 members. Moreover, any two different clubs have at most 3 members in common. Prove that there are less than 2,018 clubs in the class.

THEORY

| **Permutations** | Let S be a set of n elements. We are interested in investigating various ways in which objects from S may be selected, without replacement, to form a sequence of n elements. Each of these possible sequences is called a permutation. Formally, a **permutation** π is a bijection $\pi : [n] \to S$; $\pi(i)$ is the element that was selected at round i. (Recall that a **bijection** is a function between the elements of two sets, say A and B, where each element of A is paired with exactly one element of B, and each element of B is paired with exactly one element of A.)

The number of permutations of an n-element set (that is, the number of ways one can order n elements) is equal to

$$n! = \prod_{i=1}^{n} i = 1 \cdot 2 \cdot \ldots \cdot n . \tag{4.1}$$

One can easily prove this formula by induction. Alternatively, note that there are n ways to select the first object from S. Since the selection is done without replacements, there are $n-1$ objects left to select from; we select any of them and continue until all elements are picked. The total number of ways is then $n \cdot (n-1) \cdot \ldots \cdot 1 = n!$ and the formula (4.1) is verified.

| **Combinations** | This time k objects are selected from a set S of n elements to produce subsets without ordering (that is, unlike permutations, the order of selection does not matter). More formally, a k-**combination** of S is a subset of k distinct elements of S.

The number of k-combinations of an n-element set (provided that $1 \leq k \leq n$) is equal to the binomial coefficient

$$\binom{n}{k} = \frac{n!}{k!(n-k)!} = \frac{n(n-1)\cdots(n-k+1)}{k(k-1)\cdots 1} . \tag{4.2}$$

One can prove this formula in many ways; we provide a direct counting argument that is similar to the solution to our problem above. Select k elements, one by one, without replacement; there are $n(n-1)\cdots(n-k+1)$ ways to do it. Clearly, each subset of k elements of S can be obtained in $k!$ different ways (we know which elements are selected and we are happy with any permutation of them) so we are over-counting. The formula (4.2) holds.

[Double Counting] Let us finish with a very useful **double counting** combinatorial proof technique for showing that two expressions are equal by demonstrating that they are simply two ways of counting the same thing. For example, note that

$$\sum_{k=0}^{n} \binom{n}{k} = 2^n.$$

Indeed, on the left hand side we independently count k-elements subsets of an n-element set while the right hand side counts all subsets. Alternatively, one can use the binomial theorem (see Section 1.6) to get that

$$2^n = (1+1)^n = \sum_{k=0}^{n} \binom{n}{k} \cdot 1^k \cdot 1^{n-k} = \sum_{k=0}^{n} \binom{n}{k}.$$

As another example, note that

$$\binom{n}{k} = \binom{n-2}{k} + 2\binom{n-2}{k-1} + \binom{n-2}{k-2}.$$

In order to see it, let us color $n-2$ elements of an n-element set S red and the remaining 2 elements blue (arbitrarily). We observe that the left hand side counts k-element subsets of S. On the other hand, the right hand side independently counts k-element subsets with a given number of blue elements (that is, 0, 1, and 2, respectively).

Finally, let us show that

$$\sum_{k=0}^{n} \binom{n}{k}^2 = \binom{2n}{n}.$$

Since $\binom{n}{k} = \binom{n}{n-k}$, it is enough to show that

$$\sum_{k=0}^{n} \binom{n}{k} \cdot \binom{n}{n-k} = \binom{2n}{n}.$$

But this equality is obvious. The right hand side counts the number of n-element subsets of the set $[2n]$. The left hand side counts the same thing, where, for $0 \le k \le n$, the term $\binom{n}{k} \cdot \binom{n}{n-k}$ counts the number of subsets in which k elements are chosen from the set $[n]$ and $n-k$ elements are chosen from the set $[2n] \setminus [n]$.

SOLUTION

Let C be the set of students such that $|C| = 23$. Clubs can be represented by a family of subsets A_i of C of size 5 ($i \in [k]$, where k is the number of clubs). Since no two clubs have more than 3 members in common, each subset B of C of size 4 is contained in at most one A_i. We can then label each $B \subseteq C$ of size 4 with i if it belongs to a unique A_i, and assign it label 0 otherwise; that is, when students from B are not members of the same club. On the other hand, each A_i has clearly 5 distinct four element subsets, so 5 sets B of size 4 have label i assigned to them. Since the number of sets B of size 4 with non-zero label is at most $\binom{23}{4}$, the total number of subsets of size 4, we get that k, the number of clubs, is at most

$$\binom{23}{4}/5 = 1{,}771 < 2{,}018 \,.$$

REMARKS

The problem considered in this section is an example of a typical situation when the solution can be obtained by careful and appropriate counting technique. In our problem we first see that there are $\binom{23}{5} = 33{,}649$ sets of size 5. However, clearly not all of them can form a club as it would violate the property that clubs cannot share many members. In order to reduce the number of possible clubs, we observe that it is enough to know only 4 members of a club to uniquely identify it. This observation leads to the solution.

EXERCISES

4.3.1. The class consists of 12 people. Count in how many ways one can divide them into: 6 pairs, 4 triples, 3 quadruples, and 2 six-tuples. Which option yields the largest number of possibilities?

4.3.2. Consider an $n \times n$ square grid on which we want to place $k \le n$ chess rooks in such a way that none of them attack another rook. Count the number of ways one can do it.

4.3.3. Create all possible 4-digit numbers using digits from set $[9] = \{1, 2, 3, 4, 5, 6, 7, 8, 9\}$. Find the sum of those numbers.

4.3.4. Alice has 20 balls, all different. She first splits them into two piles and then she picks one of the piles with at least two balls, and splits it into two. She repeats this until each pile has only one ball. Find the number of ways in which she can carry out this procedure.
(Source of the problem and solution: Problem 1.8.27 from Discrete Mathematics by Lovász, Pelikán, and Vesztergombi.)

4.4 Extremal Graph Theory

SOURCE

Problem: LXVIII OM – Phase 1 – Problem 6 (modified)
Solution: our own

PROBLEM

20 boys and 20 girls attended a high-school prom. During this event, there were 98 dances. In each dance, one boy danced with one girl, and no pair danced more than once. Prove that there were two boys (say, b_1, b_2) and two girls (say, g_1, g_2) such that they all danced with one another (that is, b_1 and b_2 both danced with g_1 and g_2).

THEORY

In this section we are interested in basic **extremal graph theory** that studies extremal (maximal or minimal) graphs which satisfy some certain property. Extremality can be taken with respect to different graph invariants, such as the number of vertices, the number of edges, or the length of a longest path. Extremal graph theory officially began with Turán's theorem that we already stated (and proved) in Section 4.1.

The problem we deal with in this section is closely related to $\mathrm{ex}(n; C_4)$ defined next. The connection will be explained later.

| Turán Number | Given a class of graphs $\mathcal{F} = \{F_1, F_2, \ldots\}$, let us call a graph $\mathcal{F}$-**free** if it contains no copy of F as a subgraph for each $F \in \mathcal{F}$. Let the **Turán number**, denoted $\mathrm{ex}(n; \mathcal{F})$, be the maximal number of edges in an $\mathcal{F}$-free graph on n vertices. If the class of graphs $\mathcal{F}$ consists of a single graph, then we write $\mathrm{ex}(n; F)$ instead of $\mathrm{ex}(n; \{F\})$.

In Section 4.1 we considered $T(n, r)$, the Turán graph; that is, the complete equi-partite graph, $K_{n_1, n_2, \ldots, n_r}$ where $\sum_i n_i = n$ and $\lfloor n/r \rfloor \leq n_i \leq \lceil n/r \rceil$. By Turán's theorem we have $\mathrm{ex}(n; K_{r+1}) = e(T(n, r))$. Furthermore, $T(n, r)$ is the unique K_{r+1}-free graph that attains the extremal number. In fact, the case $\mathrm{ex}(n; K_3) = \lceil n^2/4 \rceil$ was shown earlier by Mantel.

In order to show that the bounds for the number of dances is, in some sense, best possible we need to introduce a family of graphs obtained from the projective planes. We define them now and we will explain the connection soon.

| Projective Planes | Given a set P of points and a set L of lines, we define the corresponding **incidence graph** $G(P, L)$ to be the bipartite graph whose vertices consist of the points (one partite set), and lines (the second partite set), with point $p \in P$ adjacent to line $\ell \in L$ if p lies on ℓ.

A **projective plane** consists of a set of points and lines satisfying the following axioms.

1. There is exactly one line incident with every pair of distinct points.

2. There is exactly one point incident with every pair of distinct lines.

3. There are four points such that no line is incident with more than two of them.

Finite projective planes possess $q^2 + q + 1$ points for some $q \in \mathbf{N}$ (called the order of the plane) and the same number of lines. Projective planes of order q exist for all prime powers q, and an unsettled conjecture claims that q must be a prime power for such planes to exist.

It follows immediately from the axioms that the corresponding incidence graph does not contain C_4, a cycle of length 4 (and of course any odd cycle as the graph is bipartite). It is also possible to show that this graph is $q + 1$ regular.

See Figure 4.4 for $G(P, L)$, where (P, L) is the **Fano plane** (that is, the projective plane of order 2). We note the incidence graph of the Fano plane is isomorphic to the well-known **Heawood graph**.

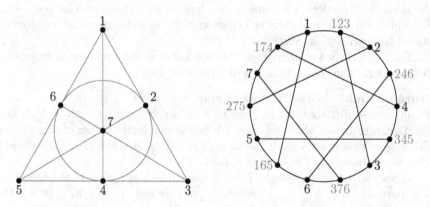

FIGURE 4.4: The Fano plane and its incidence graph.

SOLUTION

For a contradiction, suppose that there were no two boys and two girls that danced with each other. For any $i, j \in [20]$, let $x_{i,j} = 1$ if i'th boy danced with jth girl; otherwise, $x_{i,j} = 0$. Since there were 98 dances and no pair danced more than once, we have that

$$\sum_{i=1}^{20} \sum_{j=1}^{20} x_{i,j} = 98 \,.$$

Note that $g_i = \sum_{j=1}^{20} x_{i,j}$ is the number of girls that danced with the i'th boy; similarly, $b_j = \sum_{i=1}^{20} x_{i,j}$ is the number of boys that danced with the j'th girl.

Let us fix $j \in [20]$ and concentrate on the j'th girl. Our goal is to estimate $f(j)$, the number of *other* girls who danced with boys who danced with the jth girl. Clearly,

$$f(j) \leq \sum_{i=1}^{20} x_{i,j} \cdot (g_i - 1) = \left(\sum_{i=1}^{20} x_{i,j} \cdot g_i \right) - b_j . \tag{4.3}$$

Indeed, one can simply consider all boys she dances with (that is, those for which $x_{i,j} = 1$); each of them danced with $g_i - 1$ girls other than the j'th girl. In fact, the right hand side of (4.3) is not only an upper bound for $f(j)$ but equality holds. This is because since no two girls danced with the same two boys, all of these girls must be unique. Finally, since there are 19 girls other than the j'th girl, we get that $f(j) \leq 19$, or equivalently that

$$\sum_{i=1}^{20} x_{i,j} \cdot g_i \leq 19 + b_j .$$

It follows that

$$\sum_{j=1}^{20} \sum_{i=1}^{20} x_{i,j} \cdot g_i \leq \sum_{j=1}^{20} (19 + b_j) = 20 \cdot 19 + \sum_{j=1}^{20} b_j$$

$$= 380 + 98 = 478 . \tag{4.4}$$

On the other hand,

$$\sum_{j=1}^{20} \sum_{i=1}^{20} x_{i,j} \cdot g_i = \sum_{i=1}^{20} \sum_{j=1}^{20} x_{i,j} \cdot g_i = \sum_{i=1}^{20} g_i \sum_{j=1}^{20} x_{i,j}$$

$$= \sum_{i=1}^{20} g_i^2 = 20 \cdot \frac{1}{20} \sum_{i=1}^{20} g_i^2 \geq 20 \left(\frac{1}{20} \sum_{i=1}^{20} g_i \right)^2$$

$$= 20 \left(\frac{98}{20} \right)^2 = \frac{2,401}{5} = 480.2 > 478, \tag{4.5}$$

where the first inequality follows from the fact that the function $f(x) = x^2$ is convex (see Section 1.1 for more details). Inequalities (4.4) and (4.5) give us the desired contradiction.

REMARKS

In order to see the bigger picture, we will provide an alternative solution. Let us first reformulate the problem in the language of graph theory. Let B and G be the set of boys and the set of girls, respectively. Dances can be represented as bipartite graph $G = (B \cup G, E)$ where $bg \in E$ if and only if boy b danced with girl g. We know that $n = |B| = |G| = 20$ and $m = |E| = 98$. Our goal is to show that G contains C_4, a cycle of length 4.

For a contradiction, suppose that G does not contain C_4. Let

$$F = \{(b, g_1, g_2) : b \in B, bg_1 \in E, bg_2 \in E, \text{ and } g_1 \neq g_2\}.$$

Clearly,

$$\begin{aligned} |F| &= \sum_{b \in B} (\deg b)(\deg b - 1) = \sum_{b \in B} \deg^2 b - \sum_{b \in B} \deg b \\ &= \frac{1}{n}\left(\sum_{b \in B} \deg^2 b\right)\left(\sum_{b \in B} 1^2\right) - \sum_{b \in B} \deg b. \end{aligned}$$

By Cauchy-Schwarz inequality (see Section 1.7),

$$|F| \geq \frac{1}{n}\left(\sum_{b \in B}(\deg b) \cdot 1\right)^2 - \sum_{b \in B} \deg b = \frac{m^2}{n} - m,$$

as $m = \sum_{b \in B} \deg b$. On the other hand, since there is no cycle of length 4 in G, each pair of girls (g_1, g_2) is associated with at most one boy b in the family F. It follows that

$$|F| \leq n(n-1).$$

We get that

$$\frac{m^2}{n} - m - n(n-1) \leq 0 \tag{4.6}$$

and so

$$m \leq \frac{1 + \sqrt{1 + 4(n-1)}}{2/n} = \frac{n}{2}\left(1 + \sqrt{4n-3}\right).$$

Since $n = 20$, the following bound must hold: $m \leq 97.75$. We get the desired contradiction as $m = 98$.

In our problem, we assumed that the graph is bipartite but, in fact, one can easily adjust the argument for general graphs. The only difference is that $\sum_{b \in B} \deg b$ is equal to $2m$, not m. Instead of (4.6) we get

$$\frac{4m^2}{n} - 2m - n(n-1) \leq 0$$

which implies that

$$\mathrm{ex}(n, C_4) \leq \frac{2 + \sqrt{4 + 16(n-1)}}{8/n} = \frac{n}{4}\left(1 + \sqrt{4n-3}\right) = \left(\frac{1}{2} + o(1)\right)n^{3/2}.$$

On the other hand, the incidence graph of the projective plane of order q is an example of a dense graph without C_4. Indeed, $G(P, L)$ has

$$n = 2(q^2 + q + 1) = (2 + o(1))q^2$$

vertices and

$$m = (q^2 + q + 1)(q + 1) = (1 + o(1)) q^3 = \left(\frac{1}{2^{3/2}} + o(1) \right) n^{3/2}$$

edges. This construction works when q is a prime power. But, since it is known that for every integer n there exists a prime p satisfying $n \leq p \leq (1 + o(1))n$, the above estimation applies to all values of n. It follows that

$$\text{ex}(n, C_4) \geq \left(\frac{1}{2^{3/2}} + o(1) \right) n^{3/2} .$$

In fact, one can show that the upper bound is sharp; that is, there is a construction that (almost) matches this bound; that is, $\text{ex}(n, C_4) = (1 + o(1))n^{3/2}/2$.

EXERCISES

4.4.1. There is a club with 100 members where there are 1,000 pairs of friends. We want to pick a three person team from the club with one team member selected as a team leader. The procedure is that one club member first becomes a leader. The leader then chooses two followers from his/her friends and the team is formed. Show that it is possible to pick a team from the club in at least 19,000 ways.

4.4.2. Consider the following combinatorial game between two players, Builder and Painter. The game starts with the empty graph on 400 vertices. In each round, Builder presents an edge uv between two non-adjacent vertices u and v which has to be immediately colored red or blue by Painter. Show that Builder can force Painter to create a monochromatic (that is, either red or blue) path on 100 vertices in 400 rounds.
(Source of the problem and solution: well-known, classic problem related to on-line size Ramsey numbers.)

4.4.3. Consider a chess club consisting of 4^t members for some $t \in \mathbf{N}$; some of the members know each other. Show that there exist t members that all know each other, or there exist t members such that no two of them know each other.
(Source of the problem and solution: well-known, classic problem related to Ramsey numbers.)

4.5   Probabilistic Methods

SOURCE

Problem: PLMO LXX – Phase 1 – Problem 6
Solution: our own

PROBLEM

There are 100 people sitting at the round table. Each person has ordered an ice cream, either vanilla or chocolate flavoured. In total, 51 people asked for vanilla ice cream; the remaining 49 preferred chocolate one. The correct number of each flavour was prepared and placed on a table—one ice cream in front of one person. However, the waiter has forgotten who ordered which dessert so it is not guaranteed that everyone received a desert he or she ordered. Fortunately, it is possible to rotate the table and try to satisfy more customers. Prove that one can rotate the table such that at least 52 people will get what they wanted.

THEORY

$\boxed{\text{Union Bound}}$ Let us use the following elementary fact, also known as **Boole's inequality**, that we already introduced in Section 1.8. For any collection of events $A_1, \ldots A_n$,

$$\mathbf{P}\left(\bigcup_{i=1}^{n} A_i\right) \leq \sum_{i=1}^{n} \mathbf{P}(A_i) . \tag{4.7}$$

Note that this inequality is best possible—the equality holds for disjoint events.

$\boxed{\text{Bonferroni Inequalities}}$ Moreover, let us mention that (4.7) may be generalized to find stronger upper and lower bounds. These bounds are known as **Bonferroni inequalities**. In particular,

$$\mathbf{P}\left(\bigcup_{i=1}^{n} A_i\right) \geq \sum_{1 \leq i \leq n} \mathbf{P}(A_i) - \sum_{1 \leq i < j \leq n} \mathbf{P}(A_i \cap A_j) . \tag{4.8}$$

In general, for any $j \in [n]$ we define

$$S_j := \sum_{1 \leq i_1 < \ldots < i_j \leq n} \mathbf{P}\left(A_{i_1} \cap \ldots \cap A_{i_j}\right) .$$

Then, for odd $k \in [n]$,

$$\mathbf{P}\left(\bigcup_{i=1}^{n} A_i\right) \leq \sum_{j=1}^{k} (-1)^{j-1} S_j ,$$

and for even $k \in [n]$,

$$\mathbf{P}\left(\bigcup_{i=1}^{n} A_i\right) \geq \sum_{j=1}^{k} (-1)^{j-1} S_j .$$

Boole's inequality is recovered by setting $k = 1$. When $k = n$, then equality holds and the resulting identity is in fact equivalent the well-known **inclusion–exclusion principle**:

$$\mathbf{P}\left(\bigcup_{i=1}^{n} A_i\right) = \sum_{j=1}^{n} (-1)^{j-1} S_j . \tag{4.9}$$

Linearity of Expectation Consider a finite probability space and a (real) random variable X that takes values from the set $\mathcal{X}$. The **expectation** of X is defined as

$$\mathbb{E}[X] := \sum_{x \in \mathcal{X}} x \cdot \mathbf{P}(X = x) .$$

For example, roll a fair die once and define X to be the number rolled. Clearly, $\mathcal{X} = [6]$ and

$$\mathbb{E}[X] = \sum_{x \in \mathcal{X}} x \cdot \mathbf{P}(X = x) = \sum_{x=1}^{6} x \cdot \frac{1}{6} = \frac{7}{2} .$$

An important and very useful property of the expectation is that it is a linear operator; that is, for any sequence of random variables $X_1, \ldots, X_n$ and any sequence of constants $c_1, \ldots, c_n \in \mathbf{R}$,

$$\mathbb{E}\left[\sum_{i=1}^{n} c_i X_i\right] = \sum_{i=1}^{n} c_i \mathbb{E}[X_i] .$$

For example, if you roll two fair dice, the expected sum is equal to $2 \cdot (7/2) = 7$.

In order to show how the expected value can be calculated for some more complex random variables, we consider the following experiment (that we believe is natural and interesting on its own). Assume you are given a 24-card deck (that is, the deck consisting of 9-10-J-Q-K-A for each of the four suits).

Let us first consider the following experiment. Draw one card from this deck at random. If it is an Ace, then you finish the game; otherwise, you need to put the card back into the deck and restart the experiment. Our goal is to find the expected number of draws till you draw an Ace. It is clear that the process finishes after i rounds with probability $p_i := \frac{1}{6} \left(\frac{5}{6}\right)^{i-1}$. (In fact, the number of rounds is a random variable following the **geometric distribution** with parameter $p = 1/6$.) It follows that the expected number of draws is equal to

$$\sum_{i=1}^{\infty} i \cdot p_i = \sum_{i=1}^{\infty} i \cdot \frac{1}{6} \left(\frac{5}{6}\right)^{i-1} = \sum_{i=0}^{\infty} \sum_{j=i}^{\infty} \frac{1}{6} \left(\frac{5}{6}\right)^{j} = \sum_{i=0}^{\infty} \left(\frac{5}{6}\right)^{i} = 6,$$

where the formula for a sum of a geometric series was used twice. However, assuming that the expected value exists and is equal to E, we can alternatively compute it in a simpler way. Observe that we either draw an Ace in the first round (which happens with probability $1/6$) or 'lose' one draw and then repeat the identical process (this happens with probability $5/6$). It follows that

$$E = \frac{1}{6} + \frac{5}{6} \left(1 + E\right),$$

which immediately implies that $E = 6$. Let us stress that in order for this reasoning to be correct, we had to assume that the expected value of the random variable we were interested in exists (that is, it is finite).

Let us now change slightly the setting, and assume that cards are drawn at random *without* replacement. For this variant, it is obvious that the expected value exists, as it must be less than 21 (we have 24 cards and 4 aces, so at the worst case the process finishes at the end of round 21). In order to find the expectation, one could write down the sum over all possible values for the length of the process (as we did above) but in this case it would be even more cumbersome. Fortunately, it is much simpler to use the other approach which is justified as we are guaranteed that the expected value exists. Let us denote by E_n the expected number of draws from an n-card deck that contains 4 aces till we hit an ace. Clearly, $E_4 = 1$. Now, arguing as before we get that

$$E_n = \frac{4}{n} + \frac{n-4}{n} \left(1 + E_{n-1}\right) = 1 + \frac{n-4}{n} E_{n-1}.$$

It is clear that $E_n = (n+1)/5$ satisfies this recursion, and so for a 24-card deck the expected value is $(24+1)/5 = 5$. Finally, let us observe that the waiting time without replacement is smaller than with replacement. This is what one should intuitively expect as each unsuccessful draw in the variant without replacement increases the probability that we finish in the next round whereas in the other variant it remains the same.

$\boxed{\text{Probabilistic Method}}$ The **probabilistic method** is an example of non-constructive argument for proving the existence of a prescribed kind of mathematical object (see Section 4.1 for more on non-constructive arguments). It works by showing that if one randomly chooses objects, the probability that the result is of the prescribed kind is strictly greater than zero. The conclusion is that the desired object exists; indeed, the probability would have to be equal to 0 otherwise. Similarly, showing that the probability is (strictly) less than 1 can be used to prove the existence of an object that does *not* satisfy the prescribed properties. Let us stress the fact that, although the proof uses probability, the final conclusion is deterministic, without any possible error.

Here is an example of a classical result. Suppose that m basketball teams compete in a tournament and any two teams play each other exactly once. The organizers would like to select n teams and give them prizes at the end of the tournament. It is clearly a challenging task to select the best teams and it would certainly be embarrassing if the organizers ended up selecting n teams but there is another team, without the prize, that beat all n teams that had won a prize. It seems that the organizers should be safe. It feels unlikely that any selection of n teams will have some unselected team better than all of them. Perhaps surprisingly, our intuition is wrong—it is quite possible that this will be the case, at least if m is large enough. Constructing such tournament explicitly (deterministically) is not easy. On the other hand, one can easily use the probabilistic method to show that such tournament exists.

Indeed, for any fixed m, the results of all $\binom{m}{2}$ games are chosen randomly (and uniformly and independently). Now, for a given set A of n teams, the probability that there is another team that beats all teams in A is $(1/2)^n$. Hence, the probability that there is no team better than all teams in A is equal to $(1 - 1/2^n)^{m-n}$. The same formula holds for another set of n teams, say set B. Clearly, there are some correlations between the corresponding events; for example, the fact that there is a team that beats all teams in A increases the probability that there is a team that beats all teams in B, provided that $A \cap B \neq \emptyset$. However, by the union bound (see (4.7)), the probability that there is at least one set of n teams for which there is no better team is at most

$$q(n, m) := \binom{m}{n} \left(1 - \frac{1}{2^n}\right)^{m-n}.$$

If $q(n, m) < 1$, then we are guaranteed that there exists a tournament with m teams such that no n teams can be awarded without another team beating all of them.

Clearly, for any fixed $n \in \mathbf{N}$, one can find $m = m(n)$ large enough such that $q(n, m) < 1$ as $\binom{m}{n}$ grows polynomially and $(1 - 2^{-n})^{m-n}$ decreases exponentially as a function of m. In particular, $q(3, 91) < 1$ and $q(10, 102653) < 1$. Note also that for any natural numbers $m \geq n$, $\binom{m}{n} \leq (em/n)^n$ and for any $x \in \mathbf{R}$, $(1 + x) \leq \exp(x)$. It follows that

$$q(n, m) \leq \left(\frac{em}{n}\right)^n \exp\left(-\frac{m-n}{2^n}\right) = \exp\left(n + n\ln(m/n) - \frac{m-n}{2^n}\right)$$

$$= \exp\left(n + n(n\ln 2 + \ln n) - n^2 + \frac{n}{2^n}\right) < 1$$

provided $m = 2^n n^2$ and $n \geq 12$.

Let us now switch gears and discuss another elementary probabilistic method. It is obvious that one can use the expectation of random variable X to estimate the minimum and maximum value X can take. In other words, we are guaranteed that there exist $x_1, x_2 \in \mathcal{X}$ such that

$$x_1 \leq \mathbb{E}[X] \quad \text{and} \quad x_2 \geq \mathbb{E}[X].$$

Indeed, it follows immediately from the fact that

$$\mathbb{E}[X] \;=\; \sum_{x \in \mathcal{X}} x \cdot \mathbf{P}(X = x) \;\leq\; \sum_{x \in \mathcal{X}} x_{\max} \cdot \mathbf{P}(X = x)$$

$$=\; x_{\max} \sum_{x \in \mathcal{X}} \mathbf{P}(X = x) \;=\; x_{\max} \,,$$

where $x_{\max}$ is the maximum value in $\mathcal{X}$. Similarly, $\mathbb{E}[X] \geq x_{\min}$, where $x_{\min}$ is the minimum value in $\mathcal{X}$. Surprisingly, this naive method can be used to prove many non-trivial statements.

To illustrate the method, consider any n finite sets $A_1, \ldots, A_n$. Then one can pick some of them such that at least half of the underlying elements are repeated an odd number of times. In order to see this, let us pick each A_i with probability $1/2$, independently for all i. Note that for any $a \in A := \bigcup_{i \in [n]} A_i$, the probability that x is repeated an odd number of times is equal to $1/2$. Indeed, for each set A_i that a belongs to, we toss a fair coin to decide if A_i is picked or not. Regardless of the current state of the process, the last A_i that we need to consider causes a to be repeated an odd number of times with probability $1/2$. It follows that the expected number of elements that are repeated an odd number of times is equal to $|A|/2$. By the probabilistic method, we are guaranteed that it is possible to pick some sets so that the number of elements that are repeated odd number of times is at least $|A|/2$.

Here is another example, this time from graph theory. We will show that in every graph $G = (V, E)$, one can partition V, the vertex set, into V_1 and V_2 such that the number of edges with one endpoint in V_1 and another in V_2 is at least $|E|/2$. Indeed, construct a random set $V_1 \subseteq V$ by putting each vertex of V in V_1 independently, with probability $1/2$. Let $V_2 := V \setminus V_1$. For a given edge $e \in E$, let X_e denote the indicator random variable that e has exactly one endpoint in V_1; that is, $X_e = 1$ if e has the desired property and $X_e = 0$ otherwise. Clearly,

$$\mathbb{E}[X_e] \;=\; 1 \cdot \mathbf{P}(X_e = 1) + 0 \cdot \mathbf{P}(X_e = 0) \;=\; \mathbf{P}(X_e = 1) \;=\; \frac{1}{2}\,.$$

Note that G has $X = \sum_{e \in E} X_e$ edges with the desired property. By linearity of expectation,

$$\mathbb{E}[X] \;=\; \mathbb{E}\left[\sum_{e \in E} X_e \right] \;=\; \sum_{e \in E} \mathbb{E}[X_e] \;=\; \frac{|E|}{2},$$

and so the result holds by the probabilistic method.

In fact, it is possible to improve this result slightly by considering a random set V_1 consisting of *precisely* $\lfloor n/2 \rfloor$ vertices. The number of edges between V_1 and V_2 that we are guaranteed to have is at least $\frac{|E|}{2} \cdot \frac{|V|}{|V|-1} > \frac{|E|}{2}$ if $|V|$ is even, and at least $\frac{|E|}{2} \cdot \frac{|V|+1}{|V|} > \frac{|E|}{2}$ if $|V|$ is odd.

SOLUTION

Observe that we have 100 possible rotations of the table (including a trivial one; that is, without rotating at all). Let us number all possible configurations using numbers from 1 to 100; for example, in order to be precise, configuration $i \in [100]$ is obtained by rotating the table clockwise by i places. Consider a given configuration i, and let x_i be the number of people who wanted to get chocolate ice cream but got vanilla one. Since 49 people asked for chocolate ice cream, $49 - x_i$ people wanted chocolate ice cream and got what they wanted. Moreover, since 51 people got vanilla ice cream, $51 - x_i$ people wanted vanilla ice cream and got what they wanted. Therefore, $100 - 2x_i$ people got what they asked for. It remains to show that there exists $i \in [100]$ such that $x_i \leq 24$ as this guarantees that $100 - 2x_i \geq 52$.

We are going to use the double counting argument discussed in Section 4.3. We rotate the table investigating all 100 configurations and counting how many people in total wanted chocolate ice cream but got vanilla one. On the one hand, this is clearly equal to $\sum_{i=1}^{100} x_i$. Now we will count the same thing but this time from the perspective of any person out of 49 people who wanted chocolate ice cream. While table was rotating, this person saw precisely 51 vanilla ice creams. It follows that

$$\sum_{i=1}^{100} x_i = 49 \cdot 51 = 2,499 \,,$$

or equivalently $\frac{1}{100} \sum_{i=1}^{100} x_i = 24.99$. Since the average value is 24.99, there must be at least one i for which $x_i \leq 24.99$. Moreover, since all numbers are integers, we are guaranteed that $x_i \leq 24$, which finishes the proof.

REMARKS

The problem can be equivalently solved using the probabilistic method. Assume that one rotates the table uniformly at random; that is, each configuration $i \in [100]$ occurs with probability $1/100$. There are 49 people that asked for chocolate ice cream; let us mark them with labels $c_1, \ldots, c_{49}$. For any $j \in [49]$, let X_j be the random variable that equals 1 if c_j got vanilla ice cream, and equals 0 otherwise. (As mentioned above, such random variables are called indicators.) Clearly,

$$\mathbb{E}[X_j] = 1 \cdot \mathbf{P}(X_j = 1) + 0 \cdot \mathbf{P}(X_j = 0) = \mathbf{P}(X_j = 1) = \frac{51}{100},$$

as 51 vanilla ice creams were served.

Let us stress the fact that random variables X_j and X_k are correlated. Indeed, c_j and c_k are sitting around the table, and ice creams are placed on the table; it might happen that the fact that $X_j = 1$ affects the probability that $X_k = 1$. Fortunately, the linearity of expectation holds for *any* sequence of random variables. We get that

$$\mathbb{E}\left[\sum_{j=1}^{49} X_j\right] = \sum_{j=1}^{49} \mathbb{E}[X_j] = 49 \cdot \frac{51}{100} = 24.99.$$

By the probabilistic method, we get that one can rotate the table so that $\sum_{j=1}^{49} X_j \le 24.99$ and we are done.

EXERCISES

4.5.1. Let $k \in \mathbf{N}$ and fix $N = N(k) := \lfloor 2^{k/2} \rfloor$. Show that it is possible to partition set $X := [N] = \{1, 2, \ldots, N\}$ into two subsets A and B such that neither A nor B contains an arithmetic progression of length k.
(Source of the problem and solution: well-known, classic problem related to Van der Waerden numbers.)

4.5.2. Show that for any $n \in \mathbf{N}$ there is a tournament with n basketball teams participating in which there are at least $k = n!/2^{n-1}$ orderings $t_1, \ldots, t_n$ such that team t_i won against team t_{i+1}, for all $i \in [n-1]$.
(Source of the problem and solution: well-known, classic problem related to directed Hamilton paths.)

4.5.3. Consider a graph with T triangles. Show that it is possible to color the edges of this graph with two colors so that the number of monochromatic triangles is at most $T/4$.

4.5.4. There are 100 people invited to the party; 450 pairs of people know each other. Show that it is possible to select 10 people so that no two of them know each other.
(Source of the problem and solution: well-known, classic problem related to independent sets in a graph with a given degree sequence.)

4.6   Probability

SOURCE

Problem and solution: our own (inspired by a problem from the book "Are You Smart Enough to Work at Google?" by William Poundstone)

PROBLEM

Let $n \geq 2$ be any natural number. Take a unit stick and break it in n random places. Formally, each breaking point is chosen uniformly at random from the whole stick. Find the probability that one can create a polygon from the $n+1$ resulting pieces.

THEORY

Geometrical Probability | In order to solve the special case of our problem (when $n = 2$), we will use some basic **geometrical probability**. This field studies some basic properties of geometrical objects such as points, lines, planes, circles, spheres, focusing on some natural and fundamental concepts such as random points, random planes, random directions. Let us note that any rigorous discussion on geometrical probability would require sophisticated mathematical background (such as measure theory and integral geometry). As a result, we only scratch the surface in this book, focusing exclusively on very simple applications.

In order to give a flavour of results in this field, let us consider the **clean tile problem** that is an example of a mathematical game of chance that is concerned with dropping a circular coin at random. This game was studied by Buffon who is famous because of another game of chance he studied, the **needle problem** that is concerned with dropping a needle at random. The needle problem requires slightly more sophisticated tools so we only discuss the clean tile problem here.

In a room tiled with equal tiles of any shape a coin is thrown upward. One of the players bets that after its fall the coin will rest clean, that is, on one tile only. The second player bets that the coin will rest on the crack that separates tiles. We would like to investigate if the game is fair. Buffon himself considered tiles shaped as squares, equilateral triangles, rhombuses, and hexagons. We concentrate on squares, the easiest case. Suppose that a coin has diameter d and the floor is filed with squares, each of side ℓ for some $\ell > d$. We assume that the center of the coin lands at a random place on the floor. It is clear that the coin touches the separating crack if and only if the center is at distance less than $d/2$ from the crack—see Figure 4.5. Hence, the probability for the coin to be entirely within one of the tiles is given by the ratio between the area of the inner square and that of the outer square, that is, the first player

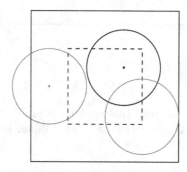

FIGURE 4.5: The clean tile problem of Buffon.

wins with probability p where

$$p := \frac{(\ell - d)^2}{\ell^2} = \left(1 - \frac{d}{\ell}\right)^2.$$

It follows that the second player wins with probability

$$1 - p = 1 - \left(1 - \frac{d}{\ell}\right)^2.$$

For the game to be fair these two probabilities must be equal, that is, the following equation has to be satisfied

$$\ell^2 - 4d\ell + 2d^2 = 0.$$

There are two solutions, $\ell = (2 \pm \sqrt{2})d$, but since $\ell > d$, the only acceptable solution is

$$\ell = (2 + \sqrt{2})d \approx 3.41\,d.$$

Smaller values of ℓ give advantage for the second player and larger values favour the first player.

| **Disjoint and Mutually Exclusive Events** | Two events A and B are said to be **disjoint** if they cannot occur at the same time, that is, $A \cap B = \emptyset$. In particular, $\mathbf{P}(A \cap B) = \mathbf{P}(\emptyset) = 0$. The simplest example of disjoint events is a coin toss. A tossed coin outcome can be either head or tails, but both outcomes cannot occur simultaneously.

Being mutually exclusive is a slightly different property of events (sets in a probability space). Two events are **mutually exclusive** if the probability of them both occurring is zero, that is, $\mathbf{P}(A \cap B) = 0$. With that definition, disjoint sets are necessarily mutually exclusive, but mutually exclusive events are not necessarily disjoint.

In order to illustrate the difference, suppose that a point is selected uniformly at random from the unit square. Each coordinate is uniformly and independently distributed from the set $[0,1]$. Let A be the event that the x coordinate is greater than or equal to the y coordinate, and B be the event that the y coordinate is greater than or equal to the x coordinate. Clearly, $\mathbf{P}(A) = \mathbf{P}(B) = 1/2$ and $A \cap B = \{(x,x) : x \in [0,1]\}$, and so the events are *not* disjoint. However, $\mathbf{P}(A \cap B) = 0$, as the area of the set $A \cap B$ is equal to 0, and so they are mutually exclusive.

Note that mutually exclusive events may not be disjoint only if in the considered probability space there exist events that have probability equal to 0. Such situations are typical when the probability space on which possible events are defined is continuous. In our example above, the space is a unit square and we defined the probability of a given event as *the area* of the corresponding sample points that are included in this event. Geometrical objects that have zero area define valid events but the associated probabilities are equal to 0. In our example, such event was the line segment from point $(0,0)$ to point $(1,1)$.

Consider now a sequence of pairwise mutually exclusive events $A_1, \ldots, A_n$. One can combine the Union Bound (4.7) and the Bonferroni inequality (4.8) to get that

$$\sum_{1 \le i \le n} \mathbf{P}(A_i) - \sum_{1 \le i < j \le n} \mathbf{P}(A_i \cap A_j) \le \mathbf{P}\left(\bigcup_{i=1}^{n} A_i\right) \le \sum_{1 \le i \le n} \mathbf{P}(A_i) ,$$

and so

$$\mathbf{P}\left(\bigcup_{i=1}^{n} A_i\right) = \sum_{i=1}^{n} \mathbf{P}(A_i) . \tag{4.10}$$

Alternatively, one can use the Inclusion–Exclusion Principle (4.9).

SOLUTION

Let us first define the problem more formally. Let A_0 and A_{n+1} be the two endpoints of the unit stick. Let $A_1, \ldots, A_n$ be the n random breaking points. Recall that, for each $i \in [n]$, A_i is chosen uniformly at random from the whole stick. It follows that X_i, the distance from A_0 to A_i, is a random variable that has a real number from the interval $[0,1]$ assigned uniformly at random. These random variables are independent and so we can generate them one by one or all the same time (simultaneously)—we will use this property at some point. Note also that with probability zero $X_i = X_j$ for some $i \ne j$, and so we may assume that such situation does not happen. As a result, we may order A_i's in an increasing order of the corresponding X_i's. Formally, let $\pi : [n] \to [n]$ be the unique permutation such that $X_{\pi(i)} < X_{\pi(i+1)}$ for any $i \in [n-1]$ (notice that π is also a random variable that depends on random variables X_i).

In order to keep the notation simple, let us fix $X_{\pi(0)} = 0$ and $X_{\pi(n+1)} = 1$. Once we break the stick, we get $n + 1$ pieces. The ith piece ($i \in [n + 1]$) has length $L_i = X_{\pi(i)} - X_{\pi(i-1)}$. Clearly, the desired polygon can be created if and only if no piece has length larger than $1/2$, that is, $L_i \leq 1/2$ for all $i \in [n+1]$. Note that the probability that some piece has length exactly $1/2$ is equal to zero and so we may or may not include this degenerate case without affecting the result.

Let A_i be the event that the ith piece is too long, that is, $L_i > 1/2$. Finding the probability that the first piece is too long is easy. Indeed, in this case one can simply look at the independent random variables X_i to get that

$$\mathbf{P}(A_1) = \mathbf{P}(L_1 > 1/2) = \mathbf{P}\left(\bigcup_{i=1}^{n}(X_i > 1/2)\right) = \left(\frac{1}{2}\right)^n.$$

Similarly, the last piece is too long with the same probability, namely, $\mathbf{P}(A_{n+1}) = 1/2^n$. But what about some middle piece? It is not clear. It feels that the situation can be different but it turns out that the distributions of all L_i's are the same. To see this we do the following trick. Instead of breaking the stick into $n + 1$ pieces by breaking it in n random places, we start with a rope that forms a circle with unit circumference (that is, we take a unit length rope and glue the two endpoints together). Now, we cut the rope in $n + 1$ random places. Again, we can do the cuts one by one or all the same time. If we do the cuts one by one, then we immediately see that the two processes are equivalent. Indeed, after the first cut the situation is exactly the same as at the beginning of the process of breaking the stick. From that point on, the two processes can be coupled together. On the other hand, if we cut the rope in $n + 1$ places simultaneously, then, by symmetry, we see that there is nothing special about L_1 or L_{n+1}. All the random variables L_i have the same distribution; in particular, $\mathbf{P}(A_i) = \mathbf{P}(L_i > 1/2) = 1/2^n$.

We need one more observation to solve our problem. Clearly, the events A_i are not independent. If the ith piece is too long, then the chance that some other piece is too long is smaller. In fact, if $i \neq j$, then the two events A_i and A_j are disjoint, that is, they cannot occur at the same time. In particular,

$$\mathbf{P}(A_i \cap A_j) = 0.$$

We use (4.10) to get

$$\mathbf{P}\left(\bigcup_{i=1}^{n+1} A_i\right) = \sum_{i=1}^{n+1}\mathbf{P}(A_i) = \frac{n+1}{2^n}.$$

It follows that the probability that one can create a polygon from the $n + 1$ pieces is equal to $1 - (n + 1)/2^n$.

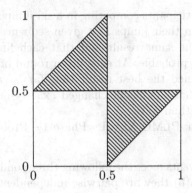

FIGURE 4.6: Breaking stick into three parts.

REMARKS

Recall that the outcome of our random experiment (that is, breaking the stick into n pieces) can be represented by n random variables X_i. Each X_i ($i \in [n]$) has a real number from the interval $[0, 1]$ assigned uniformly at random and independently. As a result, one can alternatively think about this experiment as a process of selecting a point uniformly at random from the n-dimensional unit cube. We will use this point of view to geometrically solve our problem for the specific case of breaking the stick into three parts ($n = 2$).

Let $(x, y) \in [0, 1]^2$ be a random point from the unit square (recall that $X_1 = x, X_2 = y$). If both x and y are less than $1/2$, then clearly we will not be able to construct a triangle in the original problem as the third piece will be too long. Similarly, if both are more than $1/2$, then the first piece will be too long. If $x < 1/2 < y$, then our task is doable if and only if $y - x < 1/2$, that is, the middle piece is short enough. The same argument applies to the situation when $y < 1/2 < x$. In this case, the sufficient and necessary condition for being able to achieve our task is $x - y < 1/2$. We present all four cases in Figure 4.6—shaded areas correspond to the two cases when the triangle can be constructed. It follows that the probability we are successful is equal to $1/4$ (of course, it is consistent with our general result: $1/4 = 1 - (2 + 1)/2^2$).

EXERCISES

4.6.1. Consider an urn that initially contains one white and one black ball. We repeatedly perform the following process. In a given round, one ball is drawn randomly from the urn and its color is observed. The ball is then returned to the urn, and an additional ball of the same color is added to the urn. We repeat this selection process for 50 rounds so that the urn contains 52 balls. What number of white balls is the most probable?
(Source of the problem: PLMO L – Phase 1 – Problem 11. Solution: our own.)

4.6.2. There are 65 participants competing in a ski jumping tournament. They take turns and perform their jumps in a given sequence. We assume that no two jumpers obtain the same result and that each final resulting order of participants is equally probable. At each given round of the tournament, the person that has obtained the best result thus far is called a leader. Prove that the probability that the leader changed exactly once during the whole tournament is greater than $1/16$.

(Source of the problem: PLMO XLVII – Phase 1 – Problem 11. Solution: our own.)

4.6.3. Three random events meet the following three conditions: (a) their probabilities are all equal, (b) they are pairwise independent, and (c) all of them cannot happen at the same time. What is the maximum probability that at least one of these three events holds?

(Source of the problem and solution idea: PLMO XXXV – Phase 1 – Problem 9.)

4.7 Combinations of Geometrical Objects

SOURCE

Problem and solution: PLMO LIX – Phase 3 – Problem 3

PROBLEM

Consider the set P of all points (x, y) on a plane with $x, y \in \mathbf{Z}$; that is, $P = \mathbf{Z} \times \mathbf{Z}$, the Cartesian product of $\mathbf{Z}$ and $\mathbf{Z}$. Suppose that each point in P is painted red or blue. Prove that there exists an infinite subset of P that has a center of symmetry and consists of points having the same color.

THEORY

Recall that for any two sets A and B, the **Cartesian product** $A \times B$ is the set of all ordered pairs (a, b) where $a \in A$ and $b \in B$; that is,

$$A \times B := \{(a, b) : a \in A, b \in B\} .$$

This definition extends naturally to any dimension $n \in \mathbf{N}$, the Cartesian product $A_1 \times \ldots \times A_n$, where instead of ordered pairs we deal with ordered n-tuples. Moreover, if all A_i's are the same, then we simply write A^n instead of $A \times \ldots \times A$.

$\boxed{\text{Point Reflection}}$ Let $p = (p_1, \ldots, p_n) \in \mathbf{R}^n$ be any point in n-dimensional space. For any point $a = (a_1, \ldots, a_n) \in \mathbf{R}^n$, the **reflection of a across the point p** is point

$$
\begin{aligned}
\mathrm{Ref}_P(a) &:= \left(p_1 - (a_1 - p_1), \ldots, p_n - (a_n - p_n) \right) \\
&= (2p_1 - a_1, \ldots, 2p_n - a_n) \\
&= 2p - a .
\end{aligned}
$$

In the case where $p = (0, \ldots, 0) \in \mathbf{R}^n$ is the origin, point reflection of a is simply the negation of vector a. In two dimensions, namely when $n = 2$, a point reflection is the same as a rotation of 180 degrees.

$\boxed{\text{Point Symmetry}}$ A set $S \subseteq \mathbf{R}^n$ that is invariant under a point reflection is said to possess **point symmetry**. In other words, S has point symmetry if and only if there exists point $p \in \mathbf{R}^n$ such that $S = S'$, where

$$S' := \{\mathrm{Ref}_p(a) : a \in S\} .$$

Point p is often called the **centre of the symmetry**.

Playing cards often have point symmetry, so that they look exactly the same from the top or the bottom. Some letters of the English alphabet (namely, H, I, N, O, S, X, Z) exhibit point symmetry and many geometrical figures such as circles and rectangles. Finally, graphs of many functions, such as $f(x) = x^3$ or $f(x) = 1/x$, are symmetric.

SOLUTION

Consider any coloring of P with the two colors red and blue. Towards a contradiction, suppose that there is no monochromatic infinite subset of P that has a center of symmetry. In other words, for *every* point $p \in A \times A$, where $A := \{k/2 : k \in \mathbf{Z}\}$, the set

$$P' = P'(p) := \{a \in P : \text{ both } a \text{ and } 2p - a \text{ have the same color}\}$$

is finite. In fact, we will only use this assumption for $p \in \{(0,0), (1/2, 0)\}$.

Suppose first that the center of symmetry is located at $p = (0,0)$, the origin. It follows from our assumption that there exists $M_1 \in \mathbf{N}$ such that for all $x, y \in \mathbf{Z}$ with $y \geq M_1$,

$$(x, y) \text{ and } (-x, -y) \text{ have different colors.} \qquad (4.11)$$

Similarly, if $p = (1/2, 0)$, then we are guaranteed that there exists $M_2 \in \mathbf{N}$ such that for all $x, y \in \mathbf{Z}$ with $y \geq M_2$,

$$(x + 1, y) \text{ and } (-x, -y) \text{ have different colors.} \qquad (4.12)$$

Let us now fix $y = \max\{M_1, M_2\}$. It follows immediately from (4.11) and (4.12) that for any $x \in \mathbf{Z}$, points (x, y) and $(x + 1, y)$ have the same color (since both of them have a different color than $(-x, -y)$ and there are only two colors, red and blue).

As a result, this means that all the points in $Q := \{(x, \max\{M_1, M_2\}) : x \in \mathbf{Z}\}$ have the same color. Since Q clearly has point symmetry (in fact, any point in Q is a center of the symmetry) and is infinite, we get the desired contradiction.

REMARKS

Very often problems formulated in terms of relationships of geometrical objects can be reformulated as combinatorial problems, in which geometrical properties of the considered objects form combinatorial constraints. The opposite situation may also occur; that is, sometimes combinatorial problems can be solved after rephrasing them in the language of geometry and then using geometrical tools.

In order to illustrate the power of this approach, let us consider convex n-gons. Recall that n-gon is a polygon with n sides—see Chapter 6 for more. A **convex polygon** is defined as a polygon with all its interior angles less than 180 degrees. This means that all the vertices of the polygon will point outwards, away from the interior of the shape. Assuming that there are no 3 diagonals going through the same point, let us count how many intersection points all the diagonals have.

For simplicity, let us concentrate on 8-gons and label vertices with integers from 0 to 7, starting with an arbitrary vertex and then proceeding clockwise. There are 8 diagonals from vertex i to vertex $i + 2$, $i = 0, \ldots, 7$ (using modular arithmetic), each of them intersecting $1 \cdot 5$ other diagonals. There are 8 diagonals from vertex i to vertex $i + 3$, $i = 0, \ldots, 7$; this time, each of them intersects $2 \cdot 4$ other diagonals. Finally, there are 4 diagonals from i to $i + 4$, $i = 0, \ldots, 3$ and each of them intersects $3 \cdot 3$ other diagonals. Moreover, each pair of intersecting diagonals occurs twice. Hence, the total number of intersections is equal to

$$\frac{8 \cdot 1 \cdot 5 + 8 \cdot 2 \cdot 4 + 4 \cdot 3 \cdot 3}{2} = \frac{140}{2} = 70.$$

One can repeat this argument for any value of n but it gets complicated quickly and no general formula seems to appear.

Alternatively, one can observe that each intersection point can be labelled with the set consisting of the labels of the two corresponding diagonals. For example, diagonal 15 intersects diagonal 36 at the point labelled with set $\{1, 3, 5, 6\}$. It is easy to see that no two pairs of diagonals yield the same set. On the other hand, each set of 4 labels corresponds to one intersection point. It follows that there exists a bijection from the set of points of intersections and the family of 4-element sets of set $\{0, \ldots, 7\}$ and so the two sets have the same size. Since $\binom{8}{4} = \frac{8 \cdot 7 \cdot 6 \cdot 5}{4 \cdot 3 \cdot 2} = 70$, we get an alternative way of obtaining the result. More importantly, it easily generalizes to any value of n: there are $\binom{n}{4}$ intersection points of two diagonals in a convex n-gon.

EXERCISES

4.7.1. Let P be a set of five points on a plane with the property that no three of them lie on the same line. Denote by $a(P)$ the number of obtuse triangles whose vertices lie in P. Find the minimum and the maximum value that $a(P)$ can attain over all possible sets P.

4.7.2. Every point on a circle is painted with one of three colors. Prove that there are three points on the circle that have the same color and form an isosceles triangle.
(Source of the problem and solution idea: PLMO LI – Phase 1 – Problem 4.)

4.7.3. Take a set of $n \geq 2$ points with the property that no three of them lie on the same line. We paint all line segments formed by those points in such a way that no two line segments that have a common vertex have the same color. Find the minimum number of colors for which such coloring exists.

4.8   Pigeonhole Principle

SOURCE

Problem and solution idea: LVII PLMO – Phase 1 – Problem 4

PROBLEM

During the Polish Mathematical Olympics that lasts two days, participants are solving a total of 6 problems. Each participant can get 6, 5, 2, or 0 points for the solution of each problem. During one of the competitions, the following interesting property occurred: for any two participants there were two problems for which they obtained different scores. How many participants came to this competition, at most?

THEORY

Pigeonhole Principle   The tool we introduce and use in this section is obvious but perhaps surprisingly, often an extremely powerful tool. It can be stated as follows. If one has n boxes and places more than n objects into them, then there will be at least one box that contains more than one object. In fact, one can make the following stronger statement: if k objects are placed into n boxes, then there will be at least one box that contains at least $\lceil k/n \rceil$ objects.

In order to see this tool in action (in an easy scenario) let us consider the following example. We shoot 65 shots at a square target, the side of which is 80 centimeters long. Since we are pretty good at this, all of our shots hit the target. Prove that there are two bullet holes that are closer than 15 centimeters from each other.

Suppose that our target is an old 8×8 chessboard. (Formally, we say that the target is tessellated into 8×8 square grid.) There are $8 \cdot 8 = 64$ squares and the board received $65 > 64$ shots. Hence, by the pigeonhole principle, there must be a square that received at least two shots. We claim that these two shots are at distance at most 15 centimeters from each other. Indeed, since the size of each square is 10 centimeters, the distance between any two points is, by Pythagorean theorem, at most

$$\sqrt{10^2 + 10^2} = \sqrt{200} \approx 14.1 < 15 .$$

SOLUTION

First of all, let us note that the distribution of points does not matter (well, from the perspective of our problem). Hence, we may assume that each solution gets a score from the set $P = \{0, 1, 2, 3\}$ and so the performance of each participant can be represented by a vector from set $X = \{(a_1, a_2, a_3, a_4, a_5, a_6) : a_i \in P\}$. Clearly, $|P| = 4^6 = 4{,}096$.

We know that participants got unique vectors from a subset A of P that satisfy the following property: any two vectors from A differ in at least two coordinates. Our goal is to provide an upper bound for the size of A. To that end, let us observe that the number of vectors of length 5 of elements from P is $4^5 = 1{,}024$. Hence, if A contained more than 1,024 vectors, then by pigeonhole principle it would have two vectors that coincide on the first 5 coordinates and so differ on at most one coordinate (that is, possibly the last one). This shows that $|A| \leq 1{,}024$.

Now we will show that this upper bound is sharp; that is, one can construct set A of size 1,024 with the desired property. Let

$$
A = \left\{ \left(a_1, a_2, a_3, a_4, a_5, a_6 = \sum_{i=1}^{5} a_i \pmod 4 \right) : a_i \in P \text{ for } 1 \leq i \leq 5 \right\}
$$
$$
\subseteq X .
$$

In other words, A is constructed by considering all five element vectors $(a_1, \ldots, a_5)$ from P and adding $a_6 = \sum_{i=1}^{5} a_i \pmod 4$ at the very last coordinate. (Note that $a_6 \in P$.) Clearly, $|A| = 4^5 = 1{,}024$, as the last coordinate is determined by the first five coordinates.

For a contradiction, suppose that there are two vectors in A, say, $a = (a_1, \ldots, a_6)$ and $b = (b_1, \ldots, b_6)$ that differ on at most one coordinate. Clearly, by construction, a and b differ on at least one coordinate from the first five coordinates and so a and b must differ on precisely one coordinate: $a_\ell \neq b_\ell$ for some ℓ such that $1 \leq \ell \leq 5$. In particular, $a_6 = b_6$. But this implies that

$$
\sum_{i=1}^{5} a_i = a_6 = b_6 = \sum_{i=1}^{5} b_i \pmod 4,
$$

which gives $a_\ell = b_\ell \pmod 4$. Since $P = \{0, 1, 2, 3\}$, it follows that $a_\ell = b_\ell$ and so we get the contradiction which finishes the proof that A has the desired property.

REMARKS

Let us first note how one can come up with the proof that set of size 1,024 can be constructed. The key observation was that each five element vector can be associated with one of the four signatures; moreover, if two of these vectors differ on only one coordinate, then they must have different signatures. We used these signatures to define the 6th coordinate.

Let us also mention about the following three, closely related, problems: Birthday Paradox, that can be viewed as probabilistic pigeonhole principle, Coupon Collector Problem, and Birthday Attack.

Birthday Paradox Suppose that k people are selected at random. The convenient assumption is that each day of the year (including February 29) is equally probable for a birthday, independently for each person. We are interested in estimating the probability that some pair of the selected people will have the same birthday. Since there are $n = 366$ possible birthdays, by the pigeonhole principle, this probability is equal to one if $k \geq 367$. On the other hand, if $k = n = 366$, then we are not guaranteed that such pair exists but the probability that each person has a unique birthday (that we denote by $p(n, k)$) is extremely small. Indeed, it is clear that

$$p(n, k) := \frac{n \cdot (n - 1) \cdots (n - k + 1)}{n^k} = \frac{n!}{n^k (n - k)!},$$

and so $p(366, 366) \approx 5.36 \cdot 10^{-158}$. It may seem surprising that this probability is below 50% for a group as small as 23 individuals; $p(366, 23) \approx 0.49$.

Since $1 + x \leq \exp(x)$ for any $x \in \mathbf{R}$, we can estimate $p(n, k)$ as follows.

$$p(n, k) = \prod_{i=1}^{k-1} \frac{n - i}{n} = \prod_{i=1}^{k-1} \left(1 - \frac{i}{n}\right)$$

$$\leq \exp\left(-\sum_{i=1}^{k-1} \frac{i}{n}\right) = \exp\left(-\frac{k(k - 1)}{2n}\right) =: \hat{p}(n, k).$$

Moreover, in practice this upper bound is not too far from the truth value; for example, $0.492703 \approx p(366, 23) \leq \hat{p}(366, 23) \approx 0.499998$.

Coupon Collector's Problem We continue selecting k people at random but this time we would like k to be large enough so that $q(n, k)$, the probability that every single day of the year someone has a birthday, is close to one. This problem is known as the **coupon collector's problem** as the question can be reformulated as the problem of collecting n unique coupons hidden in boxes of some brand of cereals. Clearly, for any $k < n$ we have $q(n, k) = 0$. If one is extremely lucky, then the group of $k = n = 366$ people could have the desired property but the probability is very low; indeed,

$$q(366, 366) = p(366, 366) \approx 5.36 \cdot 10^{-158}.$$

The exact values for $q(n, k)$ are extremely difficult to compute (unless n and k are small or, for example, $n = k$) as they are related to the Stirling number of the second kind, the number of ways to partition a set of k objects into n non-empty subsets. Using simulations, we determined that $k = 2{,}294$ is the smallest value for which $q(366, k) > 0.5$.

On the other hand, it is possible to show that the expected number of people that need to be selected in order for the desired property to hold is

$$n \sum_{i=1}^{n} \frac{1}{i} = n H_n = n \ln n + \gamma n + \frac{1}{2} + o(1),$$

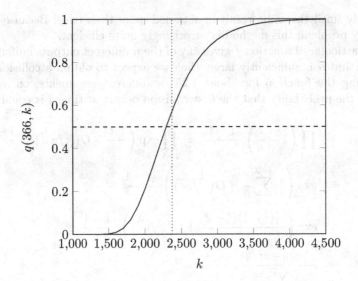

FIGURE 4.7: Plot of $q(366, k)$. Dashed horizontal line is for 50% probability, dotted vertical line denotes expected value that is roughly equal to 2,372.1245.

where H_n is the n-th harmonic number and $\gamma \approx 0.577216$ is the Euler-Mascheroni constant. In particular, for $n = 366$ it is approximately 2,372.1245. The function $q(n, k)$ has the following asymptotic behavior. If $k = n(\ln n + c_n)$ for some sequence $(c_n)_{n \in \mathbf{N}}$ of real numbers, then

$$q(n, k) \sim \begin{cases} 0 & \text{if } c_n \to -\infty \\ e^{-e^{-c}} & \text{if } c_n \to c \in \mathbf{R} \\ 1 & \text{if } c_n \to \infty. \end{cases}$$

Note that for $n = 366$ and $k = 2{,}294$ the above estimate gives $q(n, k) \approx 0.4995$, which is in line with the simulation results. So it is approximately 100 times more than what is needed for the birthday paradox case. Finally, let us highlight the following weaker but often useful statement: for any $\epsilon > 0$,

$$\lim_{n \to \infty} q\big(n, (1 - \epsilon)n \ln n\big) = 0 \quad \text{and} \quad \lim_{n \to \infty} q\big(n, (1 + \epsilon)n \ln n\big) = 1.$$

$\boxed{\text{Birthday Attack}}$ A **birthday attack** is a type of cryptographic attack that exploits the mathematics behind the birthday problem discussed above. It can be formulated as follows. Given a function $f = f(x)$, the goal of the attack is to find two different values of x, say x_1 and x_2, such that $f(x_1) = f(x_2)$. Such pair x_1, x_2 is called a collision. The method used to find a collision is simply to evaluate function f for many values of x that can be selected randomly until the same result is obtained more than once. Because of the birthday problem, this method is surprisingly quite efficient.

In particular, if function f gives any of the n different outputs uniformly at random and n is sufficiently large, then we expect to obtain a collision after evaluating the function for about $1.25\sqrt{n}$ different arguments, on average. Indeed, the probability that the first collision occurs at time i is equal to

$$
\begin{aligned}
\mathbf{P}(i) &= \prod_{t=1}^{i-2}\left(1-\frac{t}{n}\right)\cdot\frac{i-1}{n} = \prod_{t=1}^{i-2}\exp\left(-\frac{t}{n}+O((t/n)^2)\right)\cdot\frac{i-1}{n}\\
&= \exp\left(-\sum_{t=1}^{i-2}\frac{t}{n}+O(i^3/n^2)\right)\cdot\frac{i-1}{n}\\
&= \exp\left(-\frac{(i-1)(i-2)}{2n}+O(i^3/n^2)\right)\cdot\frac{i-1}{n}\\
&\sim \frac{x\exp(-x^2/2)}{\sqrt{n}},
\end{aligned}
$$

provided $i = x\sqrt{n}$ for some $x \in \mathbf{R}$. In order to calculate an asymptotic value for the probability to see a collision by time $i = x\sqrt{n}$, denoted $\mathbf{P}(\leq i)$, we need to use integrals. This part is *not* considered to be elementary mathematics so the less advanced reader can safely skip this part; we will not use this result later on. Moreover, we only provide a sketch, as a formal argument is more delicate and technical.

$$
\begin{aligned}
\mathbf{P}(\leq x\sqrt{n}) &= \sum_{i=1}^{x\sqrt{n}}\mathbf{P}(i) \sim \int_0^x z\exp(-z^2/2)\,dz\\
&= \left.-\exp(-z^2/2)\right|_0^x = 1 - \exp(-x^2/2).
\end{aligned}
$$

So one needs to evaluate function f roughly $\sqrt{2\ln 2}\,\sqrt{n} \approx 1.18\sqrt{n}$ times to get the probability close to $1/2$. Similarly, the expected number of values that need to be evaluated to get the first collision is equal to

$$
\sum_{i=1}^{\infty}i\cdot\mathbf{P}(i) \sim \int_0^{\infty}z^2\exp(-z^2/2)\,dz\cdot\sqrt{n} = \sqrt{\pi/2}\,\sqrt{n} \approx 1.25\,\sqrt{n}.
$$

EXERCISES

4.8.1 Twenty five boys and twenty five girls sit around a table. Prove that it is always possible to find a person both of whose neighbors are girls. (Source of the problem and solution: Interactive Mathematics Miscellany and Puzzles by Alexander Bogomolny, https://www.cut-the-knot.org.)

4.8.2 A person takes at least one aspirin a day for 30 days. Show that if the person takes 45 aspirin altogether, then in some sequence of consecutive days that person takes exactly 14 aspirin.

(Source of the problem and solution: Interactive Mathematics Miscellany and Puzzles by Alexander Bogomolny, https://www.cut-the-knot.org.)

4.8.3 Prove that if we take $n+1$ numbers from the set from 1 to $2n$, then in this subset there exist two numbers such that one divides the other.

4.9   Generating Functions

SOURCE

Sicherman dice – well-known problem

PROBLEM

Can you design two different dice so that their sums behave just like a pair of ordinary dice? That is, there must be two ways to roll a 3, six ways to roll a 7, one way to roll a 12, and so forth. Each die must have six sides, and each side must be labelled with a positive integer.

THEORY

| Generating Function | The (ordinary) **generating function** of a sequence $a = (a_i)_{i=0}^{\infty}$ of real numbers is defined as follows:

$$G(x) \;=\; G(a, x) \;:=\; \sum_{i=0}^{\infty} a_i x^i \;.$$

Unlike an ordinary series, this formal series is allowed to diverge, meaning that the generating function is not always a true function and the "variable" x is actually an indeterminate allowing us to perform useful algebraic manipulations.

Let us start with a simple and standard application of generating functions. The **Fibonacci sequence** $a_0, a_1, a_2, \ldots$ is defined recursively as follows:

$$a_0 = 0, a_1 = 1 \text{ and } a_{n+1} = a_n + a_{n-1} \text{ for } n \geq 1 \;.$$

Our goal is to find an explicit formula for a_n. Instead of looking for the sequence, we will look for its generating function $G(x) = \sum_{j \geq 0} a_j x^j$. Once we get it, we will try to recover the coefficient a_n in front of x^n, the nth Fibonacci number.

In order to get $G(x)$, let us take the recurrence relation $a_{n+1} = a_n + a_{n-1}$, multiply both sides of it by x^n, and sum over all values of n for which the relation is valid. We get

$$\sum_{n \geq 1} a_{n+1} x^n \;=\; \sum_{n \geq 1} a_n x^n + \sum_{n \geq 1} a_{n-1} x^n \;. \tag{4.13}$$

On the left hand side of (4.13) we have

$$\sum_{n \geq 1} a_{n+1} x^n \;=\; a_2 x + a_3 x^2 + a_4 x^3 + \ldots \;=\; \frac{G(x) - a_1 x - a_0}{x} \;=\; \frac{G(x) - x}{x}.$$

On the other hand, on the right hand side of (4.13) we have

$$\sum_{n \geq 1} a_n x^n + \sum_{n \geq 1} a_{n-1} x^n = G(x) + x G(x) .$$

It follows that

$$\frac{G(x) - x}{x} = G(x) + x G(x)$$

and so

$$G(x) = \frac{x}{1 - x - x^2} = \frac{x}{(1 - xr_+)(1 - xr_-)},$$

where $r_+ = (1 + \sqrt{5})/2$ and $r_- = (1 - \sqrt{5})/2$. Our first task is done—we have an explicit formula for the generating function of the Fibonacci sequence.

Using the partial fraction method, we get

$$\begin{aligned}
G(x) &= \frac{1}{r_+ - r_-} \left(\frac{1}{1 - xr_+} - \frac{1}{1 - xr_-} \right) \\
&= \frac{1}{\sqrt{5}} \left(\sum_{n \geq 0} r_+^n x^n - \sum_{n \geq 0} r_-^n x^n \right) \\
&= \sum_{n \geq 0} \frac{1}{\sqrt{5}} \left(r_+^n - r_-^n \right) x^n,
\end{aligned}$$

thanks to the magic of the geometric series (recall that $\sum_{n \geq 0} c^n = 1/(1 - c)$). Therefore, an explicit formula for the Fibonacci number is

$$a_n = \frac{1}{\sqrt{5}} (r_+^n - r_-^n) = \frac{1}{\sqrt{5}} \left(\left(\frac{1 + \sqrt{5}}{2} \right)^n - \left(\frac{1 - \sqrt{5}}{2} \right)^n \right).$$

In fact, note that

$$\left| \frac{1}{\sqrt{5}} \left(\frac{1 - \sqrt{5}}{2} \right)^n \right| \leq \frac{1}{\sqrt{5}} \approx 0.447 < 1/2$$

for any non-negative integer n and so

$$a_n = \left\lfloor \frac{1}{\sqrt{5}} \left(\frac{1 + \sqrt{5}}{2} \right)^n \right\rceil,$$

where $\lfloor x \rceil$ is the nearest integer to real number x.

Golden Ratio The constant that appeared in the formula for the nth Fibonacci number,

$$\phi = \frac{1 + \sqrt{5}}{2} \approx 1.618,$$

is the **Golden Ratio** that appears in mathematics surprisingly often. Perhaps even more surprising is the fact that it appears in some patterns in nature, including the spiral arrangement of leaves and other plant parts, music, architecture, or paintings. Two quantities, a and b are in the golden ratio if their ratio is the same as the ratio of their sum to the larger of the two quantities. In other words, ϕ is defined as follows:

$$\phi = \frac{a}{b} = \frac{a+b}{a},$$

where $a > b > 0$.

SOLUTION

Consider the generating function where a_k represents the number of appearances of the number k on the die. Thus, an ordinary die would be represented by the polynomial

$$\begin{aligned}
f(x) &= x + x^2 + x^3 + x^4 + x^5 + x^6 \\
&= x(x+1)(x^2 + x + 1)(x^2 - x + 1).
\end{aligned}$$

The key observation is that the result of rolling two (or more, in general) dice is represented by the product of their generating functions. Therefore, if $g(x)$ and $h(x)$ are the functions associated with the first die and the second one, respectively, then we get that

$$g(x)h(x) = f(x)^2 = \left(x(x+1)(x^2 + x + 1)(x^2 - x + 1) \right)^2.$$

There are some constraints we need to consider: we cannot have a non-zero constant term in $g(x)$ or $h(x)$ (since that would imply that some sides are labelled "0") or any negative term. It follows that we need to assign one copy of each "x" factor to $g(x)$ and $h(x)$. Moreover, $g(1) = h(1) = 6$ (the number of sides), so we need to assign one copy of each "$(x^2 + x + 1)$" and "$(x + 1)$" factor to $g(x)$ and $h(x)$ as well. It remains to distribute the two "$(x^2 - x + 1)$" factors. If we give one copy to each of $g(x)$ and $h(x)$, we get an ordinary pair of dice. Otherwise, we get

$$\begin{aligned}
g(x) &= x(x+1)(x^2 + x + 1) = x + 2x^2 + 2x^3 + x^4 \\
h(x) &= x(x+1)(x^2 + x + 1)(x^2 - x + 1)^2 = x + x^3 + x^4 + x^5 + x^6 + x^8,
\end{aligned}$$

which corresponds to two dice: $\{1, 2, 2, 3, 3, 4\}$ and $\{1, 3, 4, 5, 6, 8\}$.

REMARKS

Alternatively, one could have solved this problem using a computer in the following way. Observe first that there is only one way to get the sum to be equal to 2 or 12. This means that each dice must have exactly one 1 and both of them must have a unique maximum value (which can be different). Now, observe that this unique value must be greater than 3; otherwise, one of the dies would be $\{1, 2, 2, 2, 2, 3\}$ and the other die would have to have $1, 9$ and four numbers that are between 2 and 8 but then there are too many ways to get the sum to be equal to 11. This means that the maximum number on any die is at most 8. One can easily enumerate all such dies and check which of them meet the required criteria.

Here is a basic program written in Julia language that performs this brute-force check. We did not optimize it for speed as its run-time is under one second anyway.

```julia
using Base.Iterators

function listdies()
    ref = [1,2,3,4,5,6,5,4,3,2,1] # reference distribution
    # traverse all possible die configurations with one 1
    for d1 in product(1, 2:8, 2:8, 2:8, 2:8, 2:8)
        for d2 in product(1, 2:8, 2:8, 2:8, 2:8, 2:8)
            # filter options to avoid reporting duplicates
            if issorted(d1) && issorted(d2) && d1 <= d2
                # x will be a 6x6 matrix storing possible sums
                x = [a1 + a2 for a1 in d1, a2 in d2]
                # check if counts of sums equals what we want
                if [count(v -> v==s, x)  for s in 2:12] == ref
                    # print the result on the screen on success
                    println((d1, d2))
                end
            end
        end
    end
end
```

Now running `listdies()` ensures us that there are actually only two solutions of the problem:

```julia
julia> listdies()
((1, 2, 2, 3, 3, 4), (1, 3, 4, 5, 6, 8))
((1, 2, 3, 4, 5, 6), (1, 2, 3, 4, 5, 6))
```

In case of this problem it would be probably faster for some of the readers to write and run a program similar to the one shown above than to write out generating functions and appropriately group them. Note that in many problems it can be useful to generate the solution with the help of a computer as it might give an insight on how one can solve it or at least check if the idea is correct.

EXERCISES

4.9.1. Consider the Sicherman dice problem in which the restriction that each side is labelled with a positive integer is relaxed to any integer, not necessarily positive. Can you design more pairs of dice?

4.9.2. Solve the recurrence $x_{n+1} = x_n + 2x_{n-1}$ for $n \in \mathbf{N}$, with $x_0 = 0$ and $x_1 = 1$. Verify your solution using induction.

4.9.3. Your friend wants to play the following game with you. You toss three 6-sided fair dies and calculate the sum of outcomes. For every game you have to pay \$1. If the sum is 10 or 11 you get \$4, otherwise you get nothing. Is this game fair?

Chapter 5

Number Theory

As usual, we start the chapter with some basic definitions.

THEORY

Divisibility For any two integers a and b, we say that a **divides** b (or a **is a divisor of** b) if and only if $b/a \in \mathbf{Z}$; that is, $b = ak$ for some $k \in \mathbf{Z}$. If a divides b, then we write $a \mid b$.

For example, $5 \mid 15$ since $15/5 = 3 \in \mathbf{Z}$. On the other hand, $5 \nmid 7$. Indeed, for a contradiction suppose that $5 \mid 7$; that is, $7 = 5k$ for some $k \in \mathbf{Z}$. But this implies that $k = 7/5 \notin \mathbf{Z}$ which gives the desired contradiction.

For any $b \in \mathbf{Z}$, we have $1 \mid b$ and $b \mid b$ (since $b/1 = b \in \mathbf{Z}$ and $b/b = 1 \in \mathbf{Z}$). On the other hand, for any integer $a > 1$, $a \nmid 1$ (since $1/a \notin \mathbf{Z}$). In fact, the following more general property holds:

$$\text{for all } a, b \in \mathbf{N}, \text{ if } a > b, \text{ then } a \nmid b \tag{5.1}$$

(again, since $0 < b/a < 0$ and so $b/a \notin \mathbf{Z}$).

Let us note that divisibility is transitive; that is, if $a \mid b$ and $b \mid c$, then $a \mid c$. Indeed, suppose $a \mid b$ and $b \mid c$. It means that there exist $k, \ell \in \mathbf{Z}$ such that $b = ak$ and $c = b\ell$. Thus, $c = b\ell = (ak)\ell = a(k\ell)$ and, since $k\ell \in \mathbf{Z}$, we get that $a \mid c$.

Prime Numbers A positive integer p is said to be **prime** if and only if it has exactly two distinct positive divisors, namely, 1 and p. A positive integer p is said to be **composite** if it has more than two distinct positive divisors.

The first few primes are $2, 3, 5, 7, 11, 13, 17, 19, 23, 29, \ldots$. Note that 1 has exactly one positive divisor so it is nether prime nor composite. On the other hand, it is easy to see that every integer greater than 1 is either prime or

149

composite. Finally, let us mention that the set of prime numbers is infinite. We provide a proof of this fact in Section 4.1 as an example of a famous constructive argument due to Euclid.

Fundamental Theorem of Arithmetic   Fundamental theorem of **arithmetic** (also known as the unique factorization theorem) states that every integer $n \geq 2$ can be represented in exactly one way as a product of prime powers; that is,

$$n = \prod_{i=1}^{k} p_i^{\alpha_i} = p_1^{\alpha_1} p_2^{\alpha_2} \cdots p_k^{\alpha_k},$$

where $2 \leq p_1 < p_2 < \ldots < p_k$ are prime numbers and $\alpha_i \in \mathbf{N}$, $i \in [k]$.

Before we prove the fundamental theorem of arithmetic, let us mention that it can be alternatively stated as follows: for each integer $n \geq 2$, there exists a unique function $\alpha_n \colon \mathcal{P} \to \mathbf{N} \cup \{0\}$ from the set of all prime numbers $\mathcal{P}$ to the set of non-negative integers such that $n = \prod_{p \in \mathcal{P}} p^{\alpha_n(p)}$. This notation is often convenient; for example, it can be used to express the following well-known **Legendre's formula** that gives an expression for the exponent of the largest power of a prime p that divides the factorial $n!$:

$$\alpha_{n!}(p) = \sum_{i=1}^{\infty} \left\lfloor \frac{n}{p^i} \right\rfloor. \tag{5.2}$$

Indeed, since $n! = 1 \cdot 2 \cdot \ldots \cdot n$, we obtain at least one factor of p for each multiple of p in $[n]$, of which there are $\lfloor n/p \rfloor$. Each multiple of p^2 contributes an additional factor of p, etc. For example,

$$8! = 1 \cdot 2 \cdot 3 \cdot 4 \cdot 5 \cdot 6 \cdot 7 \cdot 8 = 2^7 \cdot 3^2 \cdot 5 \cdot 7.$$

The exponent of 2 can be computed by Legendre's formula as follows:

$$\alpha_{8!}(2) = \sum_{i=1}^{\infty} \left\lfloor \frac{8}{2^i} \right\rfloor = \left\lfloor \frac{8}{2} \right\rfloor + \left\lfloor \frac{8}{4} \right\rfloor + \left\lfloor \frac{8}{8} \right\rfloor = 4 + 2 + 1 = 7.$$

Now, let us move to the proof. The main ingredient is the following observation: every integer greater than 1 is divisible by a prime. The claim is clearly true when n is a prime, as $n \mid n$. Let us then concentrate on composite integers. For a contradiction, suppose that there are composite numbers not divisible by any prime; let us call them bad. Let N be the smallest bad number. (As we will see soon, it is convenient to concentrate on the smallest potential bad number, as it means that *no* smaller composite number is bad.) By the definition of composite numbers, N has at least 3 distinct divisors; in particular, there must be some divisor $d \mid N$ with $d \neq 1$ and $d \neq N$. By (5.1), if $d \mid N$ then $d \leq N$ and therefore $1 < d < N$. By assumption, d is not prime. Since d is a composite number less than N, d is *not* bad and so there is prime

p such that $p \mid d$ (recall that N is the smallest bad number). Since divisibility is transitive, we get that $p \mid d$ together with $d \mid N$ implies that $p \mid N$; rather, N is divisible by a prime. We get the desired contradiction which finishes the proof.

With this observation in hand we can easily prove that for any integer $n \geq 2$, there exists a function $\alpha_n \colon \mathcal{P} \to \mathbf{N} \cup \{0\}$ such that $n = \prod_{p \in \mathcal{P}} p^{\alpha_n(p)}$ (the prove of its uniqueness is slightly longer and so we skip it here). For a contradiction, suppose that there is some integer greater than 1 for which there is no such function; let n be the smallest such example. Note that n cannot be a prime number, for otherwise $\alpha_n(n) = 1$ and $\alpha_n(p) = 0$ for $p \neq n$ would be such a function (clearly, $n = n^1$). By our observation, there is a prime q such that $q \mid n$. Since n is not a prime, $q < n$. Thus, $1 < n/q < n$ and so n/q is associated with a function $\alpha_{n/q}$:

$$n/q = \prod_{p \in \mathcal{P}} p^{\alpha_{n/q}(p)}.$$

(Recall that n is the smallest example.) But it implies that

$$n = q \prod_{p \in \mathcal{P}} p^{\alpha_{n/q}(p)}$$

and so

$$\alpha_n(p) = \begin{cases} \alpha_{n/q}(p) & \text{if } p \neq q \\ \alpha_{n/q}(q) + 1 & \text{if } p = q \end{cases}$$

is the desired function associated with n which gives us a contradiction.

5.1   Greatest Common Divisors

SOURCE

Problem: PLMO XXXIV – Phase 3 — Problem 4
Solution: our own

PROBLEM

Consider any $a, b, c, d \in \mathbf{N}$ such that $ab = cd$. Prove that

$$\gcd(a, c) \cdot \gcd(a, d) \; = \; a \cdot \gcd(a, b, c, d) \,.$$

THEORY

Greatest Common Divisors The **greatest common divisor** of two integers a and b that are not both zero, denoted $\gcd(a, b)$, is the largest integer d such that $d \mid a$ and $d \mid b$. For example, $\gcd(12, 24) = 12$, $\gcd(16, 20) = 4$, $\gcd(3, 5) = 1$, and $\gcd(7, 0) = 7$. Two natural numbers a, b are said to be **relatively prime** (or **co-prime**) if $\gcd(a, b) = 1$. Finally, let us mention that this definition can be easily generalized: the greatest common divisor of k integers $a_1, a_2, \ldots, a_k$ with at least one non-zero value, denoted $\gcd(a_1, a_2, \ldots, a_k)$, is their largest common divisor.

Quotient-Remainder Theorem This useful property, which is often called the division algorithm, is in fact a simple idea that comes directly from long division. For any $n \in \mathbf{Z}$ and $d \in \mathbf{N}$, there exist unique integers q and r such that

$$n = dq + r \quad \text{and} \quad 0 \leq r < d \,.$$

We call q the **quotient** and r the **remainder** of n when divided by d.
    For example,

- if $n = 12, d = 4$, then $q = 3, r = 0$ ($12 = 4 \cdot 3 + 0$),

- if $n = 148, d = 3$, then $q = 49, r = 1$ ($148 = 3 \cdot 49 + 1$),

- if $n = -21, d = 4$, then $q = -6, r = 3$ ($-21 = 4 \cdot (-6) + 3$).

In general, $q = \lfloor n/d \rfloor$, $r = n - qd = n \mod d$. (Here, $\lfloor x \rfloor$ is the **floor** of a real number x that is defined as the largest integer not greater than x.)

There are two parts of the proof. First of all, we need to show that for any $n \in \mathbf{Z}$ and $d \in \mathbf{N}$ there exist q and r with the desired properties. In the second part, we need to show that this pair is unique. We will prove these two parts independently.

Existence: Let $n \in \mathbf{Z}$ and $d \in \mathbf{N}$. Put $q = \lfloor n/d \rfloor$ and $r = n - dq = n - d\lfloor n/d \rfloor$. Clearly, $q, r \in \mathbf{Z}$ and $n = dq + r$ so it remains to show that $0 \leq r < d$. From the definition of the floor function $\lfloor \cdot \rfloor$ it follows that $0 \leq n/d - \lfloor n/d \rfloor < 1$, so $0 \leq r/d < 1$ and the assertion follows.

Uniqueness: Let $n \in \mathbf{Z}$ and $d \in \mathbf{N}$ and take any integers q_1, q_2, r_1, r_2 such that $n = dq_1 + r_1 = dq_2 + r_2$ and $0 \leq r_1, r_2 < d$. We will show that $r_1 = r_2$ and then that also $q_1 = q_2$. For a contradiction, suppose $r_1 \neq r_2$. Without loss of generality, we may assume $r_1 < r_2$. Then, since $dq_1 + r_1 = dq_2 + r_2$, we get that $d(q_1 - q_2) = dq_1 - dq_2 = r_2 - r_1$ and so $d \mid r_2 - r_1$. But $0 \leq r_1 < r_2 < d$ and thus $d > r_2 - r_1 > 0$ which yields a contradiction. Therefore, $r_1 = r_2$. To finish the proof, it is enough to notice that the equation $dq_1 + r_1 = dq_2 + r_2$ implies that $dq_1 = dq_2$, which in turn implies that $q_1 = q_2$.

Euclidean Algorithm

The **Euclidean algorithm** (or Euclid's algorithm) is an efficient method for computing the greatest common divisor of two numbers. It is based on the following observation. Let $a, b \in \mathbf{N}$ with $a > b$. Let r be the remainder of a when divided by b ($a = bq + r$). Then,

$$\gcd(a, b) = \gcd(b, r) .$$

Clearly, $a + b < b + r$ and so one can repeat this procedure until reaching $\gcd(d, 0)$ for some $d \in \mathbf{N}$ and then the algorithm stops since $\gcd(d, 0) = d$.

In order to prove the above key observation, let $a, b \in \mathbf{N}$ with $a > b$ and let $d = \gcd(a, b), e = \gcd(b, r)$. Since $d = \gcd(a, b)$, $d \mid a$, $d \mid b$, and thus also $d \mid bq$ and $d \mid a - bq = r$. It follows that d is a common divisor of b and r. Since e is the greatest such divisor, $d \leq e$. Similarly, since $e = \gcd(b, r)$, $e \mid b$, $e \mid r$, and thus also $e \mid bq$ and $e \mid bq + r = a$. We get that e is a common divisor of a and b. Since d is the greatest such divisor, $e \leq d$. To summarize, we have shown that $d \leq e$ and $e \leq d$, so we get $d = e$.

Extended Euclidean Algorithm

Let us mention about the following straightforward but useful implication of the Euclidean algorithm. For any $a, b \in \mathbf{N}$, there exist $x, y \in \mathbf{Z}$ such that

$$ax + by = \gcd(a, b) .$$

Instead of proving this property, we will show one example which not only should convince the reader that the property above holds but also demonstrates how to actually find x, y which satisfy the desired equation.

Let us find integers x, y such that $425x + 112y = 1$. The first step is to find $\gcd(425, 112)$ using Euclidean Algorithm:

$$
\begin{aligned}
425 &= 3 \cdot 112 + 89 \quad (\Rightarrow 89 = 425 - 3 \cdot 112) \\
112 &= 1 \cdot 89 + 23 \quad (\Rightarrow 23 = 112 - 89) \\
89 &= 3 \cdot 23 + 20 \quad (\Rightarrow 20 = 89 - 3 \cdot 23) \\
23 &= 1 \cdot 20 + 3 \quad (\Rightarrow 3 = 23 - 20) \\
20 &= 6 \cdot 3 + 2 \quad (\Rightarrow 2 = 20 - 6 \cdot 3) \\
3 &= 1 \cdot 2 + 1 \quad (\Rightarrow 1 = 3 - 2) \\
2 &= 2 \cdot 1 + 0,
\end{aligned}
$$

so $\gcd(425, 112) = 1$. Now, one can reverse all the operations to get:

$$
\begin{aligned}
1 &= 3 - 2 \\
&= 3 - (20 - 6 \cdot 3) = -20 + 7 \cdot 3 \\
&= -20 + 7(23 - 20) = -8 \cdot 20 + 7 \cdot 23 \\
&= -8(89 - 3 \cdot 23) + 7 \cdot 23 = 31 \cdot 23 - 8 \cdot 89 \\
&= 31(112 - 89) - 8 \cdot 89 = -39 \cdot 89 + 31 \cdot 112 \\
&= -39(425 - 3 \cdot 112) + 31 \cdot 112 = 148 \cdot 112 - 39 \cdot 425.
\end{aligned}
$$

Hence, $x = -39$ and $y = 148$.

Finally, note that one can use the fundamental theorem of arithmetic to get that for any $a, b \in \mathbf{N}$ we have

$$
\gcd(a, b) = \prod_{p \in \mathcal{P}} p^{\min\{\alpha_a(p), \alpha_b(p)\}}. \tag{5.3}
$$

In particular, this observation shows that if one wants to compute $\gcd(a, b)$, then each prime factor can be considered independently. Let us also notice that $\gcd(a, a) = a$, $\gcd(a, 1) = 1$, $\gcd(a, b) = \gcd(b, a)$ and $\gcd(a, b, c) = \gcd(\gcd(a, b), c)$. The last inequality follows from (5.3) and the fact that for any $i, k, \ell \in \mathbf{N} \cup \{0\}$, $\min\{i, k, \ell\} = \min\{\min\{i, k\}, \ell\}$.

SOLUTION

As mentioned above, each prime factor can be considered independently. Hence, without loss of generality, we may assume that $a = p^\alpha, b = p^\beta, c = p^\gamma$, and $d = p^\delta$ for some prime p and non-negative integers α, β, γ, and δ. Our task is to show that if $\alpha + \beta = \gamma + \delta$, then

$$
\min\{\alpha, \gamma\} + \min\{\alpha, \delta\} = \alpha + \min\{\alpha, \beta, \gamma, \delta\}.
$$

We will independently consider the following three cases which will finish the proof.

Case 1: $\alpha = \min\{\alpha, \beta, \gamma, \delta\}$. Both the left hand side and the right hand side are clearly equal to 2α.

Case 2: $\beta = \min\{\alpha, \beta, \gamma, \delta\}$. Since $\alpha + \beta = \gamma + \delta$, we get $\alpha = \gamma + (\delta - \beta) \geq \gamma$. Using the same argument, we get that $\alpha \geq \delta$. It follows that the left hand side is equal to $\gamma + \delta$ whereas the right hand side is equal to $\alpha + \beta$. The sides are equal by assumption.

Case 3: $\gamma = \min\{\alpha, \beta, \gamma, \delta\}$ (the case when $\delta = \min\{\alpha, \beta, \gamma, \delta\}$ can be dealt with the same way). Arguing as before, note that $\delta \geq \alpha$ and conclude that both sides are equal to $\alpha + \gamma$.

REMARKS

Alternatively, one could solve this problem by providing a proof by induction on a. The base case is easy. Suppose that $a = 1$ and $b, c, d \in \mathbf{N}$ satisfy the required assumption, that is, $ab = cd$. Then, both the left hand side and the right hand side are equal to 1.

For the inductive step, let us fix any quadruple $a, b, c, d \in \mathbf{N}$ such that $a \geq 2$ and $ab = cd$. Let q be a prime number such that $q \mid a$. Since $ab = cd$, we get that $q \mid cd$ and so, without loss of generality, we may assume that $q \mid c$. Our inductive hypothesis is that the desired equality holds for any quadruple $a', b', c', d' \in \mathbf{N}$ satisfying $a' < a$ and $a'b' = c'd'$. In particular, since $(a/q)b = (c/q)d$ and $a/q < a$, we have

$$\gcd(a/q, c/q) \cdot \gcd(a/q, d) = (a/q) \cdot \gcd(a/q, b, c/q, d) .$$

Clearly, $\gcd(a, c) = q \cdot \gcd(a/q, c/q)$. We independently consider the following two cases:

Case 1: $\gcd(a, d) = \gcd(a/q, d)$; that is, the power of q in the (unique) factorization of d is smaller than the corresponding power for a. Hence, since $ab = cd$, the power of q in the factorization of b is smaller than the corresponding power for c; that is, $\gcd(b, c) = \gcd(b, c/q)$. It follows that

$$
\begin{aligned}
\gcd(a, c) \cdot \gcd(a, d) &= q \cdot \gcd(a/q, c/q) \cdot \gcd(a/q, d) \\
&= q \cdot (a/q) \cdot \gcd(a/q, b, c/q, d) \\
&= a \cdot \gcd(\gcd(a/q, d), \gcd(b, c/q)) \\
&= a \cdot \gcd(\gcd(a, d), \gcd(b, c)) \\
&= a \cdot \gcd(a, b, c, d) .
\end{aligned}
$$

Case 2: $\gcd(a, d) = q \cdot \gcd(a/q, d)$; that is, the power of q in the (unique) factorization of d is at least the corresponding power for a. As before, since $ab = cd$, the power of q in the factorization of b is at least the corresponding power for c; that is, $\gcd(b,c) = q \cdot \gcd(b, c/q)$. It follows that

$$
\begin{aligned}
\gcd(a,c) \cdot \gcd(a,d) &= q^2 \cdot \gcd(a/q, c/q) \cdot \gcd(a/q, d) \\
&= q^2 \cdot (a/q) \cdot \gcd(a/q, b, c/q, d) \\
&= q \cdot a \cdot \gcd(\gcd(a/q, d), \gcd(b, c/q)) \\
&= q \cdot a \cdot \gcd(\gcd(a,d)/q, \gcd(b,c)/q) \\
&= q \cdot a \cdot \gcd(\gcd(a,d), \gcd(b,c))/q \\
&= a \cdot \gcd(a,b,c,d) \, .
\end{aligned}
$$

In both cases, the desired equality holds and so the claim holds by induction.

EXERCISES

5.1.1. A positive fraction a/b is said to be in lowest terms if $\gcd(a,b) = 1$. Prove that if a positive fraction a/b is in lowest terms, then fraction

$$(a+b)/(a^2 + ab + b^2)$$

is also in lowest terms.

5.1.2. You are given two natural numbers a and b. Prove that if $a + b \mid a^2$, then $a + b \mid b^2$.

5.1.3. Consider a set A of four digit numbers whose decimal representation uses precisely two digits; moreover, both of them are non-zero. Let $f : A \to A$ be the function such that $f(a)$ flips the digits of $a \in A$ (for example, $f(1333) = 3111$). Find $n > f(n)$ for which $\gcd(n, f(n))$ is as large as possible.

5.2 Modular Arithmetic

SOURCE

Problem and solution idea: PLMO LXX – Phase 1 – Problem 10

PROBLEM

Suppose that

$$a^2 + b^2 + c^2 + d^2 = 2018! \tag{5.4}$$

for some $a, b, c, d \in \mathbf{N}$. Show that each of a, b, c, d, is greater than 10^{250}.

THEORY

$\boxed{\text{Congruence}}$ Let a, b, n be integers with $n \neq 0$. We say that a is **congruent to b modulo n**, and write $a \equiv b \pmod{n}$, if and only if $n \mid (a-b)$. For example, $127 \equiv 7 \pmod 4$, $127 \equiv 3 \pmod 4$, and $127 \equiv -1 \pmod 4$.

Clearly, the congruence is reflexive (that is, $a \equiv a \pmod n$ for any $a \in \mathbf{Z}$) and symmetric (that is, if $a \equiv b \pmod n$, then $b \equiv a \pmod n$). Moreover, it is also transitive, that is, if $a \equiv b \pmod n$, $b \equiv c \pmod n$, then $a \equiv c \pmod n$. Indeed, suppose that $a \equiv b \pmod n$ and $b \equiv c \pmod n$. Since $a \equiv b \pmod n$, $a - b = nk$ for some $k \in \mathbf{Z}$. Similarly, since $b \equiv c \pmod n$, $b - c = n\ell$ for some $\ell \in \mathbf{Z}$. Thus, $a - c = (a-b) + (b-c) = n(k+\ell)$ and $k + \ell \in \mathbf{Z}$. It follows that $a \equiv c \pmod n$.

The following straightforward observation is quite useful. Fix $n \in \mathbf{N}$ and suppose that $a \equiv b \pmod n$ and $c \equiv d \pmod n$. Then,

$$a \pm c \equiv b \pm d \pmod n, \tag{5.5}$$
$$ac \equiv bd \pmod n. \tag{5.6}$$

Since $a \equiv b \pmod n$ and $c \equiv d \pmod n$, $a - b = nk$ and $c - d = n\ell$ for some $k, \ell \in \mathbf{Z}$. Since $(a \pm c) - (b \pm d) = (a-b) \pm (c-d) = n(k \pm \ell)$ and $k \pm \ell \in \mathbf{Z}$. We get that $a \pm c \equiv b \pm d \pmod n$ and so (5.5) holds. In order to prove (5.6), we observe that $a = nk + b$, $c = n\ell + d$, and so $ac = n^2 k\ell + nkd + n\ell b + bd$. It follows that $ac - bd = n^2 k\ell + nkd + n\ell b = n(nk\ell + kd + \ell b)$ and $nk\ell + kd + \ell b \in \mathbf{Z}$ which finishes the proof.

As already mentioned, these properties are incredibly useful in computations, since when we multiply numbers in modular arithmetic (modulo n), we do not have to deal with factors larger than n. To see the power of these properties, let us answer the following question: what is the ones digit of 173^{2019}? Reformulating the question in the language of modular arithmetic, our goal is to find a (unique) digit $x \in \{0, 1, \ldots, 9\}$ such that $173^{2019} \equiv x \pmod{10}$.

Using (5.6) we notice that $173^{2019} \equiv 3^{2019}$ (mod 10) which simplifies our task. With one more trick (and applying (5.6) a few times more), we get

$$
\begin{aligned}
173^{2019} &\equiv 3^{2019} = 3^{2 \cdot 1009 + 1} = (3^2)^{1009} \cdot 3 \\
&= 9^{1009} \cdot 3 \equiv (-1)^{1009} \cdot 3 = -3 \equiv 7 \quad \text{(mod 10)}.
\end{aligned}
$$

It follows that the ones digit of 173^{2019} is 3.

To practice a bit more, let us answer a similar question for which a different argument is needed: what is the ones digit of 2^{2019}? This time we observe that

$$
\begin{aligned}
2^1 &= 2 \equiv 2 \quad \text{(mod 10)} \\
2^2 &= 2 \cdot 2 \equiv 2 \cdot 2 = 4 \quad \text{(mod 10)} \\
2^3 &= 2^2 \cdot 2 \equiv 4 \cdot 2 = 8 \quad \text{(mod 10)} \\
2^4 &= 2^3 \cdot 2 \equiv 8 \cdot 2 = 16 \equiv 6 \quad \text{(mod 10)} \\
2^5 &= 2^4 \cdot 2 \equiv 6 \cdot 2 = 12 \equiv 2 \quad \text{(mod 10)} \\
2^6 &= 2^5 \cdot 2 \equiv 2 \cdot 2 = 4 \quad \text{(mod 10)} \\
&\vdots \\
2^k &= 2^{k-1} \cdot 2 \equiv \ldots
\end{aligned}
$$

We get the following pattern $2, 4, 8, 6, 2, 4, 8, 6, 2, \ldots$. Thus we have shown that if x is even then $2^4 \cdot x \equiv x$ (mod 10). Now, since $2019 \equiv 3$ (mod 4), $2^{2019} \equiv 2^3 = 8$ (mod 10) and so the ones digit of 2^{2019} is 8.

| Multiplicative Inverse | We finish this section with a short discussion about division in modular arithmetic. The key observation is as follows:

if $\gcd(a, n) = 1$, then there exists $b \in \mathbf{Z}$ such that $ab \equiv 1 \pmod{n}$.   (5.7)

Indeed, by Extended Euclidean Algorithm, $\gcd(a, n) = 1$ implies that there exit $b, k \in \mathbf{Z}$ such that $ab + kn = 1$. Reducing both sides mod n yields the result. Let us note that b is called the **multiplicative inverse** of a mod n, denoted by a^{-1}.

One can use the Extended Euclidean Algorithm to find inverses mod n. For example, the multiplicative inverse of 5 mod 22 is 9. Indeed, $22 = 4 \cdot 5 + 2$ and $5 = 2 \cdot 2 + 1$, so $1 = 5 - 2 \cdot 2 = 5 - 2 \cdot (22 - 4 \cdot 5) = 9 \cdot 5 - 2 \cdot 22$. Thus, $9 \cdot 5 \equiv 1$ (mod 22). Let us also remark that the assumption in (5.7) that $\gcd(a, n) = 1$ is necessary; if $\gcd(a, n) > 1$, then no $b \in \mathbf{Z}$ exists. Indeed, towards a contradiction, suppose that there exists $b \in \mathbf{Z}$ such that $ab \equiv 1$ (mod n). Then $1 = ab - nk$ for some $k \in \mathbf{Z}$. But $\gcd(a, n) = d > 1$ divides the right hand side, so $d \mid 1$. We get the desired contradiction.

Here is another useful property: if $\gcd(a, n) = 1$, then

$$ax \equiv ay \pmod{n} \quad \text{implies that} \quad x \equiv y \pmod{n}.   (5.8)$$

Indeed, since $\gcd(a, n) = 1$, one can multiply both sides by a^{-1} to get

$$a^{-1}ax \equiv a^{-1}ay \pmod{n}$$
$$1x \equiv 1y \pmod{n}$$
$$x \equiv y \pmod{n}.$$

SOLUTION

For a contradiction, suppose that at least one of a, b, c, d is smaller than or equal to $10^{250} < 16^{250} = 2^{1000}$. Let α be the largest non-negative integer such that 2^α divides each of a, b, c, d; that is, using the unique factorization theorem,

$$\alpha = \min\{\alpha_a(2), \alpha_b(2), \alpha_c(2), \alpha_d(2)\}.$$

Then $a = 2^\alpha a'$, $b = 2^\alpha b'$, $c = 2^\alpha c'$, $d = 2^\alpha d'$, and at least one of a', b', c', d' is odd. Moreover, it follows from our assumption that $\alpha < 1000$.

Let us now observe that the power of 2 in the (unique) factorization of 2018! can be calculated from Legendre's formula (see (5.2)) as follows:

$$\begin{aligned}
\alpha_{2018!}(2) &= \sum_{i=1}^{\infty} \left\lfloor \frac{2018}{2^i} \right\rfloor \\
&= 1009 + 504 + 252 + 126 + 63 + 31 + 15 + 7 + 3 + 1 \\
&= 2011.
\end{aligned}$$

It is often convenient to check such computations using a computer program. Here is a simple script that implements the Legendre's formula in the Julia language and calculates the value of $\alpha_{2018!}(2)$:

```julia
julia> function legendre(n, d)
           s = 0
           q = 1
           while true
               q *= d
               a = div(n, q)
               if a == 0
                   return s
               end
               s += a
           end
       end
legendre (generic function with 1 method)

julia> legendre(2018, 2)
2011
```

Hence, after dividing both sides of equation (5.4) by $(2^\alpha)^2 = 2^{2\alpha}$, we get that

$$a'^2 + b'^2 + c'^2 + d'^2 = \frac{2018!}{2^{2\alpha}}. \tag{5.9}$$

Since $2011 - 2\alpha \geq 3$, the right hand side of equality (5.9) is congruent to $0 \pmod 8$. On the other hand, note that if n is odd, then $n^2 \equiv 1 \pmod 8$, and if n is even, then $n^2 \equiv 0$ or $4 \pmod 8$. Indeed, if $n = 2k + 1$ for some $k \in \mathbf{Z}$, then $n^2 = (2k + 1)^2 = 4k(k + 1) + 1 \equiv 1 \pmod 8$. Similarly, if $n = 2k$ for some $k \in \mathbf{Z}$, then $n^2 = 4k^2 \equiv 0$ or $4 \pmod 8$. It follows that at least one of a'^2, b'^2, c'^2, d'^2 is congruent to $1 \pmod 8$ and all of them are congruent to $0, 1$, or $4 \pmod 8$. It is easy to see then that the left hand side of equality (5.9) is *not* congruent to $0 \pmod 8$, and we get the desired contradiction.

REMARKS

The key idea in our solution is to notice that if the right hand side of equality (5.4) is divisible by 8, then each of a, b, c, d must be divisible by 2, as explained above. With this observation in hand, one can divide both sides of equality (5.4) by 4, and repeat the process recursively as long as the right hand side is divisible by 8. It shows that each of the four numbers must be large. The proof presented above uses this idea but avoids the recursive argument by noticing that in the unique factorization of 2018! prime number 2 is raised to a large power.

EXERCISES

5.2.1. Find all primes p for which $p^2 + 2$ is also prime.

5.2.2. You are given three consecutive natural numbers (say, a, $a + 1$, and $a + 2$) such that the middle one is a cube (that is, $a + 1 = \ell^3$ for some $\ell \in \mathbf{N}$). Prove that their product is divisible by 504.
(Source of the problem and solution idea: PLMO IX – Phase 3 – Problem 1.)

5.2.3. Prove that for any natural $n \in \mathbf{N}$ that is not divisible by 10 there exists $k \in \mathbf{N}$ such that n^k has in its decimal representation the same digit at the first and the last position.
(Source of the problem and solution idea: "Delta" monthly – January, 2017.)

5.3 Factorization

SOURCE

Problem and solution idea: PLMO LXVIII – Phase 2 – Problem 1

PROBLEM

Consider any prime number $p > 2$. Prove that there exists exactly one $n \in \mathbf{N}$ such that $n^2 + np$ is a square; that is, $n^2 + np = k^2$ for some $k \in \mathbf{N}$.

THEORY

In this section, we will use a basic but useful fact that follows immediately from the fundamental theorem of arithmetic:

$$\text{if } p \text{ is a prime number and } p \mid ab, \text{ then } p \mid a \text{ or } p \mid b. \qquad (5.10)$$

For example, we will use (5.10) to show that $\sqrt{2}$ is irrational. (Recall that a real number q is **rational** if $q = a/b$ for some $a, b \in \mathbf{Z}$ and $b \neq 0$; a real number r is **irrational** if it is not rational.) For a contradiction, suppose that there are integers $a, b, b \neq 0$ such that $\sqrt{2} = a/b$. We may assume that $\gcd(a, b) = 1$ (we may write $\sqrt{2}$ in lowest terms). Now, note that $\sqrt{2} = a/b$ implies that $a = b\sqrt{2}$ which, in turn, implies that $a^2 = 2b^2$; that is, $2 \mid a^2$. By (5.10), $2 \mid a^2$ implies that $2 \mid a$; that is, $a = 2k$ for some $k \in \mathbf{Z}$. Now, $a^2 = 2b^2$ implies $4k^2 = 2b^2$ that, in turn, implies $2 \mid b^2$. Using (5.10) again, we get $2 \mid b$. Since $2 \mid a$ and $2 \mid b$, $\gcd(a, b) \geq 2 > 1$ and we get the desired contradiction.

Another useful consequence of (5.10) is that if $p = ab$ for some prime number p and two integers a, b, then there are only four possibilities for a pair (a, b), namely, $(p, 1)$, $(-p, -1)$, $(1, p)$, and $(-1, -p)$. Alternatively, if one insists that a, b are natural numbers and $a \leq b$, then $(a, b) = (1, p)$. Moreover, this observation can be easily generalized. For example, if $pq = abc$ for two prime numbers p, q ($p \leq q$) and three natural numbers a, b, c ($a \leq b \leq c$), then the triple (a, b, c) must be either $(1, p, q)$ or $(1, 1, pq)$.

SOLUTION

Suppose that $n^2 + np = k^2$ for some $k \in \mathbf{N}$; that is,

$$np = k^2 - n^2 = (k - n)(k + n).$$

Let $x = \gcd(k, n)$, let $k' = k/x \in \mathbf{N}$, and let $n' = n/x \in \mathbf{N}$. Then,

$$n'p = np/x = x(k' - n')(k' + n').$$

Since $\gcd(n', k') = 1$, we have $\gcd(n', k' - n') = 1$ and $\gcd(n', k' + n') = 1$. It follows that n' has to divide x, that is, $x' = x/n' \in \mathbf{N}$ and so

$$p = x'(k' - n')(k' + n'). \qquad (5.11)$$

After noting that p is a prime and $k' - n' < k' + n'$, we get that the product on the right hand side of (5.11) is unique: $x' = 1$, $k' - n' = 1$, and $k' + n' = p$. It follows that $k' = (p+1)/2$, $n' = (p-1)/2$, and so $k = k'x = k'n' = (p^2 - 1)/4$, $n = n'x = n'^2 = (p-1)^2/4$ is the unique solution of the system, provided that both n and k are natural numbers.

In order to finish the proof, we will use the fact that p is a prime greater than 2; in particular, it is odd: $p = 2\ell + 1$ for some $\ell \in \mathbf{N}$. We get that $k = (p^2 - 1)/4 = (4\ell^2 + 4\ell)/4 = \ell^2 + \ell \in \mathbf{N}$ and $n = (p-1)^2/4 = (2\ell)^2/4 = \ell^2 \in \mathbf{N}$, and the proof is finished.

REMARKS

Alternatively one could notice that $n^2 + np = k^2$ is equivalent to

$$p^2 = (2n + p - 2k)(2n + p + 2k) .$$

But this implies that $2n + p - 2k = 1$ and $2n + p + 2k = p^2$. From this we get that

$$1 + p^2 = (2n + p - 2k) + (2n + p + 2k) = 4n + 2p$$

and so $n = (p-1)^2/4$.

EXERCISES

5.3.1. You are given two integers, a and b, and a prime $p > 2$. Prove that if $p \mid a + b$ and $p \mid a^2 + b^2$, then $p^2 \mid a^2 + b^2$.
(Source of the problem and solution idea: PLMO LXIX – Phase 1 – Problem 1.)

5.3.2. You are given four integers a, b, c, and d. Prove that if $a - c \mid ab + cd$, then $a - c \mid ad + bc$.

5.3.3. Consider any natural number $n \geq 2$. Prove that $n^{12} + 64$ has at least four different non-trivial natural factors; that is, $n^{12} + 64 = a \cdot b \cdot c \cdot d$ for some $a, b, c, d \in \mathbf{N}$ such that $1 < a < b < c < d < n^{12} + 64$.

5.4 Fermat's Little Theorem and Euler's Theorem

SOURCE

Problem and solution idea: PLMO LXV – Phase 3 – Problem 5

PROBLEM

Find all $x, y \in \mathbf{N}$ such that $2^x + 17 = y^4$.

THEORY

Fermat's Little Theorem | **Fermat's little theorem** is one of the fundamental results of elementary number theory. We state it as follows: if p is a prime and n is an integer, then

$$n^p \equiv n \pmod{p},$$

or equivalently $p \mid n^p - n$. Before we prove this fact, let us mention that it is often stated in the following form: if p is a prime and n is an integer not divisible by p, then

$$n^{p-1} \equiv 1 \pmod{p},$$

or equivalently $p \mid n^{p-1} - 1$. It is easy to show that both statements are equivalent, that is, if one of them is true then the other one must also be true.

Fix any prime number p. We will prove Fermat's little theorem by induction on n. The base case is trivial: $1^p \equiv 1 \pmod{p}$. For the inductive step, suppose that $k^p \equiv k \pmod{p}$ for some $k \in \mathbf{N}$. Our goal is to show that $(k+1)^p \equiv k+1 \pmod{p}$.

Let us first note that

$$(k+1)^p = \sum_{i=0}^{p} \binom{p}{i} k^i = 1 + \sum_{i=1}^{p-1} \binom{p}{i} k^i + k^p.$$

Then, observe that for any $i \in \mathbf{Z}$, $1 \le i \le p-1$, $p \mid \binom{p}{i}$ or, equivalently, $\binom{p}{i} \equiv 0 \pmod{p}$. Indeed,

$$\binom{p}{i} = \frac{p(p-1)\cdots(p-i+1)}{i(i-1)\cdots 1};$$

p divides the numerator but not the denominator. It follows that $(k+1)^p \equiv k^p+1 \pmod{p}$. Finally, by inductive hypothesis, we know that $k^p \equiv k \pmod{p}$ and so $(k+1)^p \equiv k+1 \pmod{p}$, as needed. The proof is finished.

Let us also mention that the converse of Fermat's little theorem is not generally true. Counter-examples, that is, composite numbers q for which $n^q \equiv n \pmod{q}$ for all integers n, are called **Carmichael numbers**. The smallest Carmichael number is 561. Indeed, $561 = 3 \cdot 11 \cdot 17$ so 561 is composite. One can show that $a^{561} \equiv a \pmod{561}$ for any a by independently showing that $a^{561} \equiv a \pmod 3$, $a^{561} \equiv a \pmod{11}$, and $a^{561} \equiv a \pmod{17}$.

Euler's Totient Function   Our next task is to show an important generalization of Fermat's little theorem. However, before we do it, we need to introduce a closely related function and discuss a few of its useful properties. **Euler's totient function** $\phi : \mathbf{N} \to \mathbf{N}$ is defined as follows: for each $q \in \mathbf{N}$, $\phi(q)$ is equal to the number of integers between 1 and q (including both 1 and q) that are co-prime with q.

As promised, we discuss next a few interesting properties of $\phi(q)$.

Property 1. If q is a prime number, then $\phi(q) = q - 1$. This fact follows directly from the definition of prime numbers.

Property 2. If $q = p^k$ for some prime number p and $k \in \mathbf{N}$, then

$$\phi(q) \;=\; p^k - p^{k-1} \;=\; q\left(1 - \frac{1}{p}\right).$$

To see this observe that there are exactly p^{k-1} numbers between 1 and q that are of the form $i \cdot p$, where $i \in [p^{k-1}]$; these are the only numbers in that interval that are *not* co-prime to q.

Property 3. Here is another useful fact:

$$\text{if } p \text{ and } q \text{ are co-prime, then } \phi(pq) \;=\; \phi(p) \cdot \phi(q). \qquad (5.12)$$

In order to show (5.12) we need to make one additional observation. Let us fix any p and q that are co-prime and consider a function $f : [pq] \to [p] \times [q]$ defined as follows: for any $d \in [pq]$, $f(d) = (r_p(d), r_q(d))$, where $r_p(d)$ is the remainder when d is divided by p and $r_q(d)$ is the remainder when d is divided by q. We claim that f is a **bijection**; that is,

a) f is **one-to-one**: if $d_1 \neq d_2$, then $f(d_1) \neq f(d_2)$, and

b) f is **onto**: for any $(r_p, r_q) \in [p] \times [q]$, there exists $d \in [pq]$ such that $f(d) = (r_p, r_q)$.

We will show first that f is one-to-one. To get a contradiction, suppose that that

$$f(d_1) \;=\; (r_p(d_1), r_q(d_1)) \;=\; (r_p(d_2), r_q(d_2)) \;=\; f(d_2)$$

for some $d_1 \neq d_2$. Due to the symmetry, without loss of generality we may assume that $d_1 > d_2$. Since $d_1 \equiv r_p(d_1) = r_p(d_2) \equiv d_2 \pmod p$ and $d_1 \equiv r_q(d_1) = r_q(d_2) \equiv d_2 \pmod q$, we have $p \mid d_1 - d_2$ and $q \mid d_1 - d_2$. Since p and q are co-prime, $pq \mid d_1 - d_2$. Finally, since $1 \leq d_2 < d_1 \leq pq$, we get that

$1 \le d_1 - d_2 < pq$ and so $pq \mid d_1 - d_2$ is only possible if $d_1 = d_2$. We get the desired contradiction.

Now, notice that the codomain of function f is of size $\|[p]\| \cdot \|[q]\| = pq$, exactly the same as the size of its domain. Hence, function f is not only a one-to-one function but it is also onto (and so it is a bijection).

We are finally ready to show (5.12). Notice that d is co-prime to pq if and only if it is co-prime to p and to q. Moreover, by the Euclidean algorithm, d is co-prime to p if and only if $r_p(d)$ is co-prime to p. Similarly, d is co-prime to q if and only if $r_q(d)$ is co-prime to q. The desired conclusion follows immediately from our observation that function f is a bijection. Indeed, we have that $\phi(pq)$, the number of values of d in $[pq]$ that are co-prime to pq, is equal to the number of pairs (r_p, r_q) where r_p is co-prime to p and r_q is co-prime to q. The number of such pairs is $\phi(p) \cdot \phi(q)$ and the proof of this property is finished.

Property 4. Combining Properties 2 and 3 we get:

$$\phi(q) = q \prod_{i=1}^{\ell} \left(1 - \frac{1}{p_i}\right),$$

where $q = \prod_{i=1}^{\ell} p_i^{k_i}$ is the unique factorization of q; in particular, (p_i) is a sequence of prime numbers and $k_i \in \mathbf{N}$ for all $i \in [\ell]$.

| Euler's Theorem | **Euler's theorem** is a generalization of Fermat's little theorem: For any integer $q \ge 2$ and any integer n that is co-prime to q, one has

$$n^{\phi(q)} \equiv 1 \pmod{q}.$$

Note that Fermat's little theorem is indeed a special case of Euler's theorem, because if q is a prime number, then $\phi(q) = q - 1$. (See Property 1 above.)

In order to prove Euler's theorem, let $r_1, r_2, \ldots, r_{\phi(q)}$ be all the numbers in $[q]$ that are co-prime with q. Let us concentrate on $n \cdot r_i$ for some $i \in [\phi(q)]$. Because both n and r_i are co-prime with q, $n \cdot r_i$ is also co-prime with q. Now, let s_i be the reminder when $n \cdot r_i$ is divided by q. Observe that s_i is also co-prime with q. Additionally, if $r_i \ne r_j$, then $s_i \ne s_j$. To see this, observe that if $s_i = s_j$, then $n \cdot r_i \equiv n \cdot r_j \pmod{q}$ and, as n is co-prime with q, it follows from (5.8) that $r_i \equiv r_j \pmod{q}$, which is not possible. This means that there is a bijection between the set $S := \{s_1, s_2, \ldots, s_{\phi(q)}\}$ and the set $R := \{r_1, r_2, \ldots, r_{\phi(q)}\}$ and so $S = R$. Hence, since $n \cdot r_i \equiv s_i \pmod{q}$ for all i, we get

$$n^{\phi(q)} \prod_{i=1}^{\phi(q)} r_i \equiv \prod_{i=1}^{\phi(q)} s_i \pmod{q}.$$

Using the fact that all r_i are co-prime with q and that $S = R$, we can use (5.8) to get the desired result.

| **Chinese Remainder Theorem** | The following result is widely used for computing with large integers, as it allows replacing a computation for which one knows a bound on the size of the result by several similar computations on small integers. The **Chinese remainder theorem** can be stated as follows in term of congruences. If the numbers in the set $\{n_1, n_2, \ldots, n_k\}$ are pairwise co-prime and $a_1, a_2, \ldots, a_k$ are any integers, then there exists an integer x such that

$$x \equiv a_1 \pmod{n_1}$$
$$\vdots$$
$$x \equiv a_k \pmod{n_k},$$

and any two such x are congruent modulo $N = \prod_{i=1}^{k} n_i$.

For each $i \in [k]$, let $N_i = N/n_i$. Note that $\gcd(N_i, n_i) = 1$ for all $i \in [k]$. Let N_i^{-1} be an inverse of N_i modulo n_i; that is, $N_i N_i^{-1} \equiv 1 \pmod{n_i}$. Let

$$\hat{x} = a_1 N_1 N_1^{-1} + a_2 N_2 N_2^{-1} + \ldots + a_k N_k N_k^{-1}.$$

Clearly, $\hat{x}$ is a simultaneous solution to all of the congruences. Indeed, for any $j \neq i$, $N_j \equiv 0 \pmod{n_i}$ and so $\hat{x} \equiv a_i N_i N_i^{-1} \equiv a_i \pmod{n_i}$, as required. Moreover, since the moduli $n_1, n_2, \ldots, n_k$ are pairwise relatively prime, any two simultaneous solutions to the system must be congruent modulo N. Hence, the full solution is $x \equiv \hat{x} \pmod{N}$.

Let us note that the proof is constructive; that is, it gives us an explicit formula for the solution. For example, let us solve the following system of congruences:

$$x \equiv 1 \pmod 5$$
$$x \equiv 2 \pmod 7$$
$$x \equiv 3 \pmod 9$$
$$x \equiv 4 \pmod{11}.$$

Note that the moduli are pairwise relatively prime, as required by the Chinese remainder theorem. Using the notation from the proof, we have $N = 5 \cdot 7 \cdot 9 \cdot 11 = 3465$, $N_1 = N/5 = 693$, $N_2 = N/7 = 495$, $N_3 = N/9 = 385$, and $N_4 = N/11 = 315$. A small calculation gives $N_1^{-1} = 2$ ($693 \cdot 2 \equiv 3 \cdot 2 = 6 \equiv 1 \pmod 5$), $N_2^{-1} = 3$ ($495 \cdot 3 \equiv 5 \cdot 3 = 15 \equiv 1 \pmod 7$), $N_3^{-1} = 4$ ($385 \cdot 4 \equiv 7 \cdot 4 = 28 \equiv 1 \pmod 9$), and $N_4^{-1} = 8$ ($315 \cdot 8 \equiv 7 \cdot 8 = 56 \equiv 1 \pmod{11}$). Hence, one solution is

$$\hat{x} = 1 \cdot 693 \cdot 2 + 2 \cdot 495 \cdot 3 + 3 \cdot 385 \cdot 4 + 4 \cdot 315 \cdot 8 = 19056.$$

The full solution is

$$x \equiv 19056 \equiv 1731 \pmod{N}.$$

In particular, $x = 1731$ is the smallest positive integer solution to the system.

SOLUTION

Suppose that $x, y \in \mathbf{N}$ are such that $2^x + 17 = y^4$. Observe that

$$y^{16} - 16^x = (y^8 - 4^x)(y^8 + 4^x) = (y^4 - 2^x)(y^4 + 2^x)(y^8 + 4^x).$$

In particular, we get that $17 = y^4 - 2^x \mid y^{16} - 16^x$ or, alternatively, that $y^{16} \equiv 16^x \pmod{17}$. Observe now that $16^x \equiv (-1)^x \pmod{17}$. On the other hand, Fermat's little theorem implies that $y^{16} \equiv 1 \pmod{17}$ if y is *not* divisible by 17 and otherwise we get that $y^{16} \equiv 0 \pmod{17}$. This implies that y is not divisible by 17 and, more importantly, that x is even. Therefore, after letting $k = x/2 \in \mathbf{N}$, we can re-write the equation as follows:

$$17 = y^4 - 2^{2k} = (y^2 - 2^k)(y^2 + 2^k).$$

Since 17 is a prime number, this equation holds only if $y^2 + 2^k = 17$ and $y^2 - 2^k = 1$. It follows that

$$2^k = \frac{(y^2 + 2^k) - (y^2 - 2^k)}{2} = \frac{17 - 1}{2} = 8$$

and

$$y^2 = \frac{(y^2 + 2^k) + (y^2 - 2^k)}{2} = \frac{17 + 1}{2} = 9,$$

which gives us $x = 2k = 6$ and $y = 3$. One can easily verify that indeed $2^6 + 17 = 3^4$.

REMARKS

It is easy to observe that the decomposition $17 = (y^2 - 2^k)(y^2 + 2^k)$ would yield an easy solution to the problem. Hence, the main difficulty in this problem is to show that x is even. We present one way to show it but, alternatively, one can use an argument similar to the one from Section 5.2. We observe that for $x = 1, 2, 3, 4, 5, 6, 7, 8$

$$2^x \equiv 2, 4, 8, 16, 15, 13, 9, 1 \pmod{17}.$$

We close the cycle of length 8 which implies that for any $x \in \mathbf{N}$, $2^{x+8} \equiv 2^x$ and so $\{2, 4, 8, 16, 15, 13, 9, 1\}$ are the only possible remainders when the left hand side of the equation is divided by 17.

On the other hand, the only possible reminders when y^4, the right hand side of the equation, is divided by 17 are 0, 1, 4, 13, and 16, which can be verified by enumerating the reminders of $0^4, 1^4, \ldots, 16^4$ when divided by 17. Here is a line of code written in Julia language that performs this calculation and shows the result:

```
julia> Set(mod(i ^ 4, 17) for i in 0:16)
Set([4, 16, 0, 13, 1])
```

Comparing possible remainders on both sides, we conclude that the only chance that the remainders match is when x is even.

EXERCISES

5.4.1. Prove that for all $a \in \mathbf{N}$, we have $35 \mid a^{64} - a^4$.

5.4.2. Prove that for any odd integer n, we have that $n \mid \prod_{i=1}^{n} \sum_{j=0}^{i} 2^j$.

5.4.3. Find the last two digits in the decimal representation of 7^{123}.

5.5 Rules of Divisibility

SOURCE

Problem and solution: PLMO LXIX – Phase 2 – Problem 2

PROBLEM

Suppose that $n \in \mathbf{N}$ is such that $n \equiv 4 \pmod 8$. Let

$$1 = k_1 < k_2 < \ldots < k_m = n$$

be all positive divisors of n. Prove that if natural number $i < m$ is not divisible by 3, then $k_{i+1} \leq 2k_i$.

THEORY

When discussing rules for divisibility, it is convenient to write an integer n we are concerned with in base 10 as follows:

$$n = 10^k a_k + 10^{k-1} a_{k-1} + \cdots + 10 a_1 + a_0$$

with $a_k \neq 0$ and $0 \leq a_i \leq 9$ for $0 \leq i \leq k$. Here are some standard rules of divisibility:

1. $2 \mid n$ if and only if $2 \mid a_0$ (the last digit is even),

2. $5 \mid n$ if and only if $5 \mid a_0$ (the last digit is 0 or 5),

3. $4 \mid n$ if and only if $4 \mid 10a_1 + a_0$ (the number formed by the two rightmost digits is divisible by 4),

4. $8 \mid n$ if and only if $8 \mid 100a_2 + 10a_1 + a_0$ (the number formed by the three rightmost digits is divisible by 8),

5. $3 \mid n$ if and only if $3 \mid \sum_{i=0}^{k} a_i$ (the sum of digits is divisible by 3),

6. $9 \mid n$ if and only if $9 \mid \sum_{i=0}^{k} a_i$ (the sum of digits is divisible by 9).

The proofs are straightforward so we only show the rule for 3 in order to illustrate the argument. Note that $10 \equiv 1 \pmod 3$, so for any $i \in \mathbf{N}$ we get $10^i \equiv 1^i \equiv 1 \pmod 3$. Thus

$$
\begin{aligned}
3 \mid n \quad &\Leftrightarrow \quad n \equiv 0 \pmod 3 \\
&\Leftrightarrow \quad 10^k a_k + 10^{k-1} a_{k-1} + \cdots + 10 a_1 + a_0 \equiv 0 \pmod 3 \\
&\Leftrightarrow \quad a_k + a_{k-1} + \cdots + a_0 \equiv 0 \pmod 3,
\end{aligned}
$$

that is, the sum of its digits is divisible by 3.

SOLUTION

Instead of proving the original conditional statement (if $3 \nmid i$, then $k_{i+1} \leq 2k_i$) we will prove its contrapositive (if $k_{i+1} > 2k_i$, then $3 \mid i$), as both statements are logically equivalent.

Let us first note that since $n \equiv 4 \pmod 8$, n is divisible by 4 but not by 8. As a result, each divisor of n is of the form $2^s(2\ell+1)$, $s \in \{0, 1, 2\}$, $\ell \in \mathbf{N} \cup \{0\}$. Consider any $i \in \mathbf{N}$ such that $i < m$ and assume that $k_{i+1} > 2k_i$. Our goal is to show that $3 \mid i$; however, in fact, we will show something stronger, namely, that the set $\{k_1, k_2, \ldots, k_i\}$ can be partitioned into $i/3$ sets, each of the form $\{d, 2d, 4d\}$ for some odd positive integer d.

We will show that if $d \leq k_i$ is an odd divisor of n, then both $2d$ and $4d$ are not only divisors of n (which is obvious) but also $2d < 4d \leq k_i$. This will finish the proof as it implies that the desired partition exists. Consider any odd divisor $d \leq k_i$ of n. By our assumption, $2d \leq 2k_i < k_{i+1}$ but this implies that $2d \leq k_i$. Repeating the argument one more time we get $4d = 2(2d) \leq 2k_i < k_{i+1}$ which gives $4d \leq k_i$. The desired property holds and we are done.

REMARKS

The key observation that leads to the solution is to notice that if $k_{i+1} > 2k_i$, then k_{i+1} must be odd. Indeed, if k_{i+1} were even, then $k_{i+1}/2 > k_i$ would be an integer that also divides n but there is no divisor of n between k_i and k_{i+1}. From this observation it follows that all divisors of the form d, $2d$ and $4d$, where d is an odd divisor of n, must be either less than k_{i+1} or all of them are greater than or equal to k_{i+1}.

EXERCISES

5.5.1. Decide if there exists $k \in \mathbf{N}$ with the property that in the decimal representation of 2^k each of the 10 digits $(0, 1, 2, \ldots, 9)$ is present the same number of times.
(Source of the problem and solution: LXX – Phase 1 – Problem 1.)

5.5.2. Find the minimum of $|20^m - 9^n|$ over all natural numbers m and n.
(Source of the problem and solution idea: LXIV – Phase 1 – Problem 5.)

5.5.3. Given $m, n, d \in \mathbf{N}$, prove that if $m^2 n + 1$ and $mn^2 + 1$ are divisible by d, then $m^3 + 1$ and $n^3 + 1$ are also divisible by d.

5.6 Remainders

SOURCE

Problem and solution idea: PLMO XLIX – Phase 3 – Problem 4

PROBLEM

Let the sequence (a_i) be defined recursively as follows: $a_1 = 1$ and for any $n \in \mathbf{N} \setminus \{1\}$, $a_n = a_{n-1} + a_{\lfloor n/2 \rfloor}$. Prove that there are infinitely many terms in the sequence (a_i) that are divisible by 7.

THEORY

Let p be any prime number. For given integers a, b, and k, let

$$S = \{a + ib : i \in \mathbf{Z}, k + 1 \le i \le k + p\}.$$

If $p \mid b$, then clearly all numbers in S give the same remainder when divided by p, namely, the remainder when a is divided by p. On the other hand, if p does *not* divide b, then all numbers in S yield unique remainders when divided by p. Indeed, for a contradiction, suppose that there exist i, j, $k + 1 \le i < j \le k + p$ such that $a + ib \equiv a + jb \pmod{p}$. It follows that $p \mid (j - i)b$ and so $p \mid (j - i)$ as p does not divide b. But $0 < j - i < p$ so it is impossible that $p \mid (j - i)$. We get the desired contradiction and so the property holds.

SOLUTION

Let us first observe that $a_2 = a_1 + a_1 = 2$, $a_3 = a_2 + a_1 = 3$, $a_4 = a_3 + a_2 = 5$, $a_5 = a_4 + a_2 = 7$. So there exists at least one term in the sequence that is divisible by 7.

For a contradiction, suppose that that there are only finitely many terms that are divisible by 7; let a_k be the largest one. Let us concentrate on the following seven consecutive terms of the sequence:

$$a_{4k-3}$$
$$a_{4k-2} = a_{4k-3} + a_{2k-1}$$
$$a_{4k-1} = a_{4k-3} + 2a_{2k-1}$$
$$a_{4k} = a_{4k-3} + 2a_{2k-1} + a_{2k} = a_{4k-3} + 3a_{2k-1} + a_k$$
$$a_{4k+1} = a_{4k-3} + 3a_{2k-1} + a_k + a_{2k} = a_{4k-3} + 4a_{2k-1} + 2a_k$$
$$a_{4k+2} = a_{4k-3} + 4a_{2k-1} + 2a_k + a_{2k+1} = a_{4k-3} + 5a_{2k-1} + 4a_k$$
$$a_{4k+3} = a_{4k-3} + 5a_{2k-1} + 4a_k + a_{2k+1} = a_{4k-3} + 6a_{2k-1} + 6a_k.$$

Since a_k is divisible by 7, the remainders of these consecutive terms when divided by 7 are the same as the corresponding reminders of numbers in

$$S = \{a_{4k-3} + ia_{2k-1} : i \in \mathbf{Z}, 0 \le i \le 6\}.$$

Since prime number 7 does not divide a_{2k-1}, it follows from the observation we made in the theory part that all numbers in S yield unique remainders when divided by 7. Hence, one of them is equal to 0 and so one of the consecutive terms we consider is divisible by 7. We get the desired contradiction and the proof is finished.

REMARKS

Since our task was to prove that there are infinite number of values of $n \in \mathbf{N}$ for which $7 \mid a_n$, it is clear that one should consider the recurrence $a_n = a_{n-1} + a_{\lfloor n/2 \rfloor}$ but concentrate on reminders when a_n in divided by 7.

One natural idea that has a chance to lead to a solution is to consider seven consecutive numbers and hope that all of them have different reminders when divided by 7. Since the recursive definition for a_n involves $a_{\lfloor n/2 \rfloor}$ and it has to be applied twice in order to reduce the number of terms to at most three, one should consider indexes of the form $n = 4k + r$ for some seven consecutive values of r.

The key observation is to notice the pattern that implies that the terms involving a_{2k-1} yield unique reminders when divided by 7, provided that a_{2k-1} is not divisible by 7. Hence, the same is true for the seven consecutive terms of the form $n = 4k + r$, provided that additionally a_k is divisible by 7. This leads to the proof by contradiction that assumes that a_k is the largest number divisible by 7.

EXERCISES

5.6.1. Find all $x, y \in \mathbf{N}$ such that $2^x + 5^y$ is a square.
(Source of the problem and solution idea: General Mathematics Vol. 15, No. 4 (2007), 145–148.)

5.6.2. Prove that for any two sequences, $(x_i)_{i=1}^{2011}$ and $(y_i)_{i=1}^{2011}$, of natural numbers, $\prod_{i=1}^{2011}(2x_i^2 + 3y_i^2)$ is *not* a square.
(Source of the problem and solution idea: PLMO LXII – Phase 3 – Problem 3.)

5.6.3. Consider any integer $n \geq 2$ and any subset S of the set $N := \{0, 1, 2, \ldots, n - 1\}$ that has more than $\frac{3}{4}n$ elements. Prove that there exist integers a, b, c such that the remainders when numbers a, b, c, $a + b$, $a + c$, $b + c$, $a + b + c$ are divided by n are all in S.
(Source of the problem and solution idea: PLMO LXI – Phase 3 – Problem 1.)

5.7 Aggregation

SOURCE

Problem and solution idea: PLMO XLII – Phase 1 – Problem 11 (modified)

PROBLEM

Let $n > 1,000$ be any integer. For $i \in [n]$, let r_i be the reminder when 2^n is divided by i. Prove that $\sum_{i=1}^{n} r_i > 3.5n$.

THEORY

When a long sequence of numbers is considered, it is often the case that some patters naturally emerge, either in the sequence itself or in the sum or the product of its terms. Here are some simple observations:

$$\sum_{i=0}^{n-1} (a + i \cdot b) = n \cdot a + \frac{n(n-1)}{2} \cdot b$$

that implies that if n is odd or b is even, then $n \mid \sum_{i=0}^{n-1}(a + i \cdot b)$. Similarly,

$$\sum_{i=0}^{n-2} b^i = \frac{b^{n-1} - 1}{b - 1}$$

is divisible by n if n is a prime number and $b - 1$ is not divisible by n. In fact, more general property holds. For any prime number q and $n = a(q-1) - 1$ for some $a \in \mathbf{N}$, the sum $\sum_{i=0}^{n-2} b^i$ is divisible by q, provided that q does not divide $b - 1$. Indeed, we recover the previous property for $a = 1$. In our problem, we will need to consider slightly more complex aggregation scheme of this flavour.

SOLUTION

Clearly, for any odd number $i \geq 3$, 2^n is not divisible by i and so $r_i \geq 1$. Since there are $\lceil n/2 \rceil$ positive odd numbers, we immediately get

$$\sum_{i=1}^{n} r_i \geq \lceil n/2 \rceil - 1 .$$

In order to get the desired lower bound of $3.5n$, we need to use a slightly more delicate argument.

Let us first note that for any number $i \in [n]$, there exist unique $a, b \in \mathbf{N} \cup \{0\}$, such that

$$i = 2^a \cdot (2b + 1) .$$

If $b = 0$, then $i = 2^a$ divides 2^n and so $r_i = 0$. As a result, we focus on numbers for which $b \geq 1$. We will say that a number $i \in [n]$ is of "type a"

if the corresponding representation is $2^a(2b+1)$ and $b \geq 1$. Since there are $\lceil \lfloor n/2^a \rfloor / 2 \rceil$ odd numbers between 1 and $\lfloor n/2^a \rfloor$, there are

$$\left\lceil \frac{\lfloor n/2^a \rfloor}{2} \right\rceil - 1 \geq \frac{\lfloor n/2^a \rfloor}{2} - 1 \geq \frac{n}{2^{a+1}} - 3/2$$

numbers of "type a."

Let us now concentrate on any number of "type a": $i = 2^a(2b+1) \in [n]$, $b \geq 1$. The key observation is that

$$r_i = 2^n - \left\lfloor \frac{2^n}{i} \right\rfloor \cdot i = \left(\frac{2^n}{i} - \left\lfloor \frac{2^n}{i} \right\rfloor \right) \cdot i$$

$$= \left(\frac{2^{n-a}}{2b+1} - \left\lfloor \frac{2^{n-a}}{2b+1} \right\rfloor \right) \cdot 2^a(2b+1)$$

$$\geq \frac{1}{2b+1} \cdot 2^a(2b+1) = 2^a,$$

since $b \neq 0$. It follows that

$$\sum_{i=1}^{n} r_i \geq \sum_{a=0}^{a_{\max}} \left(\frac{n}{2^{a+1}} - 3/2 \right) 2^a = \sum_{a=0}^{a_{\max}} \left(\frac{n}{2} - \frac{3}{2} \cdot 2^a \right),$$

where $a_{\max} = \lfloor \log_2(n/3) \rfloor$. Since $n > 1,000$, $a_{\max} \geq 8$ and so

$$\sum_{i=1}^{n} r_i \geq \sum_{a=0}^{8} \left(\frac{n}{2} - \frac{3}{2} \cdot 2^a \right) = \frac{9}{2}n - \frac{1533}{2} > 3.5n.$$

REMARKS

Let us mention that our argument above easily gives a slightly stronger asymptotic lower bound. Indeed,

$$\sum_{i=1}^{n} r_i \geq \sum_{a=0}^{a_{\max}} \left(\frac{n}{2} - \frac{3}{2} \cdot 2^a \right) = (a_{\max} + 1) \cdot \frac{n}{2} + O(2^{a_{\max}})$$

$$= (\log_2 n + O(1)) \cdot \frac{n}{2} + O(n) = \frac{n \log_2 n}{2} + O(n) \sim \frac{n \log_2 n}{2}.$$

A heuristic argument *suggests* that r_i is expected to be around $i/2$ and so it is natural to conjecture that $\sum_{i=1}^{n} r_i$ is asymptotically equal to $\sum_{i=1}^{n} i/2 \sim n^2/4$.

EXERCISES

5.7.1. Prove that if the sum of positive divisors of some natural number n is odd, then either n is a square or $n/2$ is a square.

5.7.2. Find all natural numbers n for which there exist $2n$ pairwise different numbers $a_1, a_2, \ldots, a_n, b_1, b_2, \ldots, b_n$ such that $\sum_{i=1}^{n} a_i = \sum_{i=1}^{n} b_i$ and $\prod_{i=1}^{n} a_i = \prod_{i=1}^{n} b_i$.
(Source of the problem and solution idea: PLMO LXI – Phase 1 – Problem 10.)

5.7.3. Call a natural number *white* if it is equal to 1 or is a product of an even number of prime numbers; otherwise, call it *black*. Is there an integer for which the sum of its white divisors is equal to the sum of its black divisors?
(Source of the problem and solution idea: PLMO LVIII – Phase 3 – Problem 2.)

5.8   Equations

SOURCE

Problem and solution idea: PLMO XLVII – Phase 2 – Problem 5

PROBLEM

Find all pairs of integers (x, y) that satisfy the following equation:

$$x^2(y - 1) + y^2(x - 1) = 1 .$$

THEORY

Solving an equation or a system of equations when solutions are restricted to the set of integers usually involves techniques that we investigated in earlier problems. Most common applications are rules of divisibility and factorization that will also be used in the current problem. We dedicate a separate place for these type of equations as they are well-known, interesting, and worth knowing about.

Diophantine Equations | A **Diophantine equation** is a polynomial equation whose solutions are restricted to integers. These types of equations are named after the ancient Greek mathematician Diophantus. A **linear Diophantine equation** is a first-degree equation of this type, that is, equation of the form

$$Ax + By = C,$$

where A, B, C are given integers. Although the practical applications of Diophantine analysis have been somewhat limited in the past, this kind of analysis has become much more important in the digital age. In particular, they play an important role in the theory of public-key cryptography.

In order to warm up, we concentrate on the simplest and best-understood equations, namely, linear equations. Let us first note that not all linear Diophantine equations have a solution. For example, $10x + 5y = 3$ does not have a solution as for any pair of two integers x and y, the left hand side of this equation is divisible by 5 whereas the right hand side is not. Fortunately, there is a formal process to determine whether the equation has a solution or not. Indeed, finding all solutions to linear Diophantine equations involves finding an initial solution, and then altering that solution in some way to find the remaining solutions. For the first task we are going to use the Bézout's Identity.

Bézout's Identity Let us recall that in Section 5.1 we used the extended Euclidean algorithm to prove that for any $A, B \in \mathbf{Z} \setminus \{0\}$, there exist $x, y \in \mathbf{Z}$ such that

$$Ax + By = \gcd(A, B).$$

Bézout's Identity generalizes this observation as follows. There exist integers x and y which satisfy $Ax + By = C$ if and only if $\gcd(A, B) \mid C$. In other words, the integers of the form $Ax + By$ are exactly the multiples of $\gcd(A, B)$.

Using this observation, one can determine if solutions exist or not by calculating the greatest common divisor of the coefficients of the variables, and then determining if the constant term can be divided by that greatest common divisor. As already mentioned, if solutions do exist, then there is an efficient method to find an initial solution—the extended Euclidean algorithm. With this one solution at hand, we can find all integer solutions; there are infinitely many of them.

We will show that if $(x, y) = (\hat{x}, \hat{y})$ is an integer solution of the Diophantine equation $Ax + Bx = C$, then all integer solutions to the equation are of the form

$$(x, y) = \left(\hat{x} + m\frac{B}{\gcd(A, B)}, \hat{y} - m\frac{A}{\gcd(A, B)} \right)$$

for some integer m. Indeed, we have

$$A\left(\hat{x} + m\frac{B}{\gcd(A, B)} \right) + B\left(\hat{y} - m\frac{A}{\gcd(A, B)} \right) = A\hat{x} + B\hat{y} = C,$$

which shows these are indeed solutions to the equation. On the other hand, given any solution (x, y), we have

$$Ax + By = A\hat{x} + B\hat{y}$$
$$A(x - \hat{x}) = -B(y - \hat{y})$$
$$\frac{A}{\gcd(A, B)}(x - \hat{x}) = -\frac{B}{\gcd(A, B)}(y - \hat{y}).$$

Since $\frac{A}{\gcd(A,B)}$ and $\frac{B}{\gcd(A,B)}$ are relatively prime, there exists an integer m such that $x - \hat{x} = m\frac{B}{\gcd(A,B)}$ and $y - \hat{y} = -m\frac{A}{\gcd(A,B)}$. This shows that there are no more solutions.

SOLUTION

Let us first observe that, due to the symmetry, without loss of generality we may assume that $x \leq y$. In other words, solutions to this equation come in pairs: if $(x, y) = (a, b)$ is a solution, then so is $(x, y) = (b, a)$.

If $x = 1$, then y has to satisfy $y - 1 = 1$ and we get two solutions: $(x, y) = (2, 1)$ and $(x, y) = (1, 2)$. If $x = 2$, then $4(y - 1) + y^2 = 1$ and so after solving quadratic equation we get $y = 1$ or $y = 5$. Hence, there is another pair of solutions: $(x, y) = (2, -5)$ and $(x, y) = (-5, 2)$. We will prove that there are no other solutions to this equation.

Let us start with expanding the expression:

$$x^2 y - x^2 + y^2 x - y^2 = 1 .$$

Now, we re-write it as follows:

$$xy(x + y) = 1 + x^2 + y^2$$

and so

$$xy(x + y) + 2xy = 1 + (x + y)^2 .$$

Next,

$$xy(x + y + 2) = (x + y + 2)(x + y - 2) + 5$$

and finally

$$(xy - x - y + 2)(x + y + 2) = 5.$$

There are 4 cases to consider. We will see that none of them yields new solutions and so the proof will be finished.

1. $x + y + 2 = 1$ and $xy - x - y + 2 = 5$. Adding these two equations together we get that $xy = 2$ and $x = -1 - y$. There is no pair (x, y) satisfying these two equations as the first equation implies that x and y have the same sign whereas the second one implies that they have opposite signs.

2. $x + y + 2 = 5$ and $xy - x - y + 2 = 1$. Adding these two equations together we get $xy = 2$ and $x = 3 - y$. This gives us the equation $(3 - y)y = 2$ that yields two solutions we are already aware of, $(x, y) = (1, 2)$ and $(x, y) = (2, 1)$.

3. $x + y + 2 = -1$ and $xy - x - y + 2 = -5$. This time we get $xy = -10$ and $x = -3 - y$, so $(-3 - y)y = -10$. As before, we re-discover two solutions, $(x, y) = (2, -5)$ and $(x, y) = (5, -2)$.

4. $x + y + 2 = -5$ and $xy - x - y + 2 = -1$. We get $xy = -10$ and $x = -7 - y$, so $(-7 - y)y = -10$. This quadratic equation does not have any integer solutions.

REMARKS

The key idea of the proposed solution is to use a factorization that reduces the solution to 4 simple cases. However, it is not clear how to transform the equation to get this desired form. In order to see this, it is easier to substitute $a = x - 1$ and $b = y - 1$ to get

$$(a + 1)^2 b + (b + 1)^2 a = 1 .$$

After expanding, we obtain

$$a^2b + 2ab + b + b^2a + 2ab + a = 1$$

and so

$$ab(a + b) + 4ab + a + b = 1.$$

It follows that

$$ab(a + b + 4) + a + b = 1,$$

and now it is much easier to see that it is enough to add 4 to both sides to reach the desired factorization,

$$(ab + 1)(a + b + 4) = 5.$$

EXERCISES

5.8.1. Find all natural solutions of the following equation: $x^4 + y = x^3 + y^2$.

5.8.2. Find all pairs of natural numbers that satisfy $(x - y)^n = xy$.
(Source of the problem and solution idea: PLMO LVI – Phase 3 – Problem 1.)

5.8.3. Find all natural numbers satisfying the following system of equations:

$$\begin{aligned} a + b + c &= xyz, \\ x + y + z &= abc, \end{aligned}$$

and such that $a \geq b \geq c \geq 1$ and $x \geq y \geq z \geq 1$.
(Source of the problem and solution idea: PLMO XLIX – Phase 3 – Problem 1.)

Chapter 6

Geometry

As usual, we start the chapter with some basic definitions.

THEORY

In geometry, a **point** is a specific location on a **plane** or, in general in any
d-dimensional space; however, in this chapter we are concerned with problems
in 2-dimensions. A **line** is defined as a line of points that extends infinitely in
two directions. Points that are on the same line are called **collinear** points.
A line is defined by two different points, say A and B, and is denoted by AB.
A part of a line that has defined endpoints is called a line **segment**. Usually
it is clear from the context whether we deal with a line or a line segment so
we will use the same notation, namely AB, for both the line segment between
A and B, and the line passing A and B.

Two lines that meet in a point are called **intersecting** lines. When two
lines intersect at a **right angle** to each other, they are said to be **orthogonal**
(here a right angle is $\pi/2$) . Given line AB and point C not on the line, there
is a unique line that is orthogonal to line AB which goes through point C; this
line intersects line AB at point C' and is called the **orthogonal projection**
of C on line AB.

A **polygon** is a plane figure that is bounded by a finite sequence of straight
line segments closing in a loop to form a closed polygonal circuit. These segments are called its **edges** (or **sides**), and the points where two edges meet
are the polygon's **vertices**. An n-**gon** is a polygon with n sides; for example, a **triangle** is a 3-gon, **quadrilateral** is a 4-gon. Finally, an **equilateral
triangle** is a triangle in which all three sides are equal, and an **isosceles
triangle** is a triangle that has two sides of equal length.

6.1   Circles

SOURCE

Problem: "Exercises in geometry" (in Polish) by Waldemar Pompe – Problem 19
Solution: our own

PROBLEM

Points E and F lie on sides AB and BC of square $ABCD$ and $|BE| = |BF|$. Point S is the orthogonal projection of B on line CE. Show that angle $\sphericalangle DSF$ is the right angle.

THEORY

$\boxed{\text{Circumcircle}}$ The **circumcircle** is a triangle's circumscribed circle, that is, the unique circle that passes through each of the triangle's three vertices. The center of the circumcircle is called the **circumcenter**, and the circle's radius is called the **circumradius**. The circumcenter's position depends on the type of triangle.

- If and only if it is a **right triangle** (that is, a triangle in which one angle is the **right angle**), the circumcenter lies on one of its sides (namely, the **hypotenuse**, the longest side of a right triangle, opposite the right angle).

- If and only if a triangle is **acute** (all angles smaller than the right angle), the circumcenter lies inside the triangle.

- If and only if it is **obtuse** (has one angle larger than the right angle), the circumcenter lies outside the triangle.

$\boxed{\text{Central and Inscribed Angles}}$ A **central angle** of a circle is an angle whose vertex is the center O of the circle and whose sides, called **radii**, are line segments from O to two points on the circle. In Figure 6.1, $\sphericalangle BOC$ is a central angle and we say that it **intercepts** the arc BC. An **inscribed angle** of a circle is an angle whose vertex is a point A on the circle and whose sides are line segments, called **chords**, from A to two other points on the circle. In Figure 6.1, $\sphericalangle BAC$ is an inscribed angle that intercepts the arc BC.

Here is a very useful relation between inscribed and central angles. If an inscribed angle $\sphericalangle BAC$ and a central angle $\sphericalangle BOC$ intercept the same arc, then

$$\sphericalangle BOC = 2 \cdot \sphericalangle BAC.$$

As a result, inscribed angles which intercept the same arc are equal.

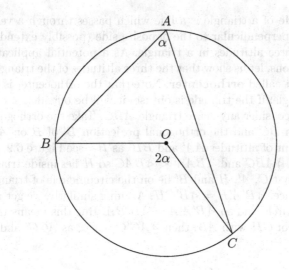

FIGURE 6.1: Relation between central and inscribed angles.

Let us stress that a central angle can be more than π. If A and O lie on the same side of line BC, then $\sphericalangle BOC$ is smaller than π (and so $\sphericalangle BAC$ is acute). On the other hand, if A and O are on different sides of BC, then it is greater than π (and so $\sphericalangle BAC$ is obtuse).

Finally, let us mention that often we are not explicitly told that we deal with inscribed angles. For example, perhaps we are given two triangles ABC and ABC' where C and C' lie on the same side of line AB. We then make a connection and notice that $\sphericalangle ACB = \sphericalangle AC'B$ if and only if they have the same circumcircle. Similarly, suppose that C and C' lie on different sides of line AB. In this case, $\sphericalangle ACB + \sphericalangle AC'B = \pi$ if and only if the two triangles have the same circumcircles.

A commonly used consequence of the above facts is that if **quadrilateral** $ABCD$ is inscribed in a circle, then its opposing angles add up to π. The other important consequence of this fact is that if we have a right triangle ABC and $\sphericalangle ACB$ is the right angle, then AB is a diameter of the circumcircle of ABC. Additionally, there are two useful facts relating circles and triangles.

- Three lines bisecting angles of a triangle intersect in one point. This point is called the **incenter** of the triangle and is a center of the **inscribed circle**; that is, the largest circle contained within the triangle.

- The three perpendicular bisectors of the sides of a triangle meet in one point, the circumcenter of the triangle.

The **altitude** of a triangle is a line which passes through a vertex of the triangle and is perpendicular to the opposite side (possibly extended). There are therefore three altitudes in a triangle. As a potential application of the above observations, let us show that the three altitudes of the triangle intersect in a single point called **orthocenter**. Note that the orthocenter is not always inside the triangle; if the triangle is obtuse, it will be outside.

Let us first consider any acute triangle ABC. Take the orthogonal projection A' of A on BC and the orthogonal projection B' of B on AC. Denote intersection point of altitudes AA' and BB' as H—see Figure 6.2. First, note that $\sphericalangle ABB' < \sphericalangle ABC$ and $\sphericalangle BAA' < \sphericalangle BAC$ so H lies inside triangle ABC. Now, observe that C, A', H and B' lie on the circumcircle of triangles $HB'C$ and $HA'C$. Hence, $\sphericalangle B'A'H = \sphericalangle B'CH$. Arguing similarly, we get that A, B', A', B lie on a circle and so $\sphericalangle B'A'A = \sphericalangle B'BA$. But this means that if C' is an intersection of CH with AB, then $\sphericalangle AC'C = \pi/2$ as ACC' and ABB' are similar.

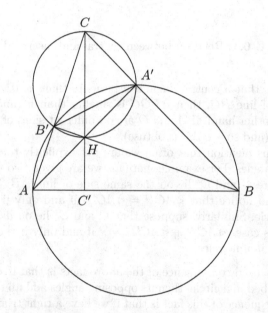

FIGURE 6.2: Orthocenter of the triangle.

The reasoning when $\sphericalangle BCA$ is not acute is identical with the only difference that H lies outside of triangle ABC, if the triangle is obtuse, or H lies on the triangle, if the triangle is right.

SOLUTION

Let F' be the point on line segment AD such that $|F'A| = |FB|$. First, note that $\sphericalangle ABF' = \sphericalangle BCE$. Hence, lines $F'B$ and CE are orthogonal and so S lies on the intersection of the two. But this means that $\sphericalangle SF'F = \sphericalangle SCF$. (In fact, these two angles are equal to the previously mentioned two but we do not need this observation here.) It follows that triangles $SF'F$ and SCF have the same circumcircle and so it contains points S, F, C, F'. Now, since $FF'DC$ is a rectangle with its three vertices (F', F, and C) lying on the circle, the fourth vertex, D, must also lie on this circle. It follows that DF is a diameter of the circle and, consequently, $\sphericalangle DSF$ is the right angle.

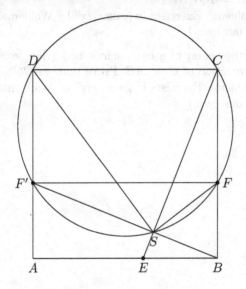

FIGURE 6.3: Illustration for Problem 6.1.

REMARKS

In geometric problems, very often drawing an additional point, line, or circle significantly helps to find a solution. In our case, introducing an auxiliary point F' turned out to be very helpful. How can one think of such a point? It is natural to extend line segment BS beyond point S and then to notice that point lying on the intersection of this line with AD forms a rectangle $AF'FB$ whose diagonal, $F'B$ contains line segment BS.

EXERCISES

6.1.1. We are given an acute triangle ABC with $\sphericalangle ACB = \pi/3$. Let A' be the orthogonal projection of A on BC, let B' be the orthogonal projection of B on AC, and let M be the middle point of line segment AB. Prove that $|A'B'| = |A'M| = |B'M|$.
(Source of the problem: "Exercises in geometry" by Waldemar Pompe – Problem 16. Solution: our own.)

6.1.2. Consider a square $ABCD$. Choose point P outside of this square such that $\sphericalangle CPB$ is the right angle. Denote by Q the intersection of AC and BD. Prove that $\sphericalangle QPC = \sphericalangle QPB$.
(Source of the problem: "Exercises in geometry" by Waldemar Pompe – Problem 15. Solution: our own.)

6.1.3. Point O is the center of a circumcircle of a triangle ABC. Point C' is the orthogonal projection of C on AB. Prove that $\sphericalangle ACC' = \sphericalangle OCB$.
(Source of the problem: "Exercises in geometry" by Waldemar Pompe – Problem 17. Solution: our own.)

6.2 Congruence

SOURCE

Problem: Student Circle – High School of Stanisław Staszic in Warsaw
Solution: our own

PROBLEM

Consider a rectangle $ABCD$. We choose point F such that triangle ABF is equilateral and AF lies inside angle $\sphericalangle BAD$. Similarly, we choose point E such that triangle BCE is equilateral and BE lies inside angle $\sphericalangle ABC$. Prove that triangle DEF is equilateral.

THEORY

Congruence │ Two figures or objects are **congruent** if they have the same shape and size, or if one has the same shape and size as the mirror image of the other. It is worth remembering that there are the following conditions for determining congruence between two triangles:

1. three sides of two triangles are equal in length (**side-side-side condition**);

2. one side and two angles are equal (**angle-side-angle** or **angle-angle-side** condition);

3. one angle and two sides associated with this angle are equal (**side-angle-side** condition);

4. one angle that is not acute and any two sides are equal.

Importantly, the exception here is when we know one acute angle and two sides of the triangle but only one of them is adjacent to the angle. In this case, there are actually two possible triangles that meet those conditions but are not congruent.

SOLUTION

Observe that angles $\sphericalangle FAD$, $\sphericalangle ECD$, and $\sphericalangle EBF$ are all equal to $\pi/6$. Also $|AD| = |EC| = |EB|$ and $|FA| = |CD| = |BF|$. This means that triangles ECD, FAD, and EBF are congruent. In particular, it implies that $|ED| = |FD| = |EF|$ and so DEF is equilateral.

REMARKS

A common strategy when solving geometry problems is to draw a picture and then try to write down all lengths and angles that one can possibly calculate (or list all relationships between them). In our solution, we simply marked the

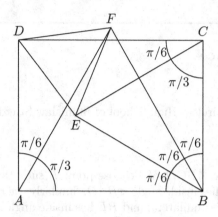

FIGURE 6.4: Illustration for Problem 6.2.

corresponding values for angles and identified sides that have equal length. When one does it, it is often easy to spot congruences. Congruence allows us to reason about unknown lengths of sides or unknown angles.

EXERCISES

6.2.1. Suppose that points P and Q lie on sides BC and CD of a square $ABCD$ such that $\sphericalangle PAQ = \pi/4$. Prove that $|BP| + |DQ| = |PQ|$.
(Source of the problem: "Exercises in geometry" by Waldemar Pompe – Problem 4. Solution: our own.)

6.2.2. Point P lies on a diagonal AC of a square $ABCD$. Points Q and R are the orthogonal projections of P on lines CD and DA, respectively. Prove that $|BP| = |RQ|$.
(Source of the problem: "Exercises in geometry" by Waldemar Pompe – Problem 1. Solution: our own.)

6.2.3. Consider an acute triangle ABC where $\sphericalangle ACB = \pi/4$. Point B' is the orthogonal projection of B on AC and point A' is the orthogonal projection of A on BC. Let H be the intersection point of AA' and BB'. Prove that $|CH| = |AB|$.
(Source of the problem: "Exercises in geometry" by Waldemar Pompe – Problem 2. Solution: our own.)

6.3 Similarity

SOURCE

Adaptation of a puzzle mentioned by Peter Winkler while visiting one of the authors of this book.

PROBLEM

You are given a triangle ABC and an n-element set S of non-overlapping disks, all having radius 1 and centers lying inside or on the triangle. Disks do not need to lie inside the triangle as long as their centers are. Moreover, by "non-overlapping" we mean that they can "touch" each other; that is, we allow the intersection of any two disks to be one point. Suppose that set S is maximal; that is, there is no disk of radius 1 such that its center lies inside or on triangle ABC and it does not overlap with any disk from S. Prove that you can completely cover triangle ABC with $4n$ disks of radius 1 (of course, this time we allow them to overlap).

THEORY

Similarities Two figures that have the same shape are said to be similar. Formally, given two figures A and B lying on the same plane, we say that they are similar if we can transform A into B only using the following two operations:

- reflection with respect to a line (notice that in particular composition of two reflections can yield us translation and rotation with respect to a point);

- scaling with respect to a point.

In particular, if two figures are **similar**, then the ratios of the lengths of their corresponding sides are equal.

For instance, any two circles or any two squares are always similar. In fact, in general, all regular n-gons are similar. If we are given two triangles, the rules of similarity are the same as rules of congruence with the only difference that the requirement that the corresponding lengths of sides are equal is replaced by equality of proportions. Finally, let us mention that if figure A has to be scaled by a factor of α to be congruent to figure B, then the ratio between the area of figure A and the area of figure B is α^2.

SOLUTION

First, let us construct another auxiliary set of n disks; this time all of them being of radius 2. For each disk D from S, we put disk E to R which has the same center as D (but radius 2, not 1). Note that R completely covers triangle

ABC. Indeed, if point X is lying inside or on the triangle but is not covered by R, then the distance from X to any of the centers is greater than 2. But this implies that a disk of radius 1, centered at X, would not overlap with any of the disks in S, contradicting the maximality of set S.

Consider now a triangle $A'B'C'$ and a set R' that are similar to triangle *ABC* and, respectively, set R but both are shrunk by a factor of two in each dimension. Clearly, triangle $A'B'C'$ is completely covered by disks of radius 1 from R'. Hence, it is enough to show that one can cover triangle *ABC* with four copies of $A'B'C'$; indeed, if this can be done, then triangle *ABC* can be covered by four copies of R'. But this is easy to do. Let D, E, and F be the midpoints of line segments AB, AC, and BC, respectively—see Figure 6.5. Now, we observe that triangle $A'B'C'$ can be partitioned into four triangles *ADE*, *BDF*, *EFC* and *DEF*, each of them congruent to $A'B'C'$. This observation finishes the proof.

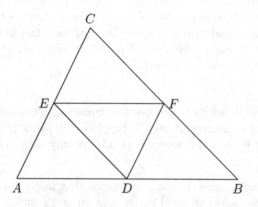

FIGURE 6.5: Illustration for Problem 6.3.

REMARKS

The key observation in our solution is that any triangle can be partitioned into four copies of themselves (scaled by a factor of two in each dimension). Clearly, triangles are not the only figures with this property.

The solution is cute but it feels that it is far from being optimal. Therefore, it is perhaps surprising that, in fact, the factor 4 is best possible! In other words, if we replace $4n$ by $\lfloor(4 - \epsilon)n\rfloor$ for some $\epsilon > 0$, then the property is no longer true, regardless how close to zero ϵ is. Formally, for any $\epsilon > 0$ there exists a counter-example, a triangle *ABC* and an n-element maximal set S of non-overlapping disks such that more than $(4 - \epsilon)n$ disks are needed to cover triangle *ABC*. To see this, we consider a very large triangle so that the boundary effects are negligible. Hence, for a moment, let us forget about the triangle and think about placing disks on the plane.

First, let us consider a tiling with regular hexagonal tiles; each hexagonal tile consists of six equilateral triangles of side lengths equal to r. We carefully choose r such that a unit disk is just a tiny bit larger than a disk inscribed into one hexagon. Since the altitude of any of the six equilateral triangles making up the hexagon is arbitrarily close to one, the radius of the disk, we may assume that r is arbitrarily close to $2/\sqrt{3}$ (but it must be a tiny bit smaller). Formally, we set $r := 2/\sqrt{3(1 + f(\epsilon))}$ for some function $f : \mathbf{R}_+ \to \mathbf{R}_+$ such that $f(\epsilon) \to 0$ as $\epsilon \to 0$, which will be determined soon. We put disks in every third tile such that their centers coincide with the corresponding centers of hexagons—see Figure 6.6. This configuration of disks just barely prevents us from adding any more disks without overlapping and so it is maximal. The fact that this is the most efficient way to prevent the addition of a non-overlapping disk is a challenging task to prove and is way beyond the scope of this book. The (limiting) ratio between the total area of all unit disks used and the area of the tiling is

$$\frac{(\pi \cdot 1^2)/3}{6(\sqrt{3}r^2/4)} = \frac{\pi\sqrt{3}(1 + f(\epsilon))}{18}.$$

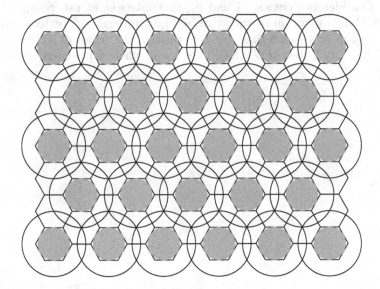

FIGURE 6.6: Maximal set of unit disks and covering with disks of radii 2.

Now, the next question is: what is the most efficient way to cover the plane by unit disks? The answer is as follows: by circumscribing the tiles in some other hexagonal tiling, this time each hexagon must have unit radius (also, as for triangles, often called circumradius). This is another difficult question that is beyond the scope of this book. In such tiling, the ratio between the total area of all unit disks used and the area of the tiling is

$$\frac{\pi}{6(\sqrt{3}/4)} = \frac{4\pi\sqrt{3}}{18}.$$

It follows that the ratio between the number of disks used in the second scenario and n, the number of disks used in the first one, is $4/(1+f(\epsilon))$. Hence, one needs more than $(4 - \epsilon)n$ disks to cover the triangle, provided that $f(\epsilon)$ is sufficiently close to zero and triangle ABC is sufficiently large so that the (finite) ratio is close to its limiting counterpart.

EXERCISES

6.3.1. You are given a rectangle that can be covered with n disks of radius r. Prove that it can be also covered by $4n$ disks of radius $r/2$.

6.3.2. You are given an acute triangle ABC. Let B' be the projection of B on AC and C' be the projection of C on AB. Show that ABC and $AB'C'$ are similar.

6.3.3. Consider two circles, o_1 and o_2, that intersect at two points, A and B. Let P be a point on o_1 such that AP goes through the center of o_1 and Q be a point on o_2 such that AQ goes through the center of o_2. Prove that if $\sphericalangle PAQ = \pi/2$, then $|PB|/|BQ| = [o_1]/[o_2]$, where $[x]$ denotes the area of figure x.

6.4   Menelaus's Theorem

SOURCE

Problem: Wojciech Guzicki Workshop
Solution: our own

PROBLEM

Consider an acute triangle ABC where $|AC| < |BC|$. Point D lies on line segment BC and $|BD| = |AC|$. Points E and F are the middle points of line segments CD and AB, respectively. Lines AC and EF intersect in point G. Prove that $|CE| = |CG|$.

THEORY

Before we state the first theorem, we need one definition. A **transversal** is a line that passes through two lines at two distinct points.

$\boxed{\text{Menelaus's Theorem}}$ Consider a triangle ABC, and a transversal line that crosses BC, AC, and AB at points D, E, and F respectively, with D, E, and F distinct from A, B, and C—see Figure 6.7. **Menelaus's theorem**

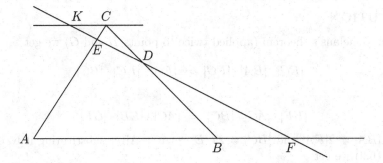

FIGURE 6.7: Menelaus's Theorem.

then states that the following relation holds:

$$|AF| \cdot |BD| \cdot |CE| = |FB| \cdot |DC| \cdot |EA| . \tag{6.1}$$

The converse is also true. If points D, E, and F are chosen on BC, AC, and AB respectively so that (6.1) holds, then D, E, and F are collinear.

In order to see this, draw the line KC parallel to AB and observe that, by similarity of the triangles, we have

$$\frac{|BD|}{|DC|} = \frac{|BF|}{|CK|}$$

and

$$\frac{|AE|}{|EC|} = \frac{|AF|}{|CK|}.$$

Now, after solving both equations for $|CK|$, we get that

$$\frac{|DC| \cdot |BF|}{|BD|} = \frac{|EC| \cdot |AF|}{|AE|},$$

which after rearrangement gives us the desired equality (6.1).

We covered the case when two of the tree intersection points lie on sides of a triangle. One can show that the theorem also holds when neither E, D nor F lies on the side of the triangle.

Ceva's Theorem   A direct consequence of Menelaus's theorem is the following **Ceva's theorem**. Given a triangle ABC and points F, D, E on sides AB, BC and, respectively, AC, the lines AD, BE and CF intersect at the same point if and only if

$$|AF| \cdot |BD| \cdot |CE| = |FB| \cdot |DC| \cdot |EA|.$$

We obtain it by writing (6.1) for triangle AFC and line BE, triangle BCF and line AD, and then dividing them side by side and rearranging the terms.

SOLUTION

From Menelaus's theorem (applied twice to points B and G) we get

$$|EC| \cdot |BA| \cdot |FG| = |FA| \cdot |EG| \cdot |BC| \qquad (6.2)$$

and

$$|EF| \cdot |AG| \cdot |BC| = |AC| \cdot |EB| \cdot |GF|. \qquad (6.3)$$

But $|BA| = 2|FA|$ and $|BC| = 2|CE| + |AC|$. After substituting it to equation (6.2) we get

$$|EC| \cdot 2|FA| \cdot (|EG| + |EF|) = |FA| \cdot |EG| \cdot (2|CE| + |AC|).$$

From this we get

$$2|EC| \cdot |EG| + 2|EC| \cdot |EF| = 2|EG| \cdot |CE| + |EG| \cdot |AC|,$$

and so

$$2|EC| \cdot |EF| = |EG| \cdot |AC|.$$

It follows that

$$|EG| = \frac{2|EC||EF|}{|AC|}$$

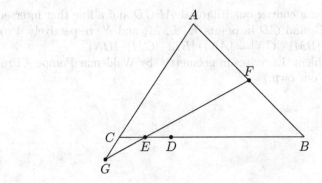

FIGURE 6.8: Illustration for Problem 6.4.

which we substitute to (6.3) to get

$$|EF|\Big(|AC| + |CG|\Big)\Big(|AC| + 2|EC|\Big)$$
$$= |AC|\Big(|AC| + |CE|\Big)\left(\frac{2|EC||EF|}{|AC|} + |EF|\right).$$

Finally, after simplification we get the desired equality; namely, $|CG| = |CE|$.

REMARKS

It is useful to remember the "arrow shaped pattern" that is obtained by two overlapping triangles. For example, coming back to our problem, Figure 6.8 contains point A which can be viewed a the "head of the arrow" consisting of two overlapping triangles, AFG and ABC. In such situations, especially when the problem concerns lengths of certain sections, Menelaus's theorem very often turns out to be useful.

EXERCISES

6.4.1. Points D, E and F lie on sides BC, CA and AB of a triangle ABC in such a way that lines AD, BE and CF intersect in a single point P. Prove that $|AF|/|FB| + |AE|/|EC| = |AP|/|PD|$.
(Source of the problem and solution idea: "Delta" monthly, March 2011 – DeltaMi – Problem 2.)

6.4.2. You are given a triangle ABC where $\sphericalangle ACB = \pi/2$. On side AC build a square $ACGH$, externally to the triangle. Similarly, on side BC build a square $CBEF$, externally to the triangle. Show that the point of intersection of AE and BH lies on the line orthogonal to AB that goes through point C.
(Source of the problem: "Exercises in geometry" by Waldemar Pompe – Problem 105. Solution: our own.)

6.4.3. You are given a convex quadrilateral $ABCD$ and a line that intersects lines DA, AB, BC, and CD in points K, L, M, and N, respectively. Prove that $|DK| \cdot |AL| \cdot |BM| \cdot |CN| = |AK| \cdot |BL| \cdot |CM| \cdot |DN|$.

(Source of the problem "Exercises in geometry" by Waldemar Pompe – Problem 106. Solution: our own.)

6.5   Parallelograms

SOURCE

Problem: "Exercises in geometry" (in Polish) by Waldemar Pompe – Problem 24

Solution: our own

PROBLEM

Point P lies inside parallelogram $ABCD$ and $\sphericalangle ABP = \sphericalangle ADP$. Show that $\sphericalangle DAP = \sphericalangle DCP$.

THEORY

Parallelogram A **parallelogram** is a quadrilateral with two pairs of parallel sides. By comparison, a quadrilateral with just one pair of parallel sides is a **trapezoid**. Hence, all parallelograms are trapezoids but the converse is not true. On the other hand, there are some special sub-families of parallelograms: **rectangle** is a parallelogram with four angles of equal size, **rhombus** is a parallelogram with four sides of equal length, and **square** is a parallelogram with four sides of equal length and angles of equal size (right angles).

There are several basic facts about parallelograms that are often useful:

- the opposite sides of a parallelogram are of equal length;

- the opposite angles of a parallelogram are equal;

- the diagonals of a parallelogram bisect each other;

- the diagonals of a parallelogram divide it into four triangles of equal area (in particular, the area of a parallelogram is twice the area of a triangle created by one of its diagonals);

- any line going through the midpoint of a parallelogram bisects the area.

SOLUTION

Draw an auxiliary point P' such that PP' is parallel to CD (and so also to AB) and for which $|CD| = |PP'|$ (and so $|AB| = |PP'|$ as well)—see Figure 6.9. Then, $\sphericalangle ABP = \sphericalangle BPP'$, as BP is a transversal that passes the two parallel lines, PP' and AB. Moreover, $\sphericalangle ADP = \sphericalangle BCP'$, as DP is parallel to CP' and AD is parallel to BC. Now, using the assumption of the problem that $\sphericalangle ABP = \sphericalangle ADP$ we get that $\sphericalangle BPP' = \sphericalangle CP'$. It follows that triangles BPP' and BCP' share one of the sides (the line segment BP') and the two angles that are opposite to BP' are equal and lie on the same side of BP'. Using the connection between central and inscribed angles discussed in Section 6.1 we deduce that one can draw a circle through points B, P, C, and P'. Therefore, $\sphericalangle P'PC = \sphericalangle P'BC$, as they are inscribed angles which intercept the same arc. But, clearly, $\sphericalangle P'PC = \sphericalangle DCP$ (as CP is a transversal that passes two parallel lines, CD and PP'), and $\sphericalangle P'BC = \sphericalangle PAD$ (as AP is parallel to BP' and AD is parallel to BC). It follows that $\sphericalangle DCP = \sphericalangle PAD$ and the proof is finished.

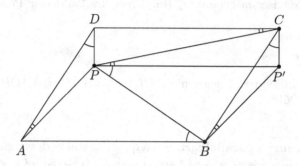

FIGURE 6.9: Illustration for Problem 6.5.

REMARKS

In this example, we see one more time how useful it is to add some auxiliary object to the figure—this time, it is point P'. The idea for adding it comes from the fact that creating another parallelogram introduces many angles that must be preserved and so many useful conditions must be satisfied.

EXERCISES

6.5.1. Consider a quadrilateral $ABCD$. Prove that the sum of distances from any point P inside this quadrilateral to the lines AB, BC, CD, and DA is constant (that is, does not depend on the choice of P) if and only if $ABCD$ is a parallelogram.

6.5.2. Consider a triangle ABC such that $|AB| = |AC|$ (that is, an isosceles triangle), AD is the height of this triangle, and E is in the middle of AD. Let F be the orthogonal projection of D on BE. Prove that $\sphericalangle AFC = \pi/2$.

6.5.3. Consider a triangle ABC. Outside of the triangle, on sides AB and AC, we built squares $ABDE$ and, respectively, $ACFG$. Let M and N be the middle points of DG and, respectively, EF. What are the possible values of the rato $|MN|/|BC|$?
(Source of the problem: "Wokół obrotów" book by Waldemar Pompe, Problem 4.22. Solution: our own.)

6.6 Power of a Point

SOURCE

Problem and solution presented in Deltoid 38 (in Polish) by Joanna Jaszuńska.

PROBLEM

Consider two externally disjoint circles A and B; that is, they are not only disjoint but also neither of them lies inside the other one. There are two lines ℓ_1 and ℓ_2 tangent to A and B selected so that they are *not* separating A and B. Line ℓ_i ($i \in \{1, 2\}$) touches circle A in point A_i and circle B in point B_i. Now consider a line $A_1 B_2$. It intersects circle A in point A_3 and and circle B in point B_3. Prove that $|A_1 A_3| = |B_2 B_3|$.

THEORY

| Tangent Line | The **tangent line** to a curve at a given point is the straight line that "just touches" the curve at that point. (A formal definition is outside the scope of this book.) As a specific example, consider a circle B and a point X outside of it. Then, there are two tangent lines, ℓ_1 and ℓ_2, to the circle; the intersection of ℓ_1 (respectively, ℓ_2) and the circle is precisely one point, B_1 (respectively, B_2).

We will show that $\sphericalangle AB_1 O = \sphericalangle AB_2 O = \pi/2$, where O is the center of the circle. Due to the symmetry, without loss of generality, it is enough to focus on showing that $\sphericalangle AB_1 O = \pi/2$. For a contradiction, suppose that $\sphericalangle AB_1 O \neq \pi/2$; that is, B_1 is *not* the orthogonal projection of O on line AB_1. This implies that there exists another point on line AB_1 that is at the same distance from O as B_1; that is, both points lie on the circle. We assumed however that there was only one point of intersection of AB_1 and the circle, and so we get the desired contradiction.

Now, we are ready to move to the main tool of this section.

| Power of a Point | We will independently consider two cases.

Case 1: Consider a circle and a point A inside it, together with any line going through A. Let the points of intersection of this line with the circle be B and C. The product $|AB| \cdot |AC|$ then does not depend on the choice of the line and is equal to the square of the radius of the circle minus the square of the distance from A to O, the center of the circle.

To see this, let us first note that if $A = O$, then the desired property is trivially true so we may assume that it is not the case. Now, let us introduce an auxiliary line going trough A and O that intersects the circle at points P and Q—see Figure 6.10. Our goal is to show that the desired property holds for all lines passing through A (including this auxiliary line OP) but, clearly, it holds for OP. Indeed, observe that

$$
\begin{aligned}
|AP| \cdot |AQ| &= (|OP| - |AO|) \cdot (|OQ| + |AO|) \\
&= (|OP| - |AO|) \cdot (|OP| + |AO|) = |OP|^2 - |AO|^2 .
\end{aligned}
$$

So it is enough to show that $|AB| \cdot |AC| = |AP| \cdot |AQ|$. Now, observe that triangles AQB and ACP are similar as $\sphericalangle QBC = \sphericalangle QPC$ and $\sphericalangle CAP = \sphericalangle QAB$ (see the connection between central and inscribed angles discussed in Section 6.1). Therefore, $|AB|/|AQ| = |AP|/|AC|$, which yields the desired property.

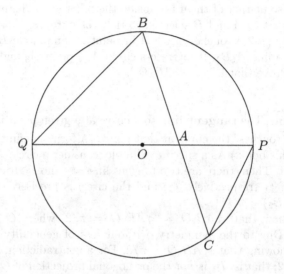

FIGURE 6.10: Power of a Point—Case 1: $|AB| \cdot |AC| = |OB|^2 - |AO|^2$.

Case 2: Now, consider a circle and a point A outside of it. Consider any line going through A that intersects with the circle. Let the points of intersection of this line with the circle be B and C. Then, the product $|AB| \cdot |AC|$ does not depend on the choice of the line and is equal to the square of the distance from A to O, the center of the circle, minus the square of the radius of the circle.

First, let us note that possibly $B = C$ (that is, the line intersects the circle at one point and so the line is, in fact, the tangent line) but this case is rather uninteresting and easy to deal with. Indeed, as argued above, in this case $\sphericalangle ABO$ is the right angle and so we get the desired property immediately.

Now assume $B \neq C$ and choose a point P on the same side of line AO as points B and C such that $\sphericalangle OPA$ is the right angle and P lies on a circle (see Figure 6.11). Since $|OB| = |OP|$, triangle BOP is an isosceles triangle and so $\sphericalangle BPO = \pi/2 - \sphericalangle BOP/2$. Since $\sphericalangle OPA$ is the right angle, $\sphericalangle BPA = \pi/2 - \sphericalangle BPO = \sphericalangle BOP/2$. But, as $\sphericalangle BCP$ is an inscribed angle and $\sphericalangle BOP$

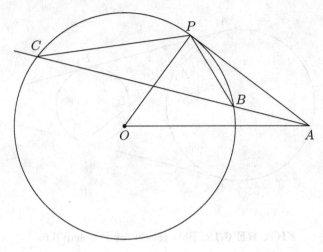

FIGURE 6.11: Power of a Point—Case 2: $|AB| \cdot |AC| = |AO|^2 - |OB|^2$.

is the central angle that intercept the same arc, $\angle BCP = \angle BOP/2$, and so $\angle BCP = \angle BPA$. This means that triangles ACP and ABP are similar. Therefore, $|AP|/|AC| = |AB|/|AP|$ and so $|AB||AC| = |AP|^2$. But, as $\angle OPA$ is the right angle, we have that $|PA|^2 = |AO|^2 - |OP|^2$, which finishes the proof.

SOLUTION

Using Power of a Point property (Case 2 above), we get

$$|A_1B_3| \cdot |A_1B_2| = |A_1O_B|^2 - |B_1O_B|^2 = |A_1B_1|^2,$$

where O_B is the center of circle B. Similarly,

$$|B_2A_3| \cdot |B_2A_1| = |B_2O_A|^2 - |A_2O_A|^2 = |A_2B_2|^2,$$

where O_A is the center of circle A. Clearly, we have $|A_1B_1| = |A_2B_2|$ as line O_AO_B is a line of symmetry of the two cycles, A, B, together with the two lines, ℓ_1 and ℓ_2. Therefore, $|A_1B_3| = |B_2A_3|$ but this means that $|A_1A_3| = |B_2B_3|$.

REMARKS

In this example we used Power of a Point property. It is a natural tool to try in situations when we have to prove facts about lengths of sections defined by a circle.

EXERCISES

6.6.1. Two circles intersect in points A and B. Point P is selected on line AB outside of the circles. Points C and D are locations where tangent lines going through point P touch both circles. Prove that $\angle PCD = \angle PDC$.

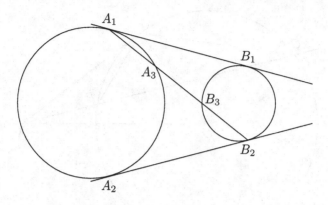

FIGURE 6.12: Illustration for Problem 6.6.

6.6.2. Consider a convex hexagon $ABCDEF$ such that $|AB| = |BC|$, $|CD| = |DE|$, and $|EF| = |FA|$. Prove that lines containing altitudes of triangles BCD, DEF, and FAB from vertices C, E, and A, respectively, intersect in one point.

6.6.3. Consider two points A and B. Take two circles o_1 and o_2 such that o_1 is tangent to AB in point A, o_2 is tangent to AB in point B, and o_1 and o_2 are externally tangent in point X. If we allow o_1 and o_2 to vary, then what is the set of points that contains all possible locations of X.

(Source of the problem: "Exercises in geometry" by Waldemar Pompe – Problem 26. Solution: our own.)

6.7 Areas

SOURCE

Problem: well-known problem
Solution: our own

PROBLEM

Let $A_1A_2A_3A_4$ be any convex quadrilateral that has area equal to 1. To simplify the notation, let $A_0 = A_4$. Let us introduce four new points: for $i \in \{1, 2, 3, 4\}$, let A_i' be the point on line A_iA_{i-1} such that A_{i-1} is the midpoint of line segment A_iA_i'. Calculate the area of $A_1'A_2'A_3'A_4'$.

THEORY

In order to be able to deal with areas of some figures, it is often the case that one needs to use a formula for the area of a triangle. Consider a triangle ABC and denote by H an orthogonal projection of C on line AB (which does not have to be on the line segment AB). The area of the triangle, denoted by $[ABC]$, is then

$$\frac{1}{2} \cdot |AB| \cdot |CH| = \frac{1}{2} bh . \tag{6.4}$$

Here $b = |AB|$ is often called the length of the base of the triangle, and $h = |CH|$ is called the altitude of the triangle. Although simple, this formula is only useful if the height can be readily found, which is not always the case. Hence, there are other formulas available. For example, the shape of the triangle is determined by the lengths of the sides. Therefore, the area can also be derived from the lengths of the sides by Heron's formula that we already used in this book (see Problem 1.9).

A direct consequence of (6.4) is that if the altitude of the triangle is fixed and one only changes the length of the base, then the area of the triangle changes proportionally. This implies that if one angle of the triangle is fixed but the two adjacent sides are rescaled by factors a and b, respectively, then the area of the triangle changes by a factor of $a \cdot b$. For a given angle $\alpha \in (0, \pi)$, this constant factor is typically defined as the area of the triangle whose sides adjacent to angle α have lengths 1 and 2. It is called the sine of angle α and is denoted by $\sin(\alpha)$. It follows that for any triangle ABC we have

$$[ABC] = \frac{1}{2} \cdot |AC| \cdot |AB| \cdot \sin(\sphericalangle BAC) = \frac{1}{2} \cdot |BA| \cdot |BC| \cdot \sin(\sphericalangle ABC)$$

$$= \frac{1}{2} \cdot |CA| \cdot |CB| \cdot \sin(\sphericalangle ACB) . \tag{6.5}$$

The Law of Sines The set of equalities in (6.5) give us immediately the following important observation known as the **law of sines**. For any triangle ABC, we have

$$\frac{|BC|}{\sin(\sphericalangle BAC)} = \frac{|AC|}{\sin(\sphericalangle ABC)} = \frac{|AB|}{\sin(\sphericalangle ACB)}.$$

This ratio is equal to the diameter of the circumscribed circle of the given triangle. Another interpretation of this observation is that every triangle with angles α, β, and γ is similar to a triangle with side lengths equal to $\sin(\alpha)$, $\sin(\beta)$, and $\sin(\gamma)$.

Let us finish with two more observations. From the discussion above, we get that $\sin(0) = \sin(\pi) = 0$, $\sin(\pi/2) = 1$, $\sin(\alpha) = \sin(\pi - \alpha)$ for any $\alpha \in (0, \pi)$, and that the sine function is increasing in the range from 0 to $\pi/2$, and decreasing in the range from $\pi/2$ to π.

Moreover, consider a triangle ABC such that $\sphericalangle ABC = \alpha$ and $\sphericalangle BAC = \pi/2$; that is, $\sphericalangle BAC$ is the right angle. Then, $[ABC] = |AB||AC|/2 = \sin(\alpha)|AB||BC|/2$, so $\sin(\alpha) = |AB|/|BC|$.

SOLUTION

We will use $[A_1 A_2 \ldots A_n]$ to denote the area of an n-gon $A_1 A_2 \ldots A_n$. Let us first observe that,

$$\begin{aligned}
[A_1' A_4 A_4'] &= 2[A_1 A_3 A_4] \\
[A_2 A_2' A_3'] &= 2[A_1 A_2 A_3].
\end{aligned}$$

Hence,

$$\begin{aligned}
[A_1' A_4 A_4'] + [A_2 A_2' A_3'] &= 2[A_1 A_3 A_4] + 2[A_1 A_2 A_3] \\
&= 2([A_1 A_3 A_4] + [A_1 A_2 A_3]) \\
&= 2[A_1 A_2 A_3 A_4] = 2.
\end{aligned}$$

Similarly, since

$$\begin{aligned}
[A_1 A_1' A_2'] &= 2[A_2 A_4 A_1] \\
[A_3 A_3' A_4'] &= 2[A_2 A_3 A_4],
\end{aligned}$$

we get $[A_1 A_1' A_2'] + [A_3 A_3' A_4'] = 2$. Combining all of these together, we conclude that

$$\begin{aligned}
[A_1' A_2' A_3' A_4'] &= [A_1 A_2 A_3 A_4] + [A_1' A_4 A_4'] + [A_2 A_2' A_3'] \\
&\quad + [A_1 A_1' A_2'] + [A_3 A_3' A_4'] = 5.
\end{aligned}$$

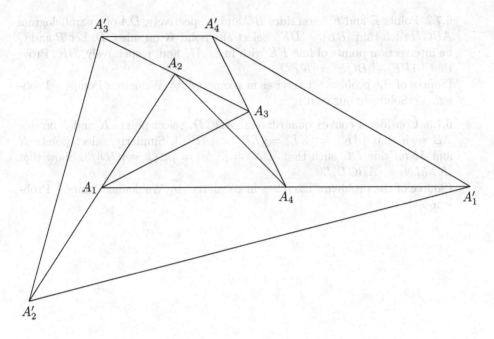

FIGURE 6.13: Illustration for Problem 6.7.

REMARKS

This problem shows that it is important to remember about the following two facts.

- Suppose that two triangles, say, ABC and DEF, are such that $|AB| = |DE|$ and $|AC| = |DF|$; moreover, $\sphericalangle BAC + \sphericalangle EDF = \pi$. (For example, triangles $A_1A_2A_3$ and $A_1A_2A'_3$ on Figure 6.13 satisfy these properties.) Then, these triangles have the same areas.

- Consider any triangle ABC and change the length of AB by a (multiplicative) factor of α, while keeping the length of AC and the angle between the two sides, namely, $\sphericalangle BAC$. (Of course, the other two angles will change as well as the length of the third side. For example, triangles $A_1A_2A'_3$ and $A'_2A_2A'_3$ have these properties with $\alpha = 2$.) Then, the area of the triangle changes by a factor of α.

EXERCISES

6.7.1. Let P be an interior point of a triangle ABC. Let lines AP, BP, and CP intersect sides BC, CA, and AB in points A', B' and, respectively, C'. Prove that $|PA|/|AA'| + |PB|/|BB'| + |PC|/|CC'| = 2$.

6.7.2. Points E and F lie on sides BC and, respectively, DA of a parallelogram $ABCD$ such that $|BE| = |DF|$. Select any point K on side CD. Let P and Q be intersection points of line FE with lines AK and, respectively, BK. Prove that $[APF] + [BQE] = [KPQ]$.

(Source of the problem: "Exercises in geometry" by Waldemar Pompe – Problem 45. Solution: our own.)

6.7.3. Consider a convex quadrilateral $ABCD$. Select points K and L on side AB such that $|AK| = |KL| = |LB| = |AB|/3$. Similarly, select points N and M on side DC such that $|DN| = |NM| = |MC| = |CD|/3$. Show that $[KLMN] = [ABCD]/3$.

(Source of the problem: "Exercises in geometry" by Waldemar Pompe – Problem 50.)

6.8 Thales' Theorem

SOURCE

Problem: "Exercises in geometry" (in Polish) by Waldemar Pompe – Problem 56 (slightly modified)
Solution: our own

PROBLEM

Given a quadrilateral $ABCD$, choose points K, L, M, N lying respectively in sides AB, BC, CD and DA such that

$$|AK|/|KB| = |CL|/|LB| = |AN|/|ND| = |CM|/|MD| .$$

Prove that $KLMN$ is a parallelogram and its area is less than or equal to the half of the area of $ABCD$.

THEORY

Thales' Theorem Let us highlight the following observation, known as **Thales' theorem** or the **intercept theorem**, about the ratios of various line segments that are created if two intersecting lines are intercepted by a pair of parallels. In fact, it is equivalent to the theorem about ratios in similar triangles. Lines A_1A_2 and B_1B_2 are parallel if and only if

$$\frac{|A_1C|}{|A_2C|} = \frac{|B_1C|}{|B_2C|} .$$

Point C may lie anywhere on the plane except for being situated on the lines A_1A_2 or B_1B_2.

SOLUTION

Denote the common ratio by α; in particular, $\alpha = |AK|/|KB|$. It follows from Thales' theorem that the three lines, NM, AD, and KL, are mutually parallel. Similarly, the three lines NK, CB, and ML are mutually parallel so $KLMN$ is a parallelogram.

In order to estimate the area of $KLMN$ (in comparison to the area of $ABCD$), note that

$$[KLMN] = [ABCD] - [AKN] - [BKL] - [CLM] - [DMN] .$$

However, since $|KB|/|AB| = |LB|/|CB| = 1/(1 + \alpha)$, the area of BKL is $1/(1 + \alpha)^2$ of the area of ABD. Considering the remaining three triangles in a similar way, we get that their total area is

$$\frac{1}{(1 + \alpha)^2} + \frac{\alpha^2}{(1 + \alpha)^2}$$

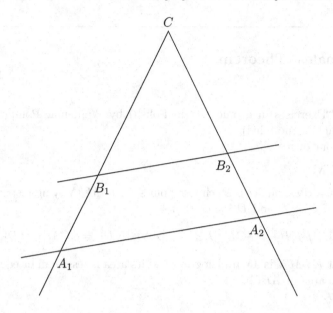

FIGURE 6.14: Thales' Theorem.

of the area of $ABCD$. Hence,

$$\frac{[KLMN]}{[ABCD]} = \frac{(1+\alpha)^2 - 1 - \alpha^2}{(1+\alpha)^2} = \frac{2\alpha}{(1+\alpha)^2}$$

$$= 2\left(\sqrt{\frac{1}{1+\alpha}\cdot\frac{\alpha}{1+\alpha}}\right)^2 \leq 2\left(\frac{1}{2}\left(\frac{1}{1+\alpha}+\frac{\alpha}{1+\alpha}\right)\right)^2 = \frac{1}{2},$$

where the inequality is obtained by the geometric-arithmetic mean inequality. The equality is obtained if $\alpha = 1$.

REMARKS

In this problem, since the ratios of the lengths of the corresponding line segments are preserved, it is natural to try to use Thales' theorem. Interestingly, one could ask a question what conclusion can be obtained if, instead, the following property holds:

$$|AK|/|KB| = |BL|/|LC| = |DM|/|MC| = |CN|/|NC| =: \alpha.$$

(Note the reversed proportions for one pair of opposing sides.) This time, $KLMN$ is not a parallelogram (in general). However, we immediately see that the area of KBL is $\alpha/(1+\alpha)^2$ fraction of the area of ABD. Similar properties are also satisfied for the three other triangles. Therefore, the area of $KLMN$ is equal to $1 - 2\alpha/(1+\alpha)^2$ of the area of $ABCD$. Arguing as

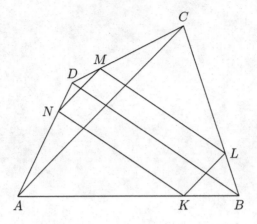

FIGURE 6.15: Illustration for Problem 6.8.

before, we get that it is at least half of the area of $ABCD$ and the equality holds when $\alpha = 1$. In fact, we get a slightly stronger property: the area of the figure obtained this way added to the area of the figure from our original problem is exactly equal to the area of $ABCD$.

EXERCISES

6.8.1. Given a parallelogram $ABCD$, consider points M and N that are in the middle of sides BC and CD, respectively. Section BD intersects with AN in point Q, and with AM in point P. Prove that $3|QP| = |BD|$.
(Source of the problem: "Exercises in geometry" by Waldemar Pompe – Problem 55. Solution: our own.)

6.8.2. Points K, L, M, and N are the middle points of sides AB, BC, CD and, respectively, DA of a parallelogram $ABCD$ whose area is equal to 1. Let P be the intersection point of KC and NB, Q be the intersection point of LD with KC, R be the intersection point of MA with LD, and, finally, S be the intersection point of NB with MA. Calculate the area of $PQRS$.
(Source of the problem: "Exercises in geometry" by Waldemar Pompe – Problem 59. Solution: our own.)

6.8.3. Points E and F are on sides AB and, respectively, AD of rhombus $ABCD$. Lines CE and CF intersect line BD in points K and L, respectively. Line EL intersects side CD in point P. Line FK intersects side BC in point Q. Prove that $|CP| = |CQ|$.
(Source of the problem: "Exercises in geometry" by Waldemar Pompe – Problem 62. Solution: our own.)

Chapter 7

Hints

In this chapter we provide hints for all exercises presented in the book.

7.1 Inequalities

1.1.1. Observe that $a + c = b + (a + c - b)$ and $a \le a + c - b \le c$.

1.1.2. Apply Jensen's inequality to function $f(k) = k^{s-1}$ and weights proportional to k.

1.1.3. Apply Jensen's inequality to function $f(x) = \sqrt{x}$.

1.2.1. Apply the arithmetic-harmonic mean inequality and observe that equality holds when $a = b = c/2 = d/4$.

1.2.2. Apply the arithmetic-geometric mean inequality to the right hand side, rearrange the terms, and finally apply the geometric-harmonic mean inequality to get the result.

1.2.3. Divide both sides by 2 and apply the arithmetic-geometric mean inequality to the left hand side.

1.3.1. Take logarithm of both sides of the inequality and directly apply the rearrangement inequality.

1.3.2. Divide both sides by abc and then apply the rearrangement inequality to the obtained inequality.

1.3.3. Take a logarithm of both sides and then apply rearrangement inequality after simplifying the expression.

1.4.1. In both cases, first invert both sides of the inequality and then apply Bernoulli's inequality, observing that $n \geq 2$.

1.4.2. Raise both sides to the power of n and then apply Bernoulli's inequality.

1.5.1. Invert both sides of the inequality and note that $n^2 - n = n \cdot (n-1)$.

1.5.2. Raise both sides of the inequality to the power of $n(n+1)$ and rearrange the obtained inequality.

1.5.3. Use the fact that $(1 + a/n)^n$ is increasing for $a \neq 0$.

1.5.4. Multiply both sides by $(n+1)/n^n$.

1.5.5. To prove the first inequality, use Bernoulli's inequality. For the second part, note that the right hand side tends to 3 and the middle term is bounded from above by e.

1.6.1. Use the binomial expansion of $(1 + x)^i$ and observe that $(1 + x)^i = 1 + ix + O(x^2)$.

1.6.2. Use the fact that for positive x: $(1 + x/n)^n > (x/n)^n$.

1.7.1. Use Cauchy-Schwarz inequality for the square roots of a, b and c.

1.7.2. Use Cauchy-Schwarz inequality for $x_1 = \sqrt{2a+1}, x_2 = \sqrt{2b+1}$, and $x_3 = \sqrt{2c+1}$, where $y_1 = y_2 = y_3 = 1$. You can alternatively use Jensen's inequality.

1.7.3. Bound function $\frac{x}{x^2+1}$ from above by a linear function passing points $(1/3, 3/10)$ and $(-3/4, -12/25)$.

1.8.1. Consider flipping a fair coin $2n$ times and calculate the probability of obtaining exactly n heads.

1.8.2. Consider the probability that in n coin tossings there are at least k heads, provided that the probability of getting a head is equal to p and, respectively, q.

1.9.1. Consider $n + 1$ points of the form $(i, \sum_{\ell=1}^{i} a_\ell), i \in \{0, 1, \ldots, n\}$.

1.9.2. Consider the area of the pentagon with each side length equal to $1/2$ with the two diagonals adjacent to the same vertex that have the same length, x.

7.2 Equalities and Sequences

2.1.1 Subtract the first equations from the third one and subtract the second equation from the fourth one to derive useful relationships between a, b, c, and d.

2.1.2 Divide the two equations.

2.1.3 Note that the system is equivalent to:

$$\begin{cases} (x - (y + z))(x - yz) = 0 \\ (y - (z + x))(y - zx) = 0 \\ (z - (x + y))(z - xy) = 0. \end{cases}$$

2.2.1 Note that $f(x) := x^3$ is an increasing function.

2.2.2 Note that $f(x) := x$, $g(x) := x^3$, and $h(x) := x^5$ are increasing functions.

2.2.3 Use the fact that at least one of the variables d, a or b attains the maximum or the minimum value from the set $\{a, b, c, d\}$.

2.3.1 Use the arithmetic-geometric mean inequality to bound the term $x^4 + 3y^2$.

2.3.2 Find an upper bound for $(x + y + z)^2$ and a lower bound for $x^2y^2 + y^2z^2 + z^2x^2$.

2.3.3 Use the substitution $a = xy$, $b = yz$, and $c = xz$.

2.4.1 Note that $x_{i+1} = x_i(4 - x_i)$, and so $x_i \in [0, 4]$ for $i \in [n]$. Use the substitution $x_1 = 4\sin^2(\alpha)$ for some $\alpha \in [0, \pi/2]$.

2.4.2 Use the identity $\cot(2x) = (\cot(x) - \tan(x))/2$ and then use the substitution $y = \tan(x)$.

2.4.3 Use the identity

$$\cot(y - x) = \frac{1 + \cot(x)\cot(y)}{\cot(x) - \cot(y)}.$$

2.5.1 For each pair of the three equations, subtract one from the other to cancel out a and one of the squares.

2.5.2 Add $x^{2009}y^{2009}$ to both sides of the equation.

2.5.3 Re-write the equations as follows: $(x_{i+1} - 6)^2 + (x_i - 8)^2 = 50$.

2.6.1 Consider the limit of x_n^3/n instead.

2.6.2 Show that $(a_n)^2/2 \leq a_{n+1} < a_n^2$.

2.6.3 Show that the sequence must diverge to ∞ or $-\infty$, unless $a = b = 0$.

2.7.1 Prove by induction that for all $\ell \in \mathbf{N} \cup \{0\}$, $a_{3\ell+1} = 1$ and $a_{3\ell+2} = -1$.

2.7.2 Note that either $a_n = 0$ for some $n \in \mathbf{N}$ or for all $n \in \mathbf{N}$, $a_{n+1}^2 + 1 = (a_{n+1} - a_n)^2$.

2.7.3 Note that $\sum_{i=1}^{n}(x_{i-1} - x_i + x_{i+1} - 1)^2 = 0$.

7.3   Functions, Polynomials, and Functional Equations

3.1.1 Using Vieta's formulas, write down the six equations involving the values we are looking for. Then, prove that $b_i \neq 0$ and that $a_1 a_2 a_3 = 1$. Using this prove that all a_i must be positive. Finally, show that all a_i are equal to 1.

3.1.2 It follows immediately from Vieta's formulas (see (3.1)) that $\sum_{i=1}^{n} x_i^2 = 0$, where x_i are roots of the polynomial $f(x)$. This implies that if all roots are real, then all of them are equal to 0.

3.1.3 Observe that we have $\frac{1}{x^4} = 9(x+2)^2$.

3.2.1 Prove that $f(0) = 0$ and then that $f(-a) = -f(a)$. Then, consider possible values of $f(2)$ depending on the arbitrarily chosen value of $f(1)$.

3.2.2 Consider the equation for $y = 0$ and $y = f(x)$ to get the relations needed to derive the solution.

3.2.3 Consider pairs of (x, y) of the form $(0,0)$, $(0, f(0))$, $(0, y)$, and $(x, f(x))$.

3.3.1 Prove that for each $x \in \mathbf{R}$ we have that $f(x) = f(x + \frac{1}{2})$.

3.3.2 Note that for $x \neq 0$ we have that $f(x) = (f(1/(1-x)) - 1)/x$.

3.3.3 Show that $f(a,b,c) = a + f(0,0,b) - f(0,0,a) + f(0,b,c) - f(0,a,b)$.

3.4.1 Since there exist $a, b \in \mathbf{Z}$ such that $P(a) = 0$ and $P(b) = 1$, we get that $(b-a) \mid (P(b) - P(a))$, and so $|b-a| = 1$. Then, define $Q(x) := P(a + (b-a)x)$ and show that $Q(f_i) = f_i$ for all $i \in \mathbf{N}$.

3.4.2 Prove that $P(0)$ is a desired integer root.

3.4.3 Prove that in order for the condition in the problem to hold we must have that $a^2 = 4b + 1$ by considering the expression

$$(2k + a - 1)P(k+1) - (2k + a + 3)P(k).$$

3.5.1 Consider a polynomial $P(q+x)+P(q-x)$ where q is any rational number and prove that it must be a constant. From this deduce that $P(x)$ must have the form $P(x) = ax + b$ for some rational numbers a and b.

3.5.2 Proof the statement by contradiction using the fact that $P(x)$ must have rational coefficients.

3.5.3 Consider polynomials $P(x) := G(x) - F(x)$ and $Q(x) := H(x) - F(x)$.

3.6.1 Re-write the polynomial in the following form:

$$f(x) = (x - a)(x - b)(2x^2 + cx + d)$$

for some $a, b, c, d \in \mathbf{R}$. Solve the resulting system of equations ensuring additionally that $ab = 2$.

3.6.2 Prove first that for sufficiently large integers x, we have that

$$P(x) < P(-x + 1) < P(x + 1).$$

Then, inspect directly the remaining cases.

3.6.3 Prove that $P(x)$ cannot have more than one term by considering $P(x) = ax^k + bx^\ell + Q(x)$, where $a, b \in \mathbf{R} \setminus \{0\}$, $\ell, k \in \mathbf{Z}$ such that $0 \le \ell < k$, and $Q(x)$ has degree less than ℓ (or $Q(x) = 0$ everywhere if $\ell = 0$). Then, consider the case when the polynomial $P(x)$ has only one term.

3.7.1 Prove first that each $P_i(x)$ has degree at most 1. Then, represent each $P_i(x)$ as follows: $P_i(x) = (a_i x + b_i)/m$, where a_i, b_i ($i \in [4]$), and m are some integers. Finally, prove that the required equality cannot be satisfied.

3.7.2 Consider the expression $r(f(p) - f(q)) + p(f(q) - f(r)) + q(f(r) - f(p))$ and observe that it is divisible by n.

3.7.3 Show that this is false for $P(x) := x^3 - 2x$.

7.4 Combinatorics

4.1.1. Greedily select pairs of members that know each other. If you stop prematurely, since there are no more pairs to select from, then select any two members that are not yet assigned, say, A_1 and A_2. Now, show that, knowing that each of them knows at least n members already assigned, it is possible to find an already selected pair of people, say, B_1 and B_2, in such a way that A_i and B_i ($i = 1, 2$) know each other. After removing the pair (B_1, B_2) and adding the pairs (A_1, B_1) and (A_2, B_2), we improve our assignment and continue the argument, if needed.

4.1.2 Start from *any* player A. Clearly, at least 6 other players played against A for the same number of rounds. If two players from this group of 6 players played the very same number of rounds, then we are done. Otherwise, there are 6 players and only two possible number of rounds for all games between them. We repeat the argument for the reduced problem.

4.1.3 Start with a person P that has at least six acquaintances. Consider then a chain of links between people, starting from this person, that oscillates between links of type *like* and *dislike*, starting with *like*. Notice that if we keep extending this chain (in any way!) it will eventually come back to P as once we enter some vertex we can always leave it. Denote this walk by W_1. If such a walk has even length, then we are finished. Otherwise, we create another walk, W_2; this time, starting from *dislike*. If its length is even, then removing it solves the problem. Otherwise, we remove both W_1 and W_2 to get the desired property.

4.2.1 Label the grid so that the bottom left cell has label $(1,1)$ and the top right one has label $(25,25)$. Put 1 in a cell with label (i,j) if $i+j$ is divisible by 3; otherwise, put 0. Observe that each block (regardless whether it is of size 1×6 or 2×3) covers precisely two 1's.

4.2.2 Show that there would have to be 9 horizontal blocks or 9 vertical ones. But this implies that there are at most 4 blocks of the other type which is not enough to cover 91 cells.

4.2.3 As usual, label the grid so that the bottom left cell has label $(1,1)$ and the top right one has label $(10,10)$. Put 1 in a cell with label (i,j) if $i+j$ is even, and 0 otherwise. How many 1's can be covered by each block?

4.3.1 Each of the four groupings can be achieved by evenly grouping some permutation of people. So it is enough to count how many permutations yield the same group.

4.3.2 First, select rows in which the rooks are going to be placed, and then select the columns for them.

4.3.3 Count first the numbers that are created. Then, observe that one can pair them in such a way that each pair has the property that the sum of digits in a given decimal position is 10.

4.3.4 Imagine the procedure backward.

4.4.1 Count the number of possible ways to select a team given an arbitrary degree distribution in the corresponding friendship graph. Then, minimize this function to get the desired lower bound.

4.4.2 Let r_t and b_t be the length of a longest red and, respectively, blue path after t rounds of the game. Show (by induction) that Builder has a strategy that increases the sum of r_t and b_t by 1 in two rounds; that is, for each $t \in \mathbf{N}$, $r_{2t} + b_{2t} \geq t$. It will follow that $\max\{r_{400}, b_{400}\} \geq 100$.

4.4.3 Select an arbitrary member of the club. We will assign him/her 1 if he/she knows at least half of the members, and 0 otherwise. Now, remove this member and all the members that do not match the chosen majority, leaving at least $4^t/2$ members. Repeat the process on the remaining subset of members. Observe that this process lasts at least $2t$ round (if there is only one member left at the beginning of round $2t$, we may assign 0 or 1 arbitrarily). But this means that at least t members have 1 assigned or at least t members have 0 assigned. The last step is to observe that this set of people satisfies the requirements of the problem.

4.5.1 Estimate the number of increasing arithmetic progressions of length k in X by $\binom{N}{2} = \frac{N(N-1)}{2} < 2^{k-1}$. Consider then a random partition of X into two subsets A and B and show that the expected number of k-element sequences in A and in B is less than 1.

4.5.2 Consider a random tournament. For each ordering, compute the probability that it has the desired property, namely, t_i won against t_{i+1}, for each $i \in [n-1]$.

4.5.3 Color edges at random, uniformly and independently. Compute the expected number of monochromatic triangles.

4.5.4 For each $i \in [100]$, let d_i be the number of acquaintances the ith person has. Notice that the average value of d_i is equal to $2 \cdot 450/100 = 9$. Take a random ordering of people. Using this ordering we investigate people, one by one, and we select a person if he or she does not know anyone already selected. Compute the expected number of people selected (function of d_i's) and then minimize it.

4.6.1 Write down the recursion for the probability of seeing exactly k white balls in an urn having n balls in total, or perform calculations for the few first rounds to make a natural conjecture that can be then proved by induction.

4.6.2 Independently consider the probability that the kth participant won the tournament of n ski jumpers and that the leader changed exactly once during the whole event.

4.6.3 Use the Inclusion–Exclusion Principle (4.9).

4.7.1 Consider two cases: a) all five points form a convex pentagon, b) one of the points lies inside the triangle formed by three other points.

4.7.2 Restrict yourself to coloring a finite number of points, namely, color 13 points that form a regular polygon inscribed in the circle. Prove that any 5 of them consist 3 that form an isosceles triangle.

4.7.3 Count first $f(n)$, the number of two element subsets of an n-element set, and then find $g(n)$, the maximum number of disjoint two element subsets of such a set. The desired upper bound is equal to $f(n)/g(n)$. Next, find a construction achieving this bound (you might want to separately consider the case when n is even and when n is odd).

4.8.1 Consider two subsets of people, those sitting in even and odd positions at the table. In one of those sets there must be at least 13 girls.

4.8.2 Consider cumulative number of aspirins taken up to and including day i. There are 30 such numbers, all distinct. Now consider those numbers increased by 14. There are also 30 such numbers, again, all distinct. There are in total 60 numbers, all of them are all less than 60 so two of them must be equal.

4.8.3 Represent each number in the selected set in the form $2^p q$, where $p \in \mathbf{N} \cup \{0\}$ and $q \in \mathbf{N}$ is an odd number.

4.9.1 Notice that by adding 1 to all sides on one die and subtracting 1 from all sides on the other die does not affect the distribution for their sum.

4.9.2 You should obtain $x_n = (2^n - (-1)^n)/3$.

4.9.3 Calculate the probability of getting 10 and 11 using $f(x)^3$, where $f(x)$ is the generating function for a fair die we have introduced in the solution. You might want to use computer to expand the resulting polynomial.

7.5   Number Theory

5.1.1 Reduce the problem to showing that $\gcd(a + b, ab) = 1$.

5.1.2 Prove that $a + b$ divides the square of $\gcd(a, b)$.

5.1.3 Consider the divisors of $n + f(n)$.

5.2.1 Observe that for any prime $p > 3$ we have $p^2 \equiv 1 \pmod 3$.

5.2.2 Observe that $504 = 8 \cdot 9 \cdot 7$. Next, independently show that the product is divisible by 8, 9, and 7.

5.2.3 For $n < 10$, the claim clearly holds for $k = 1$. For $n > 10$, one can prove that $n \equiv n^{4k+1} \pmod{10}$ for all k. Then, it is enough to prove that in the sequence $(n^{4k+1})_{k \in \mathbf{N}}$, each digit from 1 to 9 has to occur at the first position in its decimal representation.

5.3.1 Prove that $p \mid a$ and $p \mid b$ by considering $(a + b)^2 - (a^2 + b^2)$.

5.3.2 Factor the expression $(ab + cd) - (ad + bc)$.

5.3.3 Use the following observation (that often turns out to be useful)

$$a^4 + 4b^4 = (a^2 + 2b^2 - 2ab)(a^2 + 2b^2 + 2ab)$$

to get that

$$\begin{aligned}
n^{12} + 64 &= (n^6 - 4n^3 + 8)(n^6 + 4n^3 + 8) \\
&= (n^2 + 2n + 2)(n^4 - 2n^3 + 2n^2 - 4n + 4) \\
&\quad (n^2 - 2n + 2)(n^4 + 2n^3 + 2n^2 + 4n + 4).
\end{aligned}$$

Then, show that all the factors are different.

5.4.1 Use Fermat's little theorem to prove that $a^{64} - a^4$ is divisible by 5 and then that it is also divisible by 7.

5.4.2 First note that $\sum_{j=0}^{i} 2^j = 2^{i+1} - 1$ and then that 2 and n are co-prime.

5.4.3 $\phi(100) = 40$.

5.5.1 Observe that 3 does not divide 2^k.

5.5.2 Note that $|20^1 - 9^1| = 11$ and that $|20^m - 9^n| \equiv 1$ or 9 $\pmod{10}$. Then, show that $|20^m - 9^n|$ cannot be equal to 1 nor to 9.

5.5.3 Note that $d \mid m - n$ and that $m^3 + 1 = m^2(m - n) + (m^2 n + 1)$. The proof for $n^3 + 1$ holds by symmetry.

5.6.1 Separately consider the case when x is even and when x is odd.

5.6.2 Note first that one may assume that x_i and y_i are co-prime for each $i \in [2011]$. Next, consider the reminder of each term in the product when divided by 3.

5.6.3 Select any number $a \in S$ and show that it is possible to select $b \in S$ so that the reminder when dividing $a + b$ by n is in S. Then, show that having a and b fixed, one can select $c \in S$ so that $a+c$, $b+c$ and $a+b+c$ satisfy the desired conditions.

5.7.1 Consider the unique factorization of n: $n = \prod_{i=1}^{k} p_i^{\ell_i}$ for some sequence of prime numbers $2 \le p_1 < p_2 < \ldots < p_k$ and $\ell_i \in \mathbf{N}$ for $i \in [k]$. Use the fact that the sum of all positive divisors of n is equal to $\prod_{i=1}^{k} \sum_{j=0}^{\ell_i} p_i^j$ since each divisor of n has the unique representation $\prod_{i=1}^{k} p_i^{j_i}$, where $j_i \in \{0, 1, \ldots, \ell_i\}$, and two different divisors have different representations.

5.7.2 Show first that there is no solution for $n = 1$ nor for $n = 2$. Next, find solutions for $n \in \{3, 4, 5\}$. Finally, show that if there is a solution for n, then there is one for $n + 3$.

5.7.3 Denote by $D(n)$ the difference between the sum of white and the sum of black divisors of n. Prove that $D(p \cdot q) = D(p) \cdot D(q)$ when p and q are co-prime. From this conclude that it is enough to show that $D(q^k)$, where q is a prime number, is not equal to 0 to show that no such numbers exist.

5.8.1 Note that the equation can be rewritten as follows:

$$(2y - 1)^2 \;=\; (2x^2 - x)^2 - (x^2 - 1).$$

5.8.2 Consider cases $n = 1$, $n = 2$, and $n \geq 3$ separately. For the case $n = 2$, use divisibility by 3. For the case $n \geq 3$, use substitution $z = x - y$.

5.8.3 Add the two equations together and rearrange the outcome to get the following form:

$$(ab - 1)(c - 1) + (a - 1)(b - 1) + (xy - 1)(z - 1) + (x - 1)(y - 1) \;=\; 4.$$

Observe that each term at the left hand side is non-negative.

7.6   Geometry

6.1.1 Observe that points A, B', A', and B lie on a circle and that M is the center of this circle.

6.1.2 Observe that P, B, Q, and C lie on a circle.

6.1.3 Compute angles $\sphericalangle COB$ and $\sphericalangle CAB$.

6.2.1 Consider point R inside the angle $\sphericalangle PAQ$ such that $|AR| = |AB| = |AD|$ and $\sphericalangle BAP = \sphericalangle PAR$. Then, notice that $\sphericalangle DAQ = \sphericalangle QAR$.

6.2.2 Since $RPQD$ is a rectangle, $|RQ| = |PD|$.

6.2.3 Observe that $|BB'| = |CB'|$ and that $\sphericalangle B'A'H = \sphericalangle B'BA$.

6.3.1 Adapt the approach that is used for the original problem presented in this section.

6.3.2 Observe that you can inscribe $BCB'C'$ in a circle.

6.3.3 First prove that P, B, and Q are collinear. Next notice that APQ, ABQ and ABP are similar.

6.4.1 Apply Menelaus's theorem to triangle ABD and line FP, and then to triangle ACD and line EP.

6.4.2 Note that AC is parallel to BE. Then, calculate the proportion of which side AE divides side BC using Thales' theorem (see Section 6.8). Similarly, calculate the proportion in which side AC is divided by BH. Let C' be the orthogonal projection of C on AB. Finally, calculate $|AC'|/|C'B|$ and get the desired result using Ceva's theorem.

6.4.3 Add line BD to the plot and apply Menelaus's theorem twice.

6.5.1 Prove that if two half-lines, ℓ_1 and ℓ_2, are not parallel and have a common origin at point A, then any two points $B \in \ell_1$ and $C \in \ell_2$ such that $|AB| = |AC|$ have the property that all points lying on this line segment have the same total distance from lines ℓ_1 and ℓ_2.

6.5.2 Add point X such that that $ADCX$ is a rectangle. Note that B, F, E, and P are collinear and so $\sphericalangle DFX = \pi/2$. Finally, observe that points D, F, A, X, and C lie on the same circle.

6.5.3 Add point P such that $EAGP$ is a parallelogram. Observe then that $CFPE$ is also a parallelogram. Finally, note that N is the middle point of the line segment PC, and M is the middle point of the line segment PB.

6.6.1 Prove that $|PC| = |PD|$.

6.6.2 Consider three circles: circle k_1 with center in D and going through points C and E, circle k_2 with center in F and going through points E and A, and circle k_3 with a center in B and going through points A and C. Consider now three sets of points. The first one, l_{12}, has the same power with respect to k_1 and k_2. Prove that l_{12} is a line that contains the altitude of FEB going through E. Similarly, define l_{13} and l_{23}, and prove that they contain the remaining altitudes.

6.6.3 Draw a line tangent to o_1 (and so also to o_2) in point X. Denote by Y the point it intersects line AB. Now, consider power of point Y with respect to o_1 and o_2.

6.7.1 Consider the areas of triangles PAB, PBC, and PAC.

6.7.2 Note that $[ABCD]/2 = [ABEF] = [AKB]$.

6.7.3 Note first that $[DAK] = [DAB]/3$ and $[BMC] = [BDC]/3$ so $[KBMD] = 2[ABCD]/3$. Observe then that L and N bisect KB and DM.

6.8.1 Consider triangles ABQ and DQN, and then triangles APD and BMP.

6.8.2 Denote by C' the point of intersection of line CK with line AD. Observe that $|AC'| = |DA|$, and that NC' and BC are parallel. Use those facts and apply Thales' theorem to triangles $C'PN$ and BCP, and then calculate $[BCP]$. Apply the same process to triangles DQC, ARD, and BSA.

6.8.3 Show that $|DP| = |BQ| = |FD| \cdot |EB|/|BC|$. In order to show this equality for $|DP|$, use Thales' theorem for triangles FLD and BCL, and then apply it for triangles LPD and EBL. The argument for $|BQ|$ is analogous.

Chapter 8

Solutions

In this chapter we provide solutions for all exercises presented in the book.

8.1 Inequalities

Problem 1.1.1. Prove that for any $a, b, c \in \mathbf{R}$ such that $0 < a \le b \le c$,

$$\frac{1}{a} - \frac{1}{b} + \frac{1}{c} \ge \frac{1}{a + c - b}.$$

Illustrate the solution graphically. Does the same inequality hold for any function $f \colon \mathbf{R} \to \mathbf{R}$ that is convex on some connected subset of $\mathbf{R}$?

Solution. Let us observe first that $a + c = b + (a + c - b)$ and $a \le a + c - b \le c$ (see Figure 8.1 for the illustration of these observations, the length of the dashed line is equal to $(f(a) + f(c))/2 - (f(b) + f(a + c - b))/2)$. If $a = c$ (and so, in fact, $a = b = c$), then both sides are equal to $1/a$ and we are done. If $a < c$, then we note that for any convex function f (in particular, for $f(x) = 1/x$, $x > 0$) we have

$$
\begin{aligned}
f(a) + f(c) &= \frac{c-b}{c-a} f(a) + \frac{b-a}{c-a} f(a) + \frac{b-a}{c-a} f(c) + \frac{c-b}{c-a} f(c) \\
&\ge f\left(\frac{c-b}{c-a} a + \frac{b-a}{c-a} c\right) + f\left(\frac{b-a}{c-a} a + \frac{c-b}{c-a} c\right) \\
&= f(b) + f(a + c - b).
\end{aligned}
$$

Hence, not only the desired inequality holds but the same is true for any convex function f. For graphical illustration see Figure 8.1.

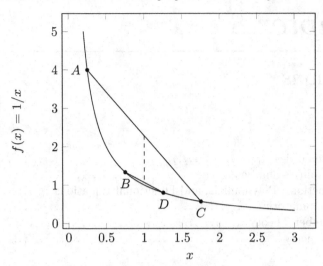

FIGURE 8.1: Illustration for Problem 1.1.1 (case $b < a + c - b$). We take $A = (a, 1/a)$, $B = (b, 1/b)$, $C = (c, 1/c)$, $D = (a + c - b, 1/(a + c - b))$.

Problem 1.1.2. Prove that for any $n \in \mathbf{N}$ and any real number $s \geq 2$, the following inequality holds:

$$\frac{\sum_{k=1}^{n} k^s}{\sum_{k=1}^{n} k} \geq \left(\frac{2}{3}n + \frac{1}{3}\right)^{s-1}.$$

Solution. It follows immediately from Jensen's inequality, applied to function $f(k) = k^{s-1}$ and $a_k = k / \sum_{i=1}^{n} i$, that

$$\frac{\sum_{k=1}^{n} k^s}{\sum_{k=1}^{n} k} = \sum_{k=1}^{n} k^{s-1} \frac{k}{\sum_{i=1}^{n} i} \geq \left(\sum_{k=1}^{n} k \frac{k}{\sum_{i=1}^{n} i}\right)^{s-1} = \left(\frac{\sum_{k=1}^{n} k^2}{\sum_{i=1}^{n} i}\right)^{s-1}$$

$$= \left(\frac{n(n+1)(2n+1)/6}{n(n+1)/2}\right)^{s-1} = \left(\frac{2}{3}n + \frac{1}{3}\right)^{s-1}.$$

(Note that $f(k) = k^{s-1}$ is convex for any $s \geq 2$.)

Problem 1.1.3. Prove that for any $x \in \mathbf{R}_+$,

$$\sqrt{x} + \sqrt{x+2} < 2\sqrt{x+1}.$$

Solution. Since $f(x) = \sqrt{x}$ is concave, we get from Jensen's inequality that for any $x \in \mathbf{R}_+$

$$\sqrt{x} + \sqrt{x+2} = 2\left(\frac{1}{2}\sqrt{x} + \frac{1}{2}\sqrt{x+2}\right) \leq 2\sqrt{\frac{1}{2}x + \frac{1}{2}(x+2)} = 2\sqrt{x+1}.$$

Since $f(x) = \sqrt{x}$ is not a linear function and we applied the inequality with $x_1 = x \neq x + 2 = x_2$, in fact sharp inequality holds.

Problem 1.2.1. Show that for any $a, b, c, d \in \mathbf{R}_+$, the following inequality holds:

$$(a + b + c + d) \left(\frac{1}{a} + \frac{1}{b} + \frac{4}{c} + \frac{16}{d} \right) \geq 64.$$

When does equality hold?

Solution. Observe that the inequality can be rewritten as follows:

$$\frac{a + b + 2\frac{c}{2} + 4\frac{d}{4}}{8} \geq \frac{8}{\frac{1}{a} + \frac{1}{b} + 2\frac{1}{c/2} + 4\frac{1}{d/4}}.$$

The right hand side of the above inequality is the harmonic mean of 8 terms, $H(a, b, c/2, c/2, d/4, d/4, d/4, d/4)$, and the left hand side is the arithmetic mean $A(a, b, c/2, c/2, d/4, d/4, d/4, d/4)$. Using arithmetic-harmonic inequality we immediately get that this inequality holds for any $a, b, c, d \in \mathbf{R}_+$, and that the equality is achieved when $a = b = c/2 = d/4$.

Problem 1.2.2. Show that for any n numbers $a_1, \ldots, a_n \in \mathbf{R}_+$, the following inequality holds:

$$\frac{a_1}{a_2 + 1} + \frac{a_2}{a_3 + 1} + \cdots + \frac{a_{n-1}}{a_n + 1} + \frac{a_n}{a_1 + 1} \geq \frac{n^2}{n + \alpha},$$

where $\alpha = \sum_{i=1}^{n} 1/a_i$.

Solution. In order to simplify the notation, let us set $a_{n+1} = a_1$. Using the arithmetic-geometric mean inequality, we get that

$$\sum_{i=1}^{n} \frac{a_i}{a_{i+1} + 1} = n \frac{\sum_{i=1}^{n} \frac{a_i}{a_{i+1} + 1}}{n} \geq n \sqrt[n]{\prod_{i=1}^{n} \frac{a_i}{a_{i+1} + 1}}.$$

However,

$$n \sqrt[n]{\prod_{i=1}^{n} \frac{a_i}{a_{i+1} + 1}} = n \sqrt[n]{\prod_{i=1}^{n} \frac{a_i}{a_i + 1}} = n \sqrt[n]{\prod_{i=1}^{n} \frac{1}{1/a_i + 1}}.$$

Finally, using the geometric-harmonic mean inequality, we get that

$$\sqrt[n]{\prod_{i=1}^{n} \frac{1}{1/a_i + 1}} \geq \frac{n}{\sum_{i=1}^{n} 1 + 1/a_i} = \frac{n}{n + \alpha},$$

and so the desired inequality holds.

Problem 1.2.3. Prove that for any $a, b \in \mathbf{R}_+$, for which $ab = 1$, we have that

$$a^m + b^m \geq 2,$$

where $m \in \mathbf{R}_+$.

Solution. It follows immediately from the arithmetic-geometric mean inequality that

$$a^m + b^m = 2\frac{a^m + b^m}{2} \geq 2\sqrt{a^m b^m} = 2.$$

Problem 1.3.1. Prove that for any $a, b \in \mathbf{R}_+$,

$$a^b b^a \leq a^a b^b.$$

Solution. The inequality we aim to prove is equivalent to the following one

$$b\log(a) + a\log(b) \leq a\log(a) + b\log(b).$$

Without loss of generality, due to the symmetry, we may assume that $a \leq b$ and so $\log(a) \leq \log(b)$. Now, the inequality above follows immediately from the rearrangement inequality.

Problem 1.3.2. Prove that for any $a, b, c \in \mathbf{R}_+$,

$$\frac{ab}{c} + \frac{bc}{a} + \frac{ca}{b} \geq a + b + c.$$

Solution. After dividing both sides by $abc \in \mathbf{R}_+$ we get the following inequality:

$$\frac{1}{c} \cdot \frac{1}{c} + \frac{1}{a} \cdot \frac{1}{a} + \frac{1}{b} \cdot \frac{1}{b} \geq \frac{1}{b} \cdot \frac{1}{c} + \frac{1}{c} \cdot \frac{1}{a} + \frac{1}{a} \cdot \frac{1}{b}.$$

Due to the symmetry, without loss of generality, we may assume that $1/a \leq 1/b \leq 1/c$. As in the previous problem, the above inequality follows immediately from the rearrangement inequality.

Problem 1.3.3. Prove that for any $a, b, c \in \mathbf{R}_+$,

$$a^a b^b c^c \geq (abc)^{(a+b+c)/3}.$$

Solution. Without loss of generality, we may assume that $a \leq b \leq c$. After taking a logarithm of both sides of the inequality we get

$$a\log(a) + b\log(b) + c\log(c) \geq \frac{a+b+c}{3}\Big(\log(a) + \log(b) + \log(c)\Big).$$

After multiplying both sides by 3 and rearranging the terms, we get

$$2a\log(a) + 2b\log(b) + 2c\log(c)$$
$$\geq (b+c)\log(a) + (a+c)\log(b) + (a+b)\log(c).$$

By rearrangement inequality, we get that

$$a\log(a) + b\log(b) + c\log(c) \geq c\log(a) + a\log(b) + b\log(c)$$

and that

$$a\log(a) + b\log(b) + c\log(c) \geq b\log(a) + c\log(b) + a\log(c).$$

These two inequalities imply immediately the desired inequality above.

Problem 1.4.1. Prove that for any integer $n > 1$,

$$\text{a)} \quad \left(\frac{n}{n+2}\right)^{n^2-n} < \frac{1}{2n-1},$$

$$\text{b)} \quad \left(\frac{n-1}{n}\right)^{n^2-1} < \frac{1}{n+2}.$$

Solution. In both parts, we first inverse both sides of the inequality and then apply Bernoulli's inequality. To show part a), that is, to show that

$$2n - 1 < \left(1 + \frac{2}{n}\right)^{n^2-n}$$

we apply Bernoulli's inequality to get that

$$\left(1 + \frac{2}{n}\right)^{n^2-n} > 1 + \frac{2}{n}(n^2 - n) = 2n - 1.$$

(Since $n \geq 2$, $n^2 - n = n(n-1) \geq 2 > 1$.) Similarly, in order to show part b), that is, to show that

$$n + 2 < \left(1 + \frac{1}{n-1}\right)^{n^2-1}$$

we again apply Bernoulli's inequality to get that

$$\left(1 + \frac{1}{n-1}\right)^{n^2-1} > 1 + \frac{1}{n-1}(n-1)(n+1) = n + 2.$$

(Since $n \geq 2$, $n^2 - 1 \geq 3 > 1$.)

Problem 1.4.2. Prove that for any real number $x > -1$ and $n \in \mathbf{N}$,

$$\sqrt[n]{1+x} \leq 1 + \frac{x}{n}.$$

Solution. Raising both sides to the power of n yields an equivalent inequality $1 + x \leq \left(1 + \frac{x}{n}\right)^n$ that follows directly from Bernoulli's inequality.

Problem 1.5.1. Prove that for any integer $n > 2$,

$$\left(\frac{n}{n+2}\right)^{n^2-n} < \frac{1}{4^{n-1}}.$$

Can the constant 4 be improved for large n?

Solution. It is enough to prove that

$$4^{n-1} < \left(1+\frac{2}{n}\right)^{n^2-n} = \left(\left(1+\frac{2}{n}\right)^n\right)^{n-1}.$$

Since for $n > 2$ we have that $(1 + 2/n)^n > (1 + 2/2)^2 = 4$, the above inequality holds. Finally, note that $\lim_{n\to\infty}(1 + 2/n)^n = e^2$, the constant 4 can be replaced by any number smaller than $e^2 \approx 7.39$. Such inequality would hold for large enough n.

Problem 1.5.2. For which $n \in \mathbf{N}$ do we have that

$$\sqrt[n]{n} > \sqrt[n+1]{n+1} ?$$

Solution. Raising both sides of the inequality to the power of $n(n + 1)$ gives us $n^{n+1} > (n + 1)^n$. After dividing both sides by n^n, we see that it is enough to show that $n > (1 + 1/n)^n$. Since $(1 + 1/n)^n \leq \exp(1/n)^n = e < 3$, this inequality holds for any $n \geq 3$. The remaining two cases can be checked by hand: it does *not* hold for $n = 1$ but it does hold for $n = 2$. We get that the inequality holds for any natural number at least 2.

Problem 1.5.3. For which $n \in \mathbf{N}$ do we have that

$$(n-1)^n(n+1)^{n+1} > n^{2n+1} ?$$

Solution. After rearranging the inequality, we get that

$$\left(1-\frac{1}{n^2}\right)^n > \frac{n}{n+1} = \frac{1}{1+1/n}$$

or, equivalently,

$$\left(1-\frac{1}{n^2}\right)^{n^2} > \frac{1}{(1+1/n)^n}.$$

Since $(1 - 1/n^2)^{n^2} \leq \exp(-1/n^2)^{n^2} = 1/e$, the left hand side is at most $1/e$. Similarly, since $(1 + 1/n)^n \leq \exp(1/n)^n = e$, the right hand side is at least $1/e$. As a result, the original inequality does *not* hold for any $n \in \mathbf{N}$.

Problem 1.5.4. For which $n \in \mathbf{N}$ do we have that

$$n^n > (n+1)^{n-1} ?$$

Solution. After multiplying both sides by $(n+1)/n^n$ we get $n+1 > (1+1/n)^n$. Since $(1 + 1/n)^n \leq \exp(1/n)^n = e < 3$, the inequality holds for any $n \geq 2$. One can directly check that it does *not* hold for $n = 1$.

Problem 1.5.5. Prove that for any $n \in \mathbf{N}$,

$$2 \leq \left(1 + \frac{1}{n}\right)^n \leq 3 \cdot \frac{n+1}{n+2} .$$

Solution. The first inequality follows immediately from Bernoulli's inequality:

$$\left(1 + \frac{1}{n}\right)^n \geq 1 + n \cdot \frac{1}{n} = 2 .$$

In order to show the second inequality, observe that $(1 + 1/n)^n \leq (\exp(1/n))^n = e < 2.72$. On the other hand, since $3(n + 1)/(n + 2)$ is increasing and tending to 3 as $n \to \infty$, the desired inequality holds for n large enough. In fact, it certainly holds for $n \geq 9$, as $3(n+1)/(n+2) = 30/11 > 2.72$ if $n = 9$. One can easily inspect the 8 missing cases $n \in [8]$ to show that the inequality holds for all $n \in \mathbf{N}$ (in fact, we get equality for $n \in \{1, 2\}$).

Problem 1.6.1. Show that for any $n \in \mathbf{N}$, there exists a non negative $x \in \mathbf{R}$ such that

$$\prod_{i=1}^n (1 + x)^i < 1 + \frac{n^2 + n + 1}{2} x .$$

Solution. Since n is fixed, there are finite number of elements in the product on the left hand side as well as in the binomial expansion of $(1 + x)^i$ for any $i \in [n]$. As a result,

$$\prod_{i=1}^n (1 + x)^i = \prod_{i=1}^n \left(1 + ix + O(x^2)\right)$$

$$= 1 + \sum_{i=1}^n ix + O(n^2) = 1 + \frac{n(n+1)}{2} x + O(x^2)$$

$$< 1 + \frac{n^2 + n}{2} x + \frac{x}{2} = 1 + \frac{n^2 + n + 1}{2} x,$$

provided that $x \in \mathbf{R}_+$ is sufficiently close to zero.

Problem 1.6.2. Prove that for any polynomial $W(x)$ and sufficiently large x we have that $(1 + x/n)^n > W(x)$, if $n \in \mathbf{N}$ is greater than the degree of W. What does it tell us about the function e^x?

Solution. Since $W(x) = O(n^{n-1})$, observe that

$$\left(1 + \frac{x}{n}\right)^n - W(x) = \frac{x^n}{n^n} + O(x^{n-1}) = \frac{x^n}{n^n}(1 + O(1/x)) > \frac{x^n}{2n^n} > 0,$$

provided that $x \in \mathbf{R}$ is sufficiently large.

As $e^x = \lim_{i \to +\infty}(1 + x/n)^n$ and the sequence $(1 + x/n)^n$ is increasing, we see that function e^x tends to infinity (as $x \to \infty$) faster than any fixed polynomial $W(x)$.

Problem 1.7.1. Prove that for $a, b, c \in \mathbf{R}_+$,

$$(a + b + c)\left(\frac{1}{a} + \frac{1}{b} + \frac{1}{c}\right) \geq 9.$$

Solution. Let us first note that

$$(a + b + c)\left(\frac{1}{a} + \frac{1}{b} + \frac{1}{c}\right) = (\sqrt{a}^2 + \sqrt{b}^2 + \sqrt{c}^2)(\sqrt{1/a}^2 + \sqrt{1/b}^2 + \sqrt{1/c}^2).$$

By Cauchy-Schwarz inequality, the right hand side is greater than or equal to

$$(\sqrt{a}\sqrt{1/a} + \sqrt{b}\sqrt{1/b} + \sqrt{c}\sqrt{1/c})^2 = 3^2 = 9.$$

As a remark, we note that the problem can be also solved by applying the arithmetic-harmonic inequality (and this approach is mentioned in the remark to the original problem presented in PLMO II – Phase 1 – Problem 6, which assumed additionally that $a + b + c = 1$).

Problem 1.7.2. Prove that for any $a, b, c \in \mathbf{R}_+$ such that $a + b + c = 1$, we have that

$$\sqrt{2a + 1} + \sqrt{2b + 1} + \sqrt{2c + 1} \leq \sqrt{15}.$$

Solution. Using Cauchy-Schwarz inequality, we get that

$$\left(\sqrt{2a + 1}\cdot 1 + \sqrt{2b + 1}\cdot 1 + \sqrt{2c + 1}\cdot 1\right)^2 \leq (2a + 1 + 2b + 1 + 2c + 1)(1 + 1 + 1) = 15,$$

which gives us the desired result. Alternatively, one can use Jensen's inequality to get that

$$\sqrt{2a + 1} + \sqrt{2b + 1} + \sqrt{2c + 1} \leq 3(\sqrt{(2a + 1 + 2b + 1 + 2c + 1)/3})$$
$$= 3\sqrt{5/3} = \sqrt{15}.$$

Problem 1.7.3. Prove that if $a, b, c \in \mathbf{R}$ are such that $a + b + c = 1$ and $\min\{a, b, c\} \geq -3/4$, then

$$\frac{a}{a^2 + 1} + \frac{b}{b^2 + 1} + \frac{c}{c^2 + 1} \leq \frac{9}{10}.$$

Does this inequality hold without the additional assumption that $\min\{a, b, c\} \geq -3/4$?

Solution. It is natural to try to bound function $f(x) := x/(x^2 + 1)$ by some linear function, that is, we search for a bound of the form

$$f(x) = \frac{x}{x^2 + 1} \leq Ax + B =: g(x)$$

for some constants A and B. Indeed, if this can be done, then the left hand side of our inequality, $f(a) + f(b) + f(c)$, would be bounded by $g(a) + g(b) + g(c) = A(a + b + c) + 3B = A + 3B$. But how can we find suitable constants A and B?

We notice that the equality holds for $a = b = c = 1/3$ and so the linear upper bound that we are searching for has to be tight for $x = 1/3$, that is,

$$f(1/3) = \frac{3}{10} = g(1/3) = \frac{A}{3} + B.$$

Moreover, the additional assumption that a, b, and c must be at least $-3/4$ suggests that the same should be true for $x = -3/4$:

$$f(-3/4) = -\frac{12}{25} = g(-3/4) = \frac{-3A}{4} + B.$$

We get the following system of equations: $A/3 + B = 3/10$ and $-3A/4 + B = -12/25$. It follows that $A = 18/25$ and $B = 3/50$. Since $A + 3B = 9/10$, the result will hold once we prove that for any $x \geq -3/4$

$$\frac{x}{x^2 + 1} \leq \frac{18}{25}x + \frac{3}{50}$$

or, equivalently, that

$$h(x) := 36x^3 + 3x^2 - 14x + 3 \geq 0$$

(One can see it after multiplying both sides of the previous inequality by $50(x^2 + 1)$.) But we already know that $g(1/3) = h(-3/4) = 0$ so $h(x)$ is divisible by $(3x - 1)(4x + 3)$ (see Section 3.6 for additional explanations of this fact). After dividing these two polynomials we find that the remaining factor is $3x - 1$ and so the inequality is equivalent to $(3x - 1)^2(4x + 3) \geq 0$, which is clearly true for any $x \geq -3/4$.

We will show now that the assumption that $\min\{a, b, c\} \geq -3/4$ is not necessary. We start with proving a few properties of function $f(x)$— for a graph of this function see a gray curve in Figure 8.2.

Property A: Function $f(x)$ is odd, that is, $f(x) = -f(x)$ for any $x \in \mathbf{R}$. Indeed,

$$f(x) = \frac{x}{x^2 + 1} = \frac{-(-x)}{(-x)^2 + 1} = -f(-x).$$

Property B: For any $x \geq 0$,

$$0 \leq f(x) \leq 1/2 = f(1).$$

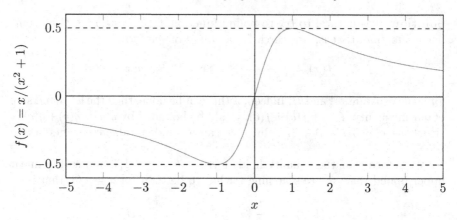

FIGURE 8.2: Function $f(x) = x/(x^2 + 1)$.

The lower bound of 0 is trivial. To see the upper bound of $1/2$, note that $x^2 - 2x + 1 = (x-1)^2 \geq 0$ which implies that $2x \leq x^2 + 1$ and so $x/(x^2 + 1) \leq 1/2$.

Property C: Function $f(x)$ is increasing on the interval $[0,1]$, that is, $f(x) < f(y)$ for any $0 \leq x < y \leq 1$. Indeed, $f(x) < f(y)$ is equivalent to $x(y^2 + 1) < y(x^2 + 1)$ which, in turn, is equivalent to $0 < (1 - yx)(y - x)$. But the last inequality is clearly true for any $0 \leq x < y \leq 1$ as $xy \in [0,1)$.

Property D: Function $f(x)$ is decreasing on the interval $[1, \infty)$, that is, $f(x) > f(y)$ for any $1 \leq x < y$. In order to see this, the same argument as before can be applied with the only difference being that now $xy \in (1, \infty)$.

Let us summarize what we have learned about function $f(x)$: it is decreasing on the interval $(-\infty, -1]$, reaching $-1/2$ at $x = -1$, then increasing on the interval $[-1, 1]$, reaching $1/2$ at $x = 1$, and finally decreasing on the interval $[1, \infty)$.

Now, let us come back to the original problem. Suppose that $c \leq b \leq a$ are any real numbers such that $a + b + c = 1$. Since we already dealt with the case $\min\{a, b, c\} \geq -3/4$, we may assume that $c < -3/4$. We independently consider 2 cases. If $c \leq -3$, then $a \geq 2$ and so

$$\frac{a}{a^2 + 1} + \frac{b}{b^2 + 1} + \frac{c}{c^2 + 1} \leq \frac{2}{2^2 + 1} + \frac{1}{1^2 + 1} + \frac{0}{0^2 + 1} = \frac{9}{10}.$$

On the other hand, if $c \in [-3, -3/4]$, then

$$\frac{a}{a^2 + 1} + \frac{b}{b^2 + 1} + \frac{c}{c^2 + 1} \leq \frac{1}{1^2 + 1} + \frac{1}{1^2 + 1} + \frac{-3}{3^2 + 1} = \frac{7}{10} < \frac{9}{10},$$

and we are done.

Problem 1.8.1. Prove that for any $n \in \mathbf{N}$,

$$\frac{1}{2n+1} \leq \frac{\binom{2n}{n}}{2^{2n}} \leq 1.$$

Can you improve these bounds for large n?

Solution. Consider flipping a fair coin $2n$ times. For any $i \in [2n]$, let $\mathbf{P}(i)$ be the probability of getting exactly i heads. Clearly, $\sum_{i=0}^{2n} \mathbf{P}(i) = 1$ (the number of heads has to be between 0 and $2n$) and for any $i \in [2n]$

$$\mathbf{P}(i) = \binom{2n}{i} \cdot \left(\frac{1}{2}\right)^{2n}.$$

(there are $\binom{2n}{i}$ ways to select i rounds when heads are obtained and each such outcome occurs with probability $(1/2)^{2n}$). Hence, our goal is to estimate $\mathbf{P}(n)$. Because of this connection, the upper bound ($\mathbf{P}(n) \leq 1$) is trivial. (Let us note that we could have obtained the same upper bound without using the above probabilistic argument by observing that $2^{2n} = (1+1)^{2n}$ and then using the binomial theorem.) In order to see the lower bound ($\mathbf{P}(i) \geq 1/(2n+1)$) we note that $\mathbf{P}(i)$, as a function of i, is maximized exactly for $i = n$; that is, for any $i \in [2n]$, $\mathbf{P}(i) \leq \mathbf{P}(n)$. To see this note that $\mathbf{P}(i) = \mathbf{P}(i-1) \cdot (2n+1-i)/i$, which is larger than one for $i \leq n$ and smaller than one otherwise.

Using this observation, it follows from the averaging argument that

$$\mathbf{P}(n) \geq \frac{1}{2n+1} \sum_{i=0}^{2n} \mathbf{P}(i) = \frac{1}{2n+1},$$

and we are done.

Finally, let us mention that better bounds can be obtained, for example, one can show that for any $n \in \mathbf{N}$,

$$\frac{1}{2\sqrt{n}} \leq \frac{\binom{2n}{n}}{2^{2n}} \leq \frac{1}{\sqrt{2n}}.$$

In fact, one can use Stirling's formula ($n! \sim \sqrt{2\pi n}(n/e)^n$) to see that

$$\frac{\binom{2n}{n}}{2^{2n}} = \frac{(2n)!}{(n!)^2 2^{2n}} \sim \frac{\sqrt{2\pi(2n)}(2n/e)^{2n}}{2\pi n(n/e)^{2n}2^{2n}} = \frac{1}{\sqrt{\pi n}}.$$

Problem 1.8.2. Prove that for $k, n \in \mathbf{N}$, such that $k \leq n$ and $p, q \in [0,1]$, such that $p < q$, we have that

$$\sum_{i=k}^{n} \binom{n}{i}\left(q^i(1-q)^{n-i} - p^i(1-p)^{n-i}\right) \geq 0.$$

Solution. In order to solve this problem, we are going to use a standard but very useful proof technique in probability theory that allows one to compare two experiments. Consider two biased coins, the first with probability p of turning up heads and the second with probability $q > p$ of turning up heads. For any fixed k, the probability p_k that the first coin produces at least k heads should be at most the probability q_k that the second coin produces at least k heads. Clearly,

$$p_k = \sum_{i=k}^{n} \binom{n}{i} p^i (1-p)^{n-i} \quad \text{and} \quad q_k = \sum_{i=k}^{n} \binom{n}{i} q^i (1-q)^{n-i}.$$

Hence, our problem reduces to showing that, indeed, $q_k \geq p_k$ for any k.

However, proving it is rather difficult with a standard counting argument. Coupling easily circumvents this problem. Let $X_1, X_2, \ldots, X_n$ be indicator random variables for heads in a sequence of n flips of the first coin. In other words, $X_i = 1$ if ith flip is a head and $X_i = 0$ otherwise. It follows that

$$p_k = \mathbf{P}\left(\sum_{i=0}^{n} X_i \geq k\right).$$

For the second coin, define a new sequence $Y_1, Y_2, \ldots, Y_n$ such that if $X_i = 1$, then $Y_i = 1$; if $X_i = 0$, then $Y_i = 1$ with probability $(q-p)/(1-p)$. Clearly, the sequence of Y_i has exactly the probability distribution of tosses made with the second coin. Indeed, for any $i \in [n]$,

$$\mathbf{P}\left(Y_i = 1\right) = \mathbf{P}\left(X_i = 1\right) + \mathbf{P}\left(X_i = 0\right) \cdot \frac{q-p}{1-p} = p + (1-p) \cdot \frac{q-p}{1-p} = q.$$

However, because of the coupling we trivially get that $X := \sum X_i \leq Y := \sum Y_i$ and so $\mathbf{P}(X \geq k) \leq \mathbf{P}(Y \geq k)$, as expected. We typically say that X is (stochastically) bounded from above by Y.

Problem 1.9.1. Prove that for all sequences of n numbers $a_1, \ldots, a_n \in \mathbf{R}$, we have that

$$\sqrt{n + \left(\sum_{i=1}^{n} a_i\right)^2} \leq \sum_{i=1}^{n} \sqrt{1 + a_i^2}.$$

When does equality hold?

Solution. Fix any sequence of numbers $a_1, \ldots, a_n \in \mathbf{R}$. Let us consider $n+1$ points, $P_0, P_1, \ldots, P_n$, on the 2-dimensional plane. For each $i \in \{0, 1, \ldots, n\}$, point P_i is defined as follows:

$$P_i := \left(i, \sum_{\ell=1}^{i} a_\ell\right).$$

(In particular, $P_0 = (0,0)$.) For any two points, P_i and P_j, let $d(P_i, P_j)$ be the distance between P_i and P_j. Clearly,

$$d(P_0, P_n) = \sqrt{(n-0)^2 + \left(\sum_{\ell=0}^{n} a_\ell - 0\right)^2} = \sqrt{n - \left(\sum_{\ell=0}^{n} a_\ell\right)^2},$$

the left hand side of our inequality. On the other hand, for each $i \in [n]$,

$$d(P_{i-1}, P_i) = \sqrt{(i-(i-1))^2 + \left(\sum_{\ell=0}^{i} a_\ell - \sum_{\ell=0}^{i-1} a_\ell\right)^2} = \sqrt{1 + a_i^2}.$$

It follows immediately from the triangle inequality that

$$d(P_0, P_n) \leq \sum_{i=1}^{n} d(P_{i-1}, P_i).$$

But this is exactly the inequality we wanted to prove and so we are done! Finally, let us note that the equality holds if and only if all the a_i are equal.

Problem 1.9.2. Prove that for any $x \in \mathbf{R}$ such that $1/4 \leq x \leq 1$, we have that

$$\frac{x}{2}\sqrt{1-x^2} + \frac{1}{16}\sqrt{16x^2 - 1} < \frac{4}{9}.$$

Solution. Consider a pentagon $ABCDE$ with each side of length $1/2$; that is, $|AB| = |BC| = |CD| = |DE| = |EA| = 1/2$. Moreover, assume that $x := |AD| = |BD|$; that is, ABD is an isosceles triangle. Since $1/2 = |AB| \leq |AD| + |BD| = 2x$, $x \geq 1/4$. On the other hand, $x = |AD| \leq |AE| + |ED| = 1$. The limiting shape when $x \to 1$ is an isosceles triangle ABD whereas if $x \to 1/4$ we get two isosceles triangles ADE and BCD (see Figure 8.3).

Fix any $x \in [1/4, 1]$ and consider the corresponding pentagon $ABCDE$ (including the two limiting scenarios, $x = 1/4$ and $x = 1$). Let us first calculate the area of our pentagon $ABCDE$ which is the sum of the areas of three isosceles triangles. The first one, ABD, has one side of length $1/2$ and two sides of length x. The remaining two, BCD and ADE, have one side of length x and two sides of length $1/2$. Since the area of isosceles triangle with base of length a and sides of equal length b is $\frac{1}{2}a^2\sqrt{b^2/a^2 - 1/4}$, we get that the area of our pentagon is equal to

$$\frac{1}{2}(1/2)^2\sqrt{x^2/(1/2)^2 - 1/4} + 2 \cdot \frac{1}{2}x^2\sqrt{(1/2)^2/x^2 - 1/4}$$

$$= \frac{1}{16}\sqrt{16x^2 - 1} + \frac{x}{2}\sqrt{1 - x^2},$$

which is exactly the left hand side of our inequality. On the other hand, the

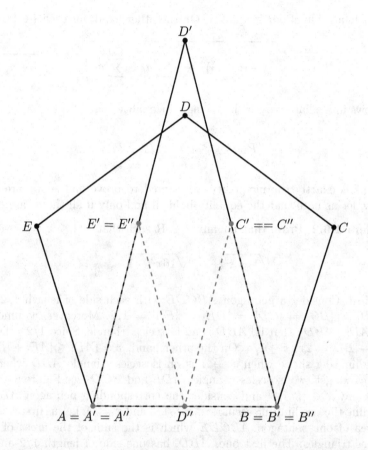

FIGURE 8.3: Illustration for Problem 1.9.2. $ABCDE$ is the regular pentagon, $A'B'C'D'E'$ is the 'limiting pentagon' when $x \to 1$ and $A''B''C''D''E''$ (dashed grey) is the 'limiting pentagon' when $x \to 1/4$.

area of our pentagon is less than or equal to the area of a regular pentagon of side length $a = 1/2$, which is equal to

$$\frac{a^2}{4}\sqrt{5(5+2\sqrt{5})} \; = \; \frac{1}{16}\sqrt{5(5+2\sqrt{5})} \; < \; 0.44 \; < \; \frac{4}{9}.$$

8.2 Equalities and Sequences

Problem 2.1.1. Solve the following system of equations, given that all variables involved are real numbers:

$$\begin{cases} a^3 + b = c \\ b^3 + c = d \\ c^3 + d = a \\ d^3 + a = b. \end{cases}$$

Solution. By subtracting the first equation from the third one, we get that

$$b - d \; = \; (c - a)(c^2 + ac + a^2 + 1).$$

Analogously, after subtracting the second equation from the fourth one, we get that

$$c - a \; = \; (d - b)(d^2 + bd + b^2 + 1)$$

and, by combining the two, we get that

$$b - d \; = \; (d - b)(d^2 + bd + b^2 + 1)(c^2 + ac + a^2 + 1). \tag{8.1}$$

Now, by considering the term $f(d) := d^2 + bd + b^2 + 1$ as a function of d, we see that $f(d) > 0$ as the discriminant of the polynomial is equal to $\Delta = b^2 - 4(b^2+1) = -3b^2 - 4 < 0$. Similarly, we get that $g(c) := c^2 + ac + a^2 + 1 > 0$. Hence, we get from (8.1) that $b = d$. A symmetric argument can be used to show that $c = a$. We can now get back to the original equations to see that $a^3 + b = a$ and $b^3 + a = b$, which implies that $a^3 = -b^3$ and so $a = -b$. As a consequence, we get that $a^3 - a = a$, and so $a \in \{0, -\sqrt{2}, \sqrt{2}\}$. It follows that there are three candidate solutions:

$$(a, b, c, d) \; \in \; \left\{ (0,0,0,0), (-\sqrt{2}, \sqrt{2}, -\sqrt{2}, \sqrt{2}), (\sqrt{2}, -\sqrt{2}, \sqrt{2}, -\sqrt{2}) \right\}.$$

We directly check that all of them satifsy the original system of equations.

Problem 2.1.2. Solve the following system of equations, given that all variables involved are real numbers:

$$\begin{cases} (x - y)(x^3 + y^3) = 7 \\ (x + y)(x^3 - y^3) = 3. \end{cases}$$

Solution. We see immediately that $x \neq y$ and $x \neq -y$. Therefore, we can divide the two equations to get that

$$\frac{(x-y)(x+y)(x^2-xy+y^2)}{(x+y)(x-y)(x^2+xy+y^2)} = \frac{7}{3},$$

and so

$$\begin{aligned} 0 &= 7(x^2+xy+y^2) - 3(x^2-xy+y^2) = 4x^2 + 10xy + 4y^2 \\ &= 2(2x+y)(2y+x). \end{aligned}$$

If $y = -2x$, then we would have to have $3x^4 = -1$ which is impossible. On the other hand, if $x = -2y$, then we get $3y^4 = 1$ and so $y = 1/\sqrt[4]{3}$ or $y = -1/\sqrt[4]{3}$. We conclude that there are two candidate solutions

$$(x,y) \in \left\{ (2/\sqrt[4]{3}, -1/\sqrt[4]{3}), (-2/\sqrt[4]{3}, 1/\sqrt[4]{3}) \right\}.$$

We directly check that all of them satisfy the original system of equations.

Problem 2.1.3. Solve the following system of equations, given that all variables involved are real numbers:

$$\begin{cases} x^2 - (y+z+yz)x + (y+z)yz = 0 \\ y^2 - (z+x+zx)y + (z+x)zx = 0 \\ z^2 - (x+y+xy)z + (x+y)xy = 0. \end{cases}$$

Solution. Let us note that the first equality can be rewritten as follows:

$$\begin{aligned} 0 &= x^2 - x(y+z) - xyz + (y+z)yz \\ &= x(x-(y+z)) + ((y+z)-x)yz \\ &= (x-(y+z))(x-yz). \end{aligned}$$

The same simplification can be done with the remaining equations to get the following equivalent system:

$$\begin{cases} (x-(y+z))(x-yz) = 0 \\ (y-(z+x))(y-zx) = 0 \\ (z-(x+y))(z-xy) = 0. \end{cases}$$

It follows that each variable is either the product or the sum of the remaining variables. We will independently consider the following four cases.

Case 1: all variables are the sums. We have $x = y+z$, $y = z+x$, and $z = x+y$. We immediately get that $x = y = z = 0$.

Case 2: two variables are the sums and one is the product. Without loss of generality, we may assume that $x = y+z$, $y = z+x$, and $z = xy$. We immediately get that $z = 0$, and this implies that $x = y = 0$, the solution we already discovered.

Case 3: one variable is the sum and two are the products. Without loss of generality, we may assume that $x = y + z$, $y = zx$, and $z = xy$. Indeed, by symmetry we can recover the whole family of solutions by permuting the solution vector (x, y, z). It follows that $x = y + z = x(z + y) = x^2$ and so $x = 0$ or $x = 1$. The case $x = 0$ leads to the solution we already discovered, namely, $(x, y, z) = (0, 0, 0)$. If $x = 1$, then $y = z = 1/2$. We conclude that there are three additional solutions, namely $(x, y, z) = (1, 1/2, 1/2)$, $(x, y, z) = (1/2, 1, 1/2)$, and $(x, y, z) = (1/2, 1/2, 1)$, by permuting the variables.

Case 4: all numbers are the products. We have $x = yz$, $y = zx$, and $z = xy$. If one variable is equal to zero, then all of them must be zero and we get the particular solution $(x, y, z) = (0, 0, 0)$ one more time. If no variable is equal to zero, then we get that $y = z(yz) = yz^2$, and so $z^2 = 1$. The symmetric arguments give us $y^2 = 1$, $x^2 = 1$, and so all of the variables are either 1 or -1. Since the value of x is determined by the value of y and z ($x = yz$), by considering all possibilities for y and z, we see that either all of the variables x, y, z are equal to 1 or precisely one of them is equal to 1. We directly check that these potential solutions are feasible, giving us the following four additional solutions: $(x, y, z) = (1, 1, 1)$, $(x, y, z) = (1, -1, -1)$, $(x, y, z) = (-1, 1, -1)$, $(x, y, z) = (-1, -1, 1)$.

Putting all of the observations together, we get the following 8 solutions:

$$(x, y, z) \in \Big\{ (0, 0, 0), (1, 1/2, 1/2), (1/2, 1, 1/2), (1/2, 1/2, 1),$$
$$(1, 1, 1), (1, -1, -1), (-1, 1, -1), (-1, -1, 1) \Big\}.$$

Problem 2.2.1. Solve the following system of equations, given that all variables involved are real numbers:

$$\begin{cases} (x + y)^3 &= 8z \\ (y + z)^3 &= 8x \\ (z + x)^3 &= 8y. \end{cases}$$

Solution. Since $f(x) := x^3$ is an increasing function, $x > z$ if and only if $(x+y)^3 > (y+z)^3$. Using this observation, we get from the first equation and the second one that $x = z$. Similarly, from the second equations and the third one we get that $x = y$. It follows that all the variables are equal. We get that $8x = (2x)^3 = 8x^3$, so $0 = 8x^3 - 8x = 8x(x-1)(x+1)$. We conclude that there are three solutions to the given systems of equations:

$$(x, y, z) \in \Big\{ (-1, -1, -1), (0, 0, 0), (1, 1, 1) \Big\}.$$

Problem 2.2.2. Solve the following system of equations, given that all variables involved are real numbers:

$$\begin{cases} x^5 = 5y^3 - 4z \\ y^5 = 5z^3 - 4x \\ z^5 = 5x^3 - 4y. \end{cases}$$

Solution. Since the system is cyclic, without loss of generality, we may assume that $x \geq y$ and $x \geq z$. We will independently consider the following two cases.

Case 1: $y \leq z$. Since $f(x) := x$ and $h(x) := x^5$ are increasing functions, we get that $5x^3 = z^5 + 4y \leq x^5 + 4z = 5y^3$, and so $x \leq y$ as $g(x) := x^3$ is also an increasing function. It follows that $x = y$ and so also $z = x$ since $y \leq z \leq x$.

Case 2: $z \leq y$. We have $5x^3 = z^5 + 4y \leq y^5 + 4x = 5z^3$, so we again get that $x = y = z$.

Since in both cases we get that $x = y = z$, variable x must satisfy the equation $x^5 = 5x^3 - 4x$ that can be rewritten as follows:

$$\begin{aligned} 0 &= x(x^4 - 5x^2 + 4) = x(x^2 - 1)(x^2 - 4) \\ &= (x + 2)(x + 1)x(x - 1)(x - 2). \end{aligned}$$

It follows that $x \in \{-2, -1, 0, 1, 2\}$. It is straightforward to check directly that the following triplets are solutions to our system:

$$(x, y, z) \in \left\{(-2, -2, -2), (-1, -1, -1), (0, 0, 0), (1, 1, 1), (2, 2, 2)\right\}.$$

Problem 2.2.3. Solve the following system of equations, given that all variables involved are *positive* real numbers:

$$\begin{cases} a^3 + b^3 + c^3 = 3d^3 \\ b^4 + c^4 + d^4 = 3a^4 \\ c^5 + d^5 + a^5 = 3b^5. \end{cases}$$

Solution. By the pigeonhole principle, at least one of the variables d, a or b attains the maximum or the minimum value from the set $\{a, b, c, d\}$. We will independently consider the following two cases.

Case 1: a, b **or** d **is the maximum.** Let us first assume that a is the maximum. We will show that $a = b = c = d$. Indeed, if this is not the case, then $3a^4 = b^4 + c^4 + d^4 < 3a^4$ which is clearly not possible. (Let us remark that here we used the fact that $f(x) := x^4$ is an increasing function on $\mathbf{R}_+$.) If b or d is the maximum, then the argument is the same but this time we need to respectively use the third or the first equation and the fact that $g(x) := x^5$ and $h(x) := x^3$ are increasing functions.

Case 2: a, b or d is the minimum. Again, we will show that $a = b = c = d$. As before, the argument is similar for each sub-case and we will present it assuming that a is the minimum. Indeed, if this is not the case that $a = b = c = d$, then $3a^4 = b^4 + c^4 + d^4 > 3a^4$ which is a contradiction.

In all the cases, we get that all the variables are equal and it is easy to check that any 4-tuple $(a, b, c, d) = (t, t, t, t)$ satisfies all the equations for each $t > 0$.

Problem 2.3.1. Solve the following equation

$$(x^4 + 3y^2)\sqrt{|x + 2| + |y|} = 4|xy^2|, \qquad (8.2)$$

provided that $x, y \in \mathbf{R}$.

Solution. Let us first note that if $(x, y) = (-2, 0)$ or $(x, y) = (0, 0)$, then both sides of (8.2) are equal to 0 and so the desired equality holds. We may then assume that $(x, y) \neq (-2, 0)$ and $(x, y) \neq (0, 0)$. In particular, the left hand side of (8.2) is non-zero.

By the arithmetic-geometric mean inequality, we get that

$$x^4 + 3y^2 \geq 4\sqrt[4]{x^4 \cdot y^2 \cdot y^2 \cdot y^2} = 4|xy|\sqrt{|y|},$$

and the equality holds only when $x^2 = |y|$. Since $(x, y) \neq (-2, 0)$, we get that $\sqrt{|x + 2| + |y|} > 0$. It follows that

$$(x^4 + 3y^2)\sqrt{|x + 2| + |y|} \geq 4|xy|\sqrt{|y|}\sqrt{|x + 2| + |y|} \geq 4|xy^2|;$$

the first equality holds when $x^2 = |y|$, and the second equality holds when $x = -2$. Since by our earlier assumption $y \neq 0$ when $x = -2$, we get that (8.2) holds only when $x = -2$ and $|y| = (-2)^2 = 4$, that is, when $(x, y) = (-2, 4)$ or $(x, y) = (-2, -4)$. We conclude that the solution is

$$(x, y) \in \left\{ (-2, -4), (-2, 0), (-2, 4), (0, 0) \right\}.$$

Problem 2.3.2. Solve the following system of equations, given that all variables involved are real numbers:

$$\begin{cases} 3(x^2 + y^2 + z^2) = 1 \\ x^2y^2 + y^2z^2 + z^2x^2 = xyz(x + y + z)^3. \end{cases}$$

Solution. Let us first note that for any $a, b \in \mathbf{R}$ we have $a^2 + b^2 \geq 2ab$, and the equality holds if and only if $a = b$. It follows that

$$\begin{aligned} (x + y + z)^2 &= x^2 + y^2 + z^2 + 2xy + 2yz + 2zx \\ &\leq 3(x^2 + y^2 + z^2) = 1, \end{aligned}$$

and the equality holds if $x = y = z$. Similarly, note that

$$(x^2y^2 + x^2z^2)/2 = x^2(y^2 + z^2)/2 \geq x^2yz,$$

and the equality holds if $x = 0$ or $y = z$. Symmetric arguments give us

$$(y^2 z^2 + y^2 x^2)/2 \;=\; y^2(z^2 + x^2)/2 \;\geq\; y^2 zx$$
$$(z^2 x^2 + z^2 y^2)/2 \;=\; z^2(x^2 + y^2)/2 \;\geq\; z^2 xy.$$

Summing the three inequalities together, we get that

$$x^2 y^2 + y^2 z^2 + z^2 x^2 \;\geq\; x^2 yz + y^2 zx + z^2 xy \;=\; xyz(x + y + z).$$

Moreover, the equality holds if and only if

$$(x = 0 \;\vee\; y = z) \;\wedge\; (y = 0 \;\vee\; z = x) \;\wedge\; (z = 0 \;\vee\; x = y).$$

For this condition to hold, clearly, if no variable is equal to 0, then $x = y = z$. It is also easy to see that it is impossible that only one variable is equal to zero. Hence, we get that the equality holds if $x = y = z$ or at least two of the three variables x, y, z are equal to 0. But this means that

$$x^2 y^2 + y^2 z^2 + z^2 x^2 \;\geq\; xyz(x + y + z) \;\geq\; xyz(x + y + z)^3,$$

where in the last step we use the fact we showed at the very beginning, namely, that $(x + y + z)^2 \leq 1$. More importantly, the condition for the equality remains the same, that is, either $x = y = z$ (since then $(x + y + z)^2 = 1$) or at least two of the three variables x, y, z are equal to 0 (since then both sides of the inequality are equal to 0). We will consider these two cases separately.

Case 1: $x = y = z$. Our system reduces to

$$\begin{cases} 9x^2 = 1 \\ 3x^4 = 27x^6, \end{cases}$$

which gives the following two solutions $(x, y, z) = (1/3, 1/3, 1/3)$ and $(x, y, z) = (-1/3, -1/3, -1/3)$.

Case 2: at least two of the three variables x, y, z are equal to 0. Without loss of generality, we may assume that $y = z = 0$ and other solutions will be obtained by permuting the variables. This time our system reduces to

$$\begin{cases} 3x^2 = 1 \\ 0 = 0. \end{cases}$$

This leads us to another six solutions of the system $(1/\sqrt{3}, 0, 0)$, $(-1/\sqrt{3}, 0, 0)$, $(0, 1/\sqrt{3}, 0)$, $(0, -1/\sqrt{3}, 0)$, $(0, 0, 1/\sqrt{3})$, and $(0, 0, -1/\sqrt{3})$.

Combining the two cases together we conclude that the solution to the system is

$$(x, y, z) \;\in\; \Big\{ (1/3, 1/3, 1/3), (-1/3, -1/3, -1/3), (1/\sqrt{3}, 0, 0), (-1/\sqrt{3}, 0, 0),$$
$$(0, 1/\sqrt{3}, 0), (0, -1/\sqrt{3}, 0), (0, 0, 1/\sqrt{3}), (0, 0, -1/\sqrt{3}) \Big\}.$$

Problem 2.3.3. Solve the following system of equations, given that all variables involved are real numbers:

$$\begin{cases} x^2y + 2 = x + 2yz \\ y^2z + 2 = y + 2zx \\ z^2x + 2 = z + 2xy. \end{cases}$$

Solution. Let us first note that if $x = 0$, then from the third equation we get that $z = 2$ and from the first one that $y = 1/2$. But this contradicts the second equation, and so $x \neq 0$. Symmetric arguments show that also $y \neq 0$ and $z \neq 0$.

Let us use the following substitution: $a = xy \neq 0$, $b = yz \neq 0$, and $c = xz \neq 0$. After multiplying the first equation by y, multiplying the second equation by 2, and adding it together we get $a^2 + 4 = a + 4c$. Symmetric operations give us the following system of equations:

$$\begin{cases} a^2 + 4 = a + 4c \\ b^2 + 4 = b + 4a \\ c^2 + 4 = c + 4b. \end{cases}$$

Due to the symmetry, without loss of generality, we may assume that a is a largest number from a, b, c and then circularly shift the solution, if needed. We get that $a^2 + 4 = a + 4c \leq a + 4a = 5a$, or alternatively that $0 \geq a^2 - 5a + 4 = (a - 1)(a - 4)$. It follows that $a \in [1, 4]$. Since function $f : [1, 4] \to [1, 4]$, $f(a) := (a^2 + 4 - a)/4$ is a bijection and $c = f(a)$, we get that also $c \in [1, 4]$. Similarly, we get that $b = f(c) \in [1, 4]$. More importantly, function $f(x)$ has the property that $1 < f(x) < x$ *unless* $x = 1$ or $x = 4$. Suppose that $1 < a < 4$. Then $1 < c = f(a) < a$, and consequently $1 < b = f(c) < c$ and $1 < a = f(b) < b$. This contradicts the fact that a is a largest value. It follows that the only possible solutions are $(a, b, c) = (1, 1, 1)$ and $(a, b, c) = (4, 4, 4)$.

Going back to the original set of equations, we see that $x = y = z$ and so there are only four potential triples (x, y, z) that satisfy the original system: $(1, 1, 1)$, $(-1, -1, -1)$, $(2, 2, 2)$, and $(-2, -2, -2)$. The last triple does *not* satisfy the original system and so the solution is

$$(x, y, z) \in \left\{ (1, 1, 1), (-1, -1, -1), (2, 2, 2) \right\}.$$

Problem 2.4.1. Let $n \geq 2$ be any natural number. Find the number of sequences $(x_1, x_2, \ldots, x_n)$ of non-negative real variables that satisfy the following system of equations: for $i \in [n]$

$$x_{i+1} + x_i^2 = 4x_i,$$

where $x_{n+1} = x_1$.

Solution. Let us first note that for each $i \in [n]$ we have

$$x_{i+1} = 4x_i - x_i^2 = x_i(4 - x_i).$$

Since x_{i+1} is non-negative, we get that $x_i \in [0, 4]$. As there exist $\alpha_i \in [0, \infty)$ such that $x_i = 4\sin^2(\alpha_i)$, it is natural to use this substitution. In fact, there are many choices for α_i for a given x_i. However, α_1 has a unique value in $[0, \pi/2]$ that satisfies $x_1 = 4\sin^2(\alpha_1) \in [0, 4]$. Let us then note that

$$\begin{aligned} 4\sin^2(\alpha_{i+1}) &= x_{i+1} = x_i(4 - x_i) = 4\sin^2(\alpha_i)(4 - 4\sin^2(\alpha_i)) \\ &= 4(2\sin(\alpha_i)\cos(\alpha_i))^2 = 4\sin^2(2\alpha_i), \end{aligned}$$

and so $x_i = 4\sin^2(2^{i-1}\alpha_1)$. In particular, since $x_1 = x_{n+1}$, we get that $\sin^2(\alpha) = \sin^2(2^n\alpha)$. It follows that $\sin(\alpha) = \sin(2^n\alpha)$ or $\sin(\alpha) = -\sin(2^n\alpha)$.

Let us first deal with a degenerate case, namely, $\alpha = 0$ that yields the following particular solution: $(x_1, x_2, \ldots, x_n) = (0, 0, \ldots, 0)$. If $\sin(\alpha) = \sin(2^n\alpha)$ for some $\alpha > 0$, then there exists $k \in \mathbf{N}$ such that either $\alpha + 2k\pi = 2^n\alpha$ or $-\alpha + \pi + 2k\pi = 2^n\alpha$. On the other hand, if $\sin(\alpha) = -\sin(2^n\alpha)$ for some $\alpha > 0$, then there exists $k \in \mathbf{N}$ such that either $\alpha + \pi + 2k\pi = 2^n\alpha$ or $-\alpha + 2k\pi = 2^n\alpha$. Combining these two observations together, we get that $(2^n + 1)\alpha = k\pi$ for some $k \in \mathbf{N}$ or $(2^n - 1)\alpha = k\pi$ for some $k \in \mathbf{N}$. Since $\alpha \in [0, \pi/2]$, including the degenerate case, we get that

$$\alpha \in \left\{ \frac{k\pi}{2^n + 1} : k \in [2^{n-1}] \right\} \cup \left\{ \frac{k\pi}{2^n - 1} : k \in [2^{n-1} - 1] \right\} \cup \{0\}.$$

Finally, note that these two first sets above are disjoint as $2^n + 1$ and $2^n - 1$ are co-prime for $n \geq 2$. It follows that there are $2^{n-1} + (2^{n-1} - 1) + 1 = 2^n$ solutions to the original set of equations.

Problem 2.4.2. Let $n \in \mathbf{N}$. Find all solutions of the equation

$$|\tan(x)^n - \cot(x)^n| = 2n|\cot(2x)|.$$

Solution. Using double-angle identities, we get that

$$\cot(2x) = \frac{\cos(2x)}{\sin(2x)} = \frac{\cos^2(x) - \sin^2(x)}{2\sin(x)\cos(x)} = \frac{\cot(x) - \tan(x)}{2}.$$

Using the substitution $y = \tan(x)$, our equality can be equivalently rewritten as follows:

$$\left| y^n - \frac{1}{y^n} \right| = n \left| \frac{1}{y} - y \right| = n \left| y - \frac{1}{y} \right|. \tag{8.3}$$

The problem is easy if $n = 1$, as then (8.3) is always satisfied, provided that $y \neq 0$. As a result, any value of $x \in \mathbf{R}$ that falls into the domain of both $\tan(x)$ and $\cot(x)$ functions, satisfies the original equation. In other words, the solution is:

$$x \in \mathbf{R} \setminus \left\{ \frac{k\pi}{2} : k \in \mathbf{Z} \right\}.$$

Suppose then that $n \geq 2$. We will independently consider the following two cases.

Case 1: $y = 1$ **or** $y = -1$. Both sides of (8.3) are equal to zero, which yields the following family of solutions: $x = (2k + 1)\pi/4$ for some $k \in \mathbf{Z}$.

Case 2: $y \neq 1$ **and** $y \neq -1$. This time $|y - 1/y| \neq 0$ and so after dividing both sides of (8.3) by $|y - 1/y|$ we get:

$$n = \left| \frac{\frac{y^{2n}-1}{y^n}}{\frac{y^2-1}{y}} \right| = \left| y^{1-n} \cdot \frac{(y^2 - 1)(y^{2n-2} + y^{2n-4} + \ldots + y^2 + 1)}{(y^2 - 1)} \right|$$

$$= \left| \sum_{i=1}^{n} y^{n+1-2i} \right| = \sum_{i=1}^{n} |y|^{n+1-2i}, \tag{8.4}$$

where the last equality holds because either all the terms y^{n+1-2i} are positive or all are negative. Using the arithmetic-geometric mean inequality, we get that

$$\frac{1}{n} \sum_{i=1}^{n} |y|^{n+1-2i} \geq \left(\prod_{i=1}^{n} |y|^{n+1-2i} \right)^{1/n} = \sqrt[n]{1} = 1,$$

where the equality holds if and only if $|y| = 1$. It follows that (8.4) holds if and only if $|y| = 1$, which are excluded in Case 2 (we already considered them in Case 1).

Combining the two cases together, we conclude that for $n \geq 2$ the solution is:

$$x \in \left\{ \frac{(2k + 1)\pi}{4} : k \in \mathbf{Z} \right\}.$$

Problem 2.4.3. For a given $a \in \mathbf{R}$, let us recursively define the following sequence: $x_0 = \sqrt{3}$ and for all non-negative integers n,

$$x_{n+1} = \frac{1 + ax_n}{a - x_n}.$$

Find all values of a for which the sequence has period equal to 8.

Solution. Let us first note that for any $a \in \mathbf{R}$, there exists $\alpha = \alpha(a) \in \mathbf{R}$ such that $a = \cot(\alpha)$. In particular, for convenience we set $\alpha = \cot^{-1}(a) \in (0, \pi)$, where $\cot^{-1}(\cdot)$ is the inverse of the cotangent function. (Note that none of the six trigonometric functions are one-to-one. They are restricted to their principal branch in order to have inverse functions. For cotangent the principal branch is $(0, \pi)$.) Let us now recursively define another sequence: $y_0 = \pi/6$ and for all non-negative integers n, $y_{n+1} = y_n - \alpha$. Of course, it means that $y_n = y_0 - n\alpha = \pi/6 - n\alpha$ for $n \in \mathbf{N}$. We will prove by induction on n that $x_n = \cot(y_n)$ for all $n \in \mathbf{N} \cup \{0\}$.

The base case ($n = 1$) is easy: $\cot(y_0) = \cot(\pi/6) = \sqrt{3} = x_0$. For the inductive step, assume that $x_n = \cot(y_n)$ for some $n \in \mathbf{N} \cup \{0\}$. Our goal is to show that $x_{n+1} = \cot(y_{n+1})$ but this follows almost immediately from the following identity:

$$\cot(y - x) = \frac{1 + \cot(x)\cot(y)}{\cot(x) - \cot(y)}.$$

Indeed, note that

$$\cot(y_{n+1}) = \cot(y_n - \alpha) = \frac{1 + \cot(\alpha)\cot(y_n)}{\cot(\alpha) - \cot(y_n)} = \frac{1 + a x_n}{a - x_n} = x_{n+1}.$$

The proof by induction is finished.

With this convenient representation at hand, let us come back to our problem of finding values of a that yield period equal to 8. (Not necessarily the fundamental period has to be equal to 8.) For the sequence $x_n = y_n$ to have period of 8, we must have $8\alpha = k\pi$ for some $k \in \mathbf{N}$, as the cotangent function has period of π. Since $\alpha \in (0, \pi)$, in fact, $k \in [7]$. It follows that all the possible values of a that satisfy the desired condition of the problem are of the form $a = \cot(k\pi/8)$ for some $k \in [7]$.

Suppose that $a = \cot(k\pi/8)$ for some $k \in [7]$. We get that for each $n \in \mathbf{N} \cup \{0\}$,

$$x_n = \cot\left(\frac{\pi}{6} - n\alpha(a)\right) = \cot\left(\frac{\pi}{6} - \frac{nk\pi}{8}\right) = \cot\left(\pi \cdot \frac{4 - 3nk}{24}\right).$$

Since $4 - 3nk$ is not divisible by 3, $(4 - 3nk)/24$ is never an integer. As a result, $\pi(4 - 3nk)/24$ always belongs to the domain of the cotangent function and so for all seven identified values of a the sequence is properly defined.

Note that in the problem we required the function to have a period of 8 but its fundamental period can be smaller. In particular we note that for $k \in \{2, 6\}$ the fundamental period of the sequence is 4 and for $k = 4$ the fundamental period of the sequence is 2.

Problem 2.5.1. For a given $a \in \mathbf{R}$, consider the following system of equations:

$$\begin{cases} x + y^2 + z^2 = a \\ x^2 + y + z^2 = a \\ x^2 + y^2 + z = a. \end{cases}$$

Find the number of real solutions (x, y, z) of this system as a function of a.

Solution. By subtracting the first equation from the second one, we get that

$$0 = x^2 - x + y - y^2 = (x - y)(x + y - 1),$$

which implies that either $x = y$ or $x = 1 - y$. Symmetric arguments may be applied to the remaining two pairs of equations. Since $1 - (1 - t) = t$, we conclude that all the solutions must have one of the following four forms: $(x, y, z) = (t, t, t)$, $(x, y, z) = (t, t, 1 - t)$, $(x, y, z) = (t, 1 - t, t)$, or $(x, y, z) = (1 - t, t, t)$, where $t \in \mathbf{R}$. Clearly, if $t = 1/2$, then only one solution should be counted, namely, $(x, y, z) = (1/2, 1/2, 1/2)$. It means that we need to be extra careful with the case $a = (1/2) + (1/2)^2 + (1/2)^2 = 1$. On the other hand, if $t \neq 1/2$, then all solutions are distinct. More importantly, by symmetry, the last three forms are associated in the following sense: if one of them is a solution, then so are the remaining two. We will independently consider the two cases.

Case 1: there is a solution of the form (t, t, t) **for some** $t \in \mathbf{R}$. It follows that $2t^2 + t - a = 0$. Since the discriminant is equal to $\Delta = 1 - 8a$, we conclude that there are no solutions of this form if $a < -1/8$ (that is, $\Delta < 0$), precisely one solution if $a = -1/8$, and two solutions if $a > -1/8$.

Case 2: there is a solution of the form $(1 - t, t, t)$ **for some** $t \in \mathbf{R}$. This time we get that $2t^2 - t + (1 - a) = 0$ and so $\Delta = 1 - 8(1 - a) = 8a - 7$. It follows that there are no solutions of this form if $a < 7/8$, precisely one solution if $a = 7/8$, and two solutions if $a > 7/8$.

Let us now come back to the special case $a = 1$ that requires more attention. If $a = 1$, then we have two solutions of the form (t, t, t) for some $t \in \mathbf{R}$, namely $(1/2, 1/2, 1/2)$ and $(-1, -1, -1)$. Moreover, there are two solutions of the form $(1 - t, t, t)$ for some $t \in \mathbf{R}$, again including $(1/2, 1/2, 1/2)$ which we do not want to count. The other one, namely $(1, 0, 0)$, yields another two solutions of the form $(t, 1 - t, t)$ and $(t, t, 1 - t)$. So there are 5 solutions for this special case:

$$(x, y, z) \in \left\{ (-1, -1, -1), (1/2, 1/2, 1/2), (1, 0, 0), (0, 1, 0), (0, 0, 1) \right\}.$$

Let us summarize our observations. The number of the solutions of our system of equations is equal to:

- $0 = 0 + 3 \cdot 0$, provided $a < -1/8$;
- $1 = 1 + 3 \cdot 0$, provided $a = -1/8$;
- $2 = 2 + 3 \cdot 0$, provided $-1/8 < a < -7/8$;
- $5 = 2 + 3 \cdot 1$, provided $a = 7/8$ or $a = 1$;
- $8 = 2 + 3 \cdot 2$, provided $a > 7/8$ and $a \neq 1$.

Problem 2.5.2. Solve the following system of equations, given that all variables involved are positive real numbers:

$$(x^{2010} - 1)(y^{2009} - 1) = (x^{2009} - 1)(y^{2010} - 1).$$

Solution. Let us first note that if $x = 1$ or $y = 1$, then the equality trivially holds. We will assume then that they are *not* equal to 1.

Let us first re-write the equation as follows:

$$x^{2010}(y^{2009} - 1) - y^{2009} = (x^{2009} - 1)y^{2010} - x^{2009}.$$

After adding $x^{2009}y^{2009}$ to both sides of the equation, we get that

$$x^{2010}(y^{2009} - 1) + (x^{2009} - 1)y^{2009} = (x^{2009} - 1)y^{2010} + x^{2009}(y^{2009} - 1),$$

that can be equivalently written as follows:

$$x^{2009}(x - 1)(y^{2009} - 1) = (x^{2009} - 1)y^{2009}(y - 1).$$

Now, after dividing both sides by $(x - 1)(y - 1)$, we get that

$$x^{2009} \sum_{i=0}^{2008} y^i = y^{2009} \sum_{i=0}^{2008} x^i.$$

(Recall that we assumed that $x \neq 1$ and $y \neq 1$.) Since x and y are both non-zero (in fact, they are both positive real numbers), this equation can be equivalently rewritten as follows:

$$\sum_{i=1}^{2009} \frac{1}{y^i} = \sum_{i=1}^{2009} \frac{1}{x^i}.$$

The final observation is that for each $i \in [2009]$, $f(t) := 1/t^i$ is a decreasing function on the domain $(0, \infty)$. It follows that $g(t) := \sum_{i=1}^{2009} 1/t^i$ is also a decreasing function on that domain. Since x and y are positive, the equation holds only if $x = y$.

We conclude that all solutions of the equation are of the form (t, t), $(1, t)$, or $(t, 1)$, where $t \in (0, \infty)$. It is straightforward to directly check that they indeed satisfy the original system.

Problem 2.5.3. Fix an integer $n \geq 2$, and consider the following system of n equations: for $i \in [n]$

$$x_{i+1}^2 + x_i^2 + 50 = 12x_{i+1} + 16x_i.$$

(As usual, we use the convention that $x_{n+1} = x_1$.) Find the number of solutions of this system, given that all variables involved are integers.

Solution. Let us start with rewriting the system as follows: for $i \in [n]$

$$(x_i - 8)^2 + (x_{i+1} - 6)^2 = 50.$$

Since both $x_i - 8$ and $x_{i+1} - 6$ are integers, we need to decompose 50 into a sum of two squares of integers. The only decompositions involving natural numbers are $1^2 + 7^2 = 50$ and $5^2 + 5^2 = 50$. It follows that for each $i \in [n]$,

$$(x_i, x_{i+1}) \in S := \Big\{ (1,5), (1,7), (3,1), (3,11), (7,-1), (7,13),$$
$$(9,-1), (9,13), (13,1), (13,11), (15,5), (15,7) \Big\}.$$

We will show now that there is no $i \in [n]$ for which $(x_i, x_{i+1}) = (1,5)$. Indeed, for a contradiction, suppose that $(x_i, x_{i+1}) = (1,5)$ for some $i \in [n]$. Then we get that $(x_{i+1}, x_{i+2}) = (5, x_{i+2}) \in S$, but there is no pair in S with the first coordinate equal to 5 (here we extended our convention and use $x_{n+2} = x_2$). We get the desired contradiction and so the pair $(1,5)$ is eliminated from the set of potential pairs. Using similar arguments one can eliminate more pairs to get that for each $i \in [n]$,

$$(x_i, x_{i+1}) \in T := \Big\{ (1,7), (7,13), (13,1) \Big\}.$$

Our next observation is that the first pair, pair (x_1, x_2), uniquely determines the sequence $(x_1, x_2, \ldots, x_n, x_{n+1})$. Moreover, the numbers form the cycle of length three. As a result, since $x_1 = x_{n+1}$, we get that the solution exists if and only if $3 \mid n$. We conclude that the system has 3 solutions if $3 \mid n$, and no solution otherwise.

Problem 2.6.1. Let x_1 be any positive real number, and for each $n \in \mathbf{N}$ let

$$x_{n+1} = x_n + \frac{1}{x_n^2}.$$

Prove that $x_n / \sqrt[3]{n}$ has a limit and then find it.

Solution. It is clear that $x_n > 0$ for all $n \in \mathbf{N}$. (One can formally prove it by induction.) It will be convenient to use the following substitution: for each $n \in \mathbf{N}$, $y_n := x_n^3 > 0$. It follows that

$$y_{n+1}^{1/3} = y_n^{1/3} + \frac{1}{y_n^{2/3}}.$$

By raising both sides to the power of 3, we get

$$y_{n+1} = y_n + 3 + \frac{3}{y_n} + \frac{1}{y_n^2}.$$

As a result, by unrolling the recursion all the way to y_1, we get that

$$y_n = y_1 + \sum_{i=1}^{n-1} \left(3 + \frac{3}{y_i} + \frac{1}{y_i^2} \right) = 3(n-1) + y_1 + \sum_{i=1}^{n-1} \left(\frac{3}{y_i} + \frac{1}{y_i^2} \right).$$

In particular, we get that $y_i > 3(i-1)$ for all $i \in \mathbf{N}$. After switching back to x_n, we get the following bounds that hold for all $n \in \mathbf{N}$:

$$a_n := \sqrt[3]{3 - \frac{3}{n}} < \frac{x_n}{\sqrt[3]{n}} = \sqrt[3]{\frac{y_n}{n}} < b_n\,,$$

where

$$b_n := \sqrt[3]{3 + \frac{-3 + x_1^3 + 3/x_1^3 + 1/x_1^6}{n} + \frac{1}{n}\sum_{i=2}^{n-1}\left(\frac{3}{3(i-1)} + \frac{1}{(3(i-1))^2}\right)}.$$

It is clear that $a_n \to \sqrt[3]{3}$ as $n \to \infty$. Hence, by sandwiching the sequence $(x_n/\sqrt[3]{n})_{n\in\mathbf{N}}$ between $(a_n)_{n\in\mathbf{N}}$ and $(b_n)_{n\in\mathbf{N}}$, to show that $x_n/\sqrt[3]{n} \to \sqrt[3]{3}$, it is enough to show that $b_n \to \sqrt[3]{3}$ (see the squeeze theorem). To that end, it is needed to show that

$$\frac{1}{n}\sum_{i=2}^{n-1}\left(\frac{3}{3(i-1)} + \frac{1}{(3(i-1))^2}\right) = \frac{1}{n}\sum_{i=1}^{n-2}\frac{1}{i} + \frac{1}{9n}\sum_{i=1}^{n-2}\frac{1}{i^2} \to 0.$$

We will independently show that $\sum_{i=1}^{\infty} 1/i^2$ is finite (and so the second term tends to 0) and that $H_n = \sum_{i=1}^{n} 1/i \leq \ln(n)$ (and so the first term tends to 0 as well—clearly, $\ln(n)$ tends to infinity much slower than n).

For the first task, let us note that for $n \geq 2$ we have

$$\sum_{i=1}^{n}\frac{1}{i^2} < 1 + \sum_{i=2}^{n}\frac{1}{i(i-1)} = 1 + \sum_{i=2}^{n}\frac{i-(i-1)}{i(i-1)}$$

$$= 1 + \sum_{i=2}^{n}\left(\frac{1}{i-1} - \frac{1}{i}\right) = 2 - \frac{1}{n}.$$

It follows that $\sum_{i=1}^{\infty} 1/i^2 = \lim_{n\to\infty}\sum_{i=1}^{n} 1/i^2$ is smaller than or equal to 2. In fact, $\sum_{i=1}^{\infty} 1/i^2 = \pi^2/6 \approx 1.6449$.

For the second task, let us recall that in Section 1.5 we showed that $e < (1 + 1/(i-1))^i$ and so $e^{1/i} < i/(i-1)$. It follows that $1/i < \ln(i) - \ln(i-1)$. Since $\ln(1) = 0$, we get that for $n \geq 2$,

$$H_n = \sum_{i=1}^{n}\frac{1}{i} < 1 + \sum_{i=2}^{n}\left(\ln(i) - \ln(i-1)\right) = 1 + \ln(n).$$

This bound is quite good as one can show that $H_n > \ln(n+1)$ (see the solution to Problem 4.6.2) and asymptotically $H_n = \ln(n) + \gamma + o(1)$, where $\gamma \approx 0.577216$ is the Euler-Mascheroni constant.

Let us mention about an alternative solution to this problem that uses the **Stolz–Cesàro theorem**. This theorem is a criterion for proving the convergence of a sequence and can be viewed as a generalization of a **L'Hôpital's rule**. Suppose that $(a_n)_{n \in \mathbf{N}}$ and $(b_n)_{n \in \mathbf{N}}$ are sequences of real numbers such that $(a_n)_{n \in \mathbf{N}}$ is increasing and tends to ∞. If the limit of $(b_{n+1} - b_n)/(a_{n+1} - a_n)$ exists, then the limit of b_n/a_n also exists and they are equal. (Let us note that the converse of this implication is not true in general.)

Let us now come back to our problem. We observe that $x_n \to \infty$ as $n \to \infty$. Then, we deal with the limit of x_n^3/n using the Stolz–Cesàro theorem:

$$\lim_{n \to \infty} \frac{x_n^3}{n} = \lim_{n \to \infty} \frac{x_{n+1}^3 - x_n^3}{(n+1) - n} = \lim_{n \to \infty} \frac{(x_n + 1/x_n^2)^3 - x_n^3}{(n+1) - n}$$

$$= \lim_{n \to \infty} \left(3 + \frac{3}{x_n^3} + \frac{1}{x_n^6} \right) = 3.$$

This immediately gives us that $x_n/\sqrt[3]{n} \to \sqrt[3]{3}$.

Problem 2.6.2. Consider the following sequence defined recursively: $a_1 = 4$ and for each $n \in \mathbf{N}$, let $a_{n+1} = a_n(a_n - 1)$. Moreover, for each $n \in \mathbf{N}$, let $b_n = \log_2(a_n)$ and $c_n = n - \log_2(b_n)$. Prove that c_n is bounded.

Solution. It is clear that sequence $(a_n)_{n \in \mathbf{N}}$ is increasing and $a_n \geq 4$ for each $n \in \mathbf{N}$. We will upper and lower bound term a_{n+1} in terms of a_n. An upper bound is easy: $a_{n+1} = a_n(a_n - 1) < a_n^2$. On the other hand, since $a_n \geq 4$, we get that $a_{n+1} = a_n(a_n - 1) = a_n^2/2 + a_n^2/2 - a_n = a_n^2/2 + a_n(a_n/2 - 1) \geq a_n^2/2$. It follows that

$$a_n^2/2 \leq a_{n+1} < a_n^2.$$

Since $a_1 = 4 = 2^2$, we get that $2^{2^{n-1}+1} \leq a_n < 2^{2^n}$. Aiming for a simpler argument, we will use slightly weaker bounds, namely, $2^{2^{n-1}} < a_n < 2^{2^n}$. It follows that $2^{n-1} < b_n < 2^n$, and so $0 < c_n < 1$.

Problem 2.6.3. You are given two numbers $a, b \in \mathbf{R}$. Let $x_1 = a$, $x_2 = b$, and for each $n \in \mathbf{N}$ let $x_{n+2} = x_{n+1} + x_n$. Show that there exist $a, b \in \mathbf{R}$, $a \neq b$, for which there are at least 2,000 distinct pairs (k, ℓ), $k < \ell$, such that $x_k = x_\ell$. On the other hand, the number of such pairs is finite even if $a = b$, unless $a = b = 0$.

Solution. Consider any pair $a, b \in \mathbf{R}$ that is different than $a = b = 0$. In particular, it is allowed that $a = b \in \mathbf{R} \setminus \{0\}$. We will first show that all but possibly a finite number of terms of the sequence are unique. This proves the second part of the problem. We will independently consider the following three cases.

Case 1: $x_i > 0$ and $x_{i+1} > 0$ for some $i \in \mathbf{N}$. It is clear that the sequence $(x_{i+n})_{n \in \mathbf{N}}$ is strictly increasing and so indeed all but finitely many terms of the sequence $(x_n)_{n \in \mathbf{N}}$ are unique.

Case 2: $x_i < 0$ and $x_{i+1} < 0$ **for some** $i \in \mathbf{N}$. We get the same conclusion as before since the sequence $(x_{i+n})_{n \in \mathbf{N}}$ is strictly decreasing.

Before we move to the last case, let us suppose that $x_i = 0$ for some $i \in \mathbf{N}$. If there are more terms x_i that are equal to 0, then we concentrate on the first one. Since the case $a = b = 0$ is excluded, we get that $x_{i+1} \neq 0$. Indeed, if $0 = x_{i+1} = x_i + x_{i-1} = 0 + x_{i-1}$ for some $i \geq 2$, then $x_{i-1} = 0$ which gives us a contradiction (x_i is the first term equal to 0). It follows that $x_{i+2} = x_{i+1} + x_i = x_{i+1}$ and we arrive in either Case 1 or Case 2. Hence, without loss of generality, we may assume that $x_i \neq 0$ for all $i \in \mathbf{N}$ and it remains to investigate oscillating sequences.

Case 3: the sequence $(x_n)_{n \in \mathbf{N}}$ **oscillates between positive and negative values.** Suppose that for some $i \in \mathbf{N}$ we have $x_i > 0$ and $x_{i+1} < 0$. But then $0 < x_{i+2} = x_{i+1} + x_i < x_i$. It follows that the sequence $(x_{i+2n})_{n \in \mathbf{N}}$ is a strictly decreasing sequence of positive numbers. Similarly, $0 > x_{i+3} = x_{i+1} + x_{i+2} > x_{i+1}$, and so the sequence $(x_{i+1+2n})_{n \in \mathbf{N}}$ is a strictly increasing sequence of negative numbers. As a result, all but finitely many terms of the sequence $(x_n)_{n \in \mathbf{N}}$ are unique. (In fact, with a slightly more delicate argument one can argue that all of them are unique.)

Before we move to the proof of the first part of the problem, let us make one remark. One can show (by induction on n) that for each $n \in \mathbf{N}$,

$$x_n = A \left(\frac{1 + \sqrt{5}}{2} \right)^n + B \left(\frac{1 - \sqrt{5}}{2} \right)^n$$

for some carefully chosen $A = A(a, b)$ and $B = B(a, b)$. (Constants A and B can be determined by considering $x_1 = a$ and $x_2 = b$.) Note that $A = B = 0$ only if $a = b = 0$ and in which case we have infinitely many pairs (k, ℓ). Otherwise, at some point the sequence $(x_n)_{n \in \mathbf{N}}$ must be strictly increasing or strictly decreasing, and so the number of pairs (k, ℓ) we are interested in is finite. This also shows that Case 3 is impossible.

Let us now come back to our problem. We will show that it is possible to select a and b such that there are at least 2,000 pairs (k, ℓ), $k < \ell$, with $x_k = x_\ell$. In order to see this, let us consider the classic Fibonacci sequence where $x_1 = x_2 = 1$. (See Section 4.9 for more on that sequence.) However, we will extend it to negative indices. Since we want to preserve that $x_{n+2} = x_{n+1} + x_n$ for each $n \in \mathbf{Z}$, we get that

$$x_n = x_{n+2} - x_{n+1}. \tag{8.5}$$

In particular, $x_0 = x_2 - x_1 = 1 - 1 = 0$, $x_{-1} = x_1 - x_0 = 1 - 0 = 1$, and $x_{-2} = x_0 - x_{-1} = 0 - 1 = -1$.

We will show by (strong) induction on i that for each $i \in \mathbf{N}$,

$$x_{-i} = -(-1)^i x_i. \tag{8.6}$$

The base case ($i = 1$ and $i = 2$) clearly holds: $x_{-1} = 1 = -(-1)^1 x_1$ and $x_{-2} = -1 = -(-1)^2 x_2$. For the inductive step, suppose that for some $i \in \mathbf{N}$, we have $x_{-i} = -(-1)^i x_i$ and $x_{-(i+1)} = -(-1)^{i+1} x_{i+1}$. Our goal is to show that $x_{-(i+2)} = -(-1)^{i+2} x_{i+2}$. It follows immediately from (8.5) that

$$
\begin{aligned}
x_{-(i+2)} &= x_{-i} - x_{-(i+1)} = -(-1)^i x_i - \left(-(-1)^{i+1} x_{i+1} \right) \\
&= -(-1)^{i+2} x_i - (-1)^{i+2} x_{i+1} = -(-1)^{i+2}(x_i + x_{i+1}) \\
&= -(-1)^{i+2} x_{i+2},
\end{aligned}
$$

as required. This finishes the proof of (8.6).

The rest of the proof is straightforward. It follows from (8.6) that for each $k \in \mathbf{N}$, $x_{2k+1} = x_{-2k-1}$ (by applying it to $i = 2k + 1$). This means that it is enough to "shift" the Fibonacci sequence, that is, take $a := x_{-3999}$ and $b := x_{-3998}$ to generate the desired sequence.

Problem 2.7.1. Find the number of infinite sequences $(a_i)_{i \in \mathbf{N}}$, such that $a_i \in \{-1, 1\}$ for all $i \in \mathbf{N}$, $a_{mn} = a_m a_n$ for all $m, n \in \mathbf{N}$, and each consecutive triple contains at least one 1 and one -1.

Solution. Let us first note that for any $n \in \mathbf{N}$ we have $a_{n^2} = a_n a_n = (a_n)^2 = 1$. In particular, $a_1 = a_4 = a_9 = 1$. We will show that $a_2 = -1$. For a contradiction, suppose that $a_2 = 1$. Then, $a_3 = -1$ (as the triple a_1, a_2, a_3 has to contain at least one -1), $a_6 = a_2 a_3 = -1$, $a_8 = a_2 a_4 = 1$, $a_7 = -1$ (as the triple a_7, a_8, a_9 has to contain at least one -1), $a_5 = 1$ (as the triple a_5, a_6, a_7 has to contain at least one 1), and $a_{10} = a_2 a_5 = 1$. But then the triple a_8, a_9, a_{10} does *not* contain any -1, and we get the desired contradiction. It follows that $a_2 = -1$.

Suppose that $a_k = a_{k+1} = x \in \{-1, 1\}$ for some $k \in \mathbf{N}$, $k \geq 2$. Using the property for consecutive triples, we get that $a_{k-1} = a_{k+2} = -x$. Then, as a result, $a_{2k-2} = a_2 a_{k-1} = -(-x) = x$ and $a_{2k+4} = a_2 a_{k+2} = x$. Similar argument gives us $a_{2k} = a_2 a_{k+2} = -x$. Using the property for consecutive triples one more time, we get that $x_{2k+1} = x$. It follows that $a_{2k-2} = a_{2k+1} = a_{2k+4}(= x)$. If $k = 3\ell + 1$ for some $\ell \in \mathbf{N}$, then we would get that $a_{6\ell} = a_{6\ell+3} = a_{6\ell+6}$ but, as a result, also $a_{2\ell} = a_{2\ell+1} = a_{2\ell+2}$, which is not possible (the corresponding triple does *not* satisfy the desired property). It follows that the following property is satisfied:

$$ a_{3\ell+1} = -a_{3\ell+2} \quad \text{for all } \ell \in \mathbf{N}. \tag{8.7} $$

We will prove, by (strong) induction on ℓ, that for all $\ell \in \mathbf{N} \cup \{0\}$, $a_{3\ell+1} = 1$ and $a_{3\ell+2} = -1$. The base case ($\ell = 0$) holds: $a_1 = 1$ and $a_2 = -1$. For the inductive step, suppose that $a_{3\ell+1} = 1$ and $a_{3\ell+2} = -1$ for all non-negative integers that are less than $\ell_0 \in \mathbf{N}$. Our goal is to show that $a_{3\ell_0+1} = 1$ and $a_{3\ell_0+2} = -1$. We will independently investigate two cases, depending on the parity of ℓ_0.

Case 1: $\ell_0 = 2m$ **for some** $m \in \mathbf{N}$. We get $a_{3\ell_0+2} = a_{6m+2} = a_2 a_{3m+1} = -a_{3m+1} = -1$ by inductive hypothesis, as $m = \ell_0/2 < \ell_0$. It follows that $a_{3\ell_0+1} = -a_{3\ell_0+2} = 1$ by (8.7).

Case 2: $\ell_0 = 2m + 1$ **for some** $m \in \mathbf{N} \cup \{0\}$. We get $a_{3\ell_0+1} = a_{6m+4} = a_2 a_{3m+2} = -a_{3m+2} = 1$ by inductive hypothesis, as $m = (\ell_0 - 1)/2 < \ell_0$. Again, by (8.7), it follows that $a_{3\ell_0+2} = -a_{3\ell_0+1} = -1$.

We showed above that the values of the sequence $(a_n)_{n \in \mathbf{N}}$ are determined when $n \equiv 1$ or $n \equiv 2 \pmod{3}$. We will now show that the value of a_3 uniquely determines the whole sequence. Indeed, note that any natural number n (not necessarily divisible by 3) is uniquely represented as follows: $n = 3^p(3q + r)$, where $p, q \in \mathbf{N} \cup \{0\}$ and $r \in \{1, 2\}$. It follows that

$$a_n = a_{3^p(3q+r)} = (a_3)^p a_{3q+r} = (a_3)^p (-1)^{r+1}.$$

Therefore, indeed, by fixing $a_3 \in \{-1, 1\}$ we uniquely define two possible sequences.

It remains to show that both sequences satisfy the two desired properties. Property (8.7) guarantees that all consecutive triples contain at least one 1 and one -1. In order to show that $a_{mn} = a_m a_n$ for all $m, n \in \mathbf{N}$, let us concentrate on any $m, n \in \mathbf{N}$. As mentioned above, we may uniquely represent m and n as follows: $m = 3^{p_1}(3q_1 + r_1)$ and $n = 3^{p_2}(3q_2 + r_2)$, where $p_1, p_2, q_1, q_2 \in \mathbf{N} \cup \{0\}$ and $r_1, r_2 \in \{1, 2\}$. It follows that

$$mn = 3^{p_1 + p_2}(9q_1 q_2 + 3(q_1 r_2 + q_2 r_1) + r_1 r_2).$$

As a result, if $r_1 = r_2$, then $mn = 3^{p_1 + p_2}(3q_3 + 1)$ for some $q_3 \in \mathbf{N} \cup \{0\}$ and so

$$\begin{aligned} a_{mn} &= (a_3)^{p_1 + p_2}(-1)^2 = (a_3)^{p_1 + p_2}(-1)^2(-1)^{r_1 + r_2} \\ &= (a_3)^{p_1}(-1)^{r_1+1}(a_3)^{p_2}(-1)^{r_2+1} = a_m a_n, \end{aligned}$$

as desired. Similarly, if $r_1 \neq r_2$, then $mn = 3^{p_1 + p_2}(3q_3 + 2)$ for some $q_3 \in \mathbf{N} \cup \{0\}$ and so

$$\begin{aligned} a_{mn} &= (a_3)^{p_1 + p_2}(-1)^3 = (a_3)^{p_1 + p_2}(-1)^3(-1)^{r_1 + r_2 - 1} \\ &= (a_3)^{p_1}(-1)^{r_1+1}(a_3)^{p_2}(-1)^{r_2+1} = a_m a_n. \end{aligned}$$

The two desired properties are satisfied and the proof is finished.

Problem 2.7.2. Let us fix any real number a. We recursively define sequence $(a_n)_{n \in \mathbf{N}}$ as follows: let $a_1 = a$ and for each $n \in \mathbf{N}$, let $a_{n+1} = (a_n - 1/a_n)/2$ if $a_n \neq 0$ and $a_{n+1} = 0$ if $a_n = 0$. Prove that this sequence has infinitely many non-positive elements and infinitely many non-negative elements.

Solution. Let us first note that if $a_N = 0$ for some $N \in \mathbf{N}$, then $a_n = 0$ for all $n \geq N$ and so the desired property is trivially satisfied. Therefore, we may assume that $a_n \neq 0$ for all $n \in \mathbf{N}$. It follows that $a_{n+1} = (a_n - 1/a_n)/2$ for all $n \in \mathbf{N}$ which can be rewritten as $2a_{n+1}a_n = a_n^2 - 1$, or equivalently as $a_{n+1}^2 + 1 = (a_{n+1} - a_n)^2$. This implies that for each $n \in \mathbf{N}$, $|a_{n+1} - a_n| > 1$.

Suppose that $a_n > 0$ for some $n \in \mathbf{N}$. We get that $a_{n+1} = (a_n - 1/a_n)/2 < a_n/2 < a_n$. Combining this with the previous observation we conclude that $a_{n+1} < a_n - 1$. As a result, for some $k > n$ we get that $a_k < 0$ (for example, it is easy to see that $k - n \leq \lceil a_n \rceil$); recall that we had assumed that $a_n \neq 0$ for all $n \in \mathbf{N}$. Similarly, if $a_n < 0$ for some $n \in \mathbf{N}$, then $a_{n+1} = (a_n - 1/a_n)/2 > a_n/2 > a_n$, and so $a_{n+1} > a_n + 1$. It follows that $a_k > 0$ for some $k > n$. The conclusion is that, regardless of the choice of $a \in \mathbf{R}$, the sequence $(a_n)_{n \in \mathbf{N}}$ must either reach zero (and stay zero forever), or oscillate infinitely many times between positive and negative values, as required.

Problem 2.7.3. Let n be any natural number such that $n \geq 3$. Find all sequences of real numbers $(x_1, x_2, \ldots, x_n)$ that satisfy the following conditions:

$$\sum_{i=1}^{n} x_i = n \quad \text{and} \quad \sum_{i=1}^{n} (x_{i-1} - x_i + x_{i+1})^2 = n,$$

where we set $x_0 = x_n$ and $x_{n+1} = x_1$.

Solution. Let us first note that

$$\sum_{i=1}^{n} (x_{i-1} - x_i + x_{i+1} - 1)^2$$

$$= \sum_{i=1}^{n} \left((x_{i-1} - x_i + x_{i+1})^2 - 2(x_{i-1} - x_i + x_{i+1}) + 1 \right)$$

$$= \sum_{i=1}^{n} (x_{i-1} - x_i + x_{i+1})^2 - 2 \sum_{i=1}^{n} (x_{i-1} - x_i + x_{i+1}) + n$$

$$= \sum_{i=1}^{n} (x_{i-1} - x_i + x_{i+1})^2 - 2 \left(\sum_{i=1}^{n} x_{i-1} - \sum_{i=1}^{n} x_i + \sum_{i=1}^{n} x_{i+1} \right) + n$$

$$= n - 2 \left(n - n + n \right) + n = 0.$$

(Recall our convention: $x_0 = x_n$ and $x_{n+1} = x_1$.) This implies that $x_{i-1} - x_i + x_{i+1} - 1 = 0$ for all $i \in [n]$, or equivalently that $x_{i+1} = 1 + x_i - x_{i-1}$. It follows that $x_{i+2} = 1 + x_{i+1} - x_i = 1 + (1 + x_i - x_{i-1}) - x_i = 2 - x_{i-1}$. As a result, $x_{i+5} = 2 - x_{i+2} = 2 - (2 - x_{i-1}) = x_{i-1}$. It follows that $x_{i+6} = x_i$ for all $i \in [n - 6]$. We will independently consider the following two cases depending whether n is divisible by 6 or not.

Case 1: 6 divides n. We will show that the values of x_1 and x_2 uniquely determine the whole sequence. Indeed, once x_1 and x_2 are fixed, we get that $x_3 = 1 + x_2 - x_1$, $x_4 = 1 + x_3 - x_2 = 2 - x_1$, $x_5 = 2 - x_2$, and $x_6 = 1 + x_5 - x_4 = 1 - x_2 + x_1$. Since $x_{i+6} = x_i$ for all $i \in [n - 6]$, the remaining values are determined. It is straightforward to check that $\sum_{i=1}^{6} x_i = 6$ and so $\sum_{i=1}^{n} x_i = n$, as desired. The second condition is forced by the fact that $x_{i-1} - x_i + x_{i+1} = 1$ for all $i \in [n]$. We conclude that in this case one can fix any values of x_1 and x_2, and these two values determine the sequence that satisfy the desired properties. These are the only sequences.

Case 2: 6 does not divide n. As before, we fix the values of x_1 and x_2. Arguing as before, we determine the remaining values of the sequence. However, since 6 does not divide n, we obtain additional constraints for x_1 and x_2. Depending on the remainder of n when divided by 6, we get one of the following conditions:

- $n \equiv 1 \pmod 6$: $\begin{cases} x_1 = x_2 \\ x_2 = 1 + x_2 - x_1 \end{cases}$

- $n \equiv 2 \pmod 6$: $\begin{cases} x_1 = 1 + x_2 - x_1 \\ x_2 = 2 - x_1 \end{cases}$

- $n \equiv 3 \pmod 6$: $\begin{cases} x_1 = 2 - x_1 \\ x_2 = 2 - x_2 \end{cases}$

- $n \equiv 4 \pmod 6$: $\begin{cases} x_1 = 2 - x_2 \\ x_2 = 1 - x_2 + x_1 \end{cases}$

- $n \equiv 5 \pmod 6$: $\begin{cases} x_1 = 1 - x_2 + x_1 \\ x_2 = x_1 \end{cases}$

In each case, the only solution is $x_1 = x_2 = 1$ and then all other values are also equal to 1. As a result, if 6 does not divide n, the only solution is a constant sequence, namely, $x_i = 1$ for all $i \in [n]$.

8.3   Functions, Polynomials, and Functional Equations

Problem 3.1.1. Find all sets of six real numbers $a_1, a_2, a_3, b_1, b_2, b_3$ with the property that for all $i \in [3]$, a_{i+1} and b_{i+1} are two different solutions of the equation $x^2 + a_i x + b_i = 0$ (here we let $a_4 = a_1$ and $b_4 = b_1$).

Solution. Let us first observe that if $b_{i+1} = 0$ for some i, then b_i is also equal to 0, as b_{i+1} is a root of $x^2 + a_i x + b_i$. Therefore, we would have that all b_is are equal to 0. But this would mean that $a_1 = -a_3 = a_2 = -a_1$, and so $a_1 = 0$. However, this contradicts the fact that $a_i \neq b_i$. As a result, we may assume that $b_i \neq 0$ for all i. Using Viete's formulas, we get that

$$a_1 + b_1 = -a_3$$
$$a_2 + b_2 = -a_1$$
$$a_3 + b_3 = -a_2$$
$$a_1 b_1 = b_3$$
$$a_2 b_2 = b_1$$
$$a_3 b_3 = b_2 .$$

After multiplying the last three equations and dividing both sides by $b_1 b_2 b_3$, we get that $a_1 a_2 a_3 = 1$. This, in particular, implies that no coefficient is equal to 0. Now, calculate b_i from the first three equations and substitute them into the last three equations to get that

$$a_1(a_3 + a_1) = a_2 + a_3$$
$$a_2(a_1 + a_2) = a_3 + a_1$$
$$a_3(a_2 + a_3) = a_1 + a_2 .$$

Since $a_1 a_2 a_3 = 1$, either all coefficients are positive or only one of them is positive. We will independently investigate both cases.

Suppose first that all a_i are positive. Due to the symmetry, without loss of generality, we may assume that a_1 is a largest coefficient; in particular, $a_1 \geq 1$. From the first equation above, we get that

$$a_2 + a_3 = a_1(a_3 + a_1) \geq a_3 + a_1 ,$$

and so $a_2 \geq a_1$. However, because of our assumption that a_1 is a largest coefficient and the fact that the inequality above is sharp when $a_1 > 1$, this is only possible when $a_1 = a_2 = 1$. But then we get that also $a_3 = 1$, which in turn implies that $b_1 = b_2 = b_3 = -2$. It is straightforward to check that, indeed, these coefficients yield the desired solution, as $x^2 + x - 2 = (x-1)(x+2)$.

Suppose now that two coefficients a_i are negative and one of them is positive. Without loss of generality, we may assume that $a_1 > 0$, $a_2 < 0$, and $a_3 < 0$. From the first equation, we have that

$$a_1^2 + (a_1 - 1)a_3 = a_2 .$$

Note that if $a_1 \leq 1$, then $a_1^2 = a_2 + (1 - a_1)a_3 < 0$ which is impossible. It follows that $a_1 > 1$. From the same equation we have that $a_1(a_3 + a_1) = a_2 + a_3 < 0$ which implies that $a_3 + a_1 < 0$ and so $a_3 < -a_1 < -1$. Using this inequality and the fact that $a_2 + a_3 < 0$, it follows from the third equation that

$$a_1 + a_2 = a_3(a_2 + a_3) > -(a_2 + a_3),$$

or equivalently that $a_1 + a_3 > -2a_2$. Since $a_2 < 0$, we get that $a_1 + a_3 > 0$, which is a contradiction, as above we showed that $a_3 + a_1 < 0$.

Combining the two cases together we get that the only solution is $a_1 = a_2 = a_3 = 1$ and $b_1 = b_2 = b_3 = -2$.

Problem 3.1.2. Let $n \geq 3$ be an integer. Prove that the polynomial

$$f(x) = x^n + \sum_{i=0}^{n-3} a_i x^i$$

has n real roots if and only if all a_i are equal to 0.

Solution. Trivially, if all a_i are equal to 0, then $f(x) = x^n$ has n real roots, all of them equal to zero. Suppose now that $f(x)$ has n real roots: x_i, $i \in [n]$. It follows from (3.1) that

$$\sum_{i=1}^{n} x_i^2 = \frac{a_{n-1}^2 - 2 \cdot a_n \cdot a_{n-2}}{a_n^2}.$$

Since $a_{n-1} = a_{n-2} = 0$, we get that $\sum_{i=1}^{n} x_i^2 = 0$. If all x_i are real, then we get that $x_i = 0$ for all i. As a result, the considered polynomial is $f(x) = x^n$ (all a_i are equal to 0), and the proof is finished.

Problem 3.1.3. Let x_1, x_2, x_3 be the roots of the equation $3x^3 + 6x^2 - 1 = 0$. Find the value of $\sum_{i=1}^{3} \frac{1}{x_i^4}$.

Solution. Let us first re-write the given equation as follows:

$$1 = 3x^3 + 6x^2 = 3x^2(x + 2).$$

After squaring both sides and then dividing both sides by x^4, we get that $\frac{1}{x^4} = 9(x + 2)^2$. It follows that

$$\sum_{i=1}^{3} \frac{1}{x_i^4} = 9 \sum_{i=1}^{3} (x_i + 2)^2 = 9 \sum_{i=1}^{3} (x_i^2 + 4x_i + 4)$$

$$= 108 + 9 \sum_{i=1}^{3} x_i^2 + 36 \sum_{i=1}^{3} x_i.$$

From Vieta's formulas and their consequence (3.1), we get that

$$\sum_{i=1}^{3} x_i = \frac{-6}{3} = -2$$

and

$$\sum_{i=1}^{3} x_i^2 = \frac{6^2 - 2 \cdot 3 \cdot 0}{3^2} = 4.$$

Therefore, the sum we are looking for is equal to $108 + 9 \cdot 4 + 36 \cdot (-2) = 72$.

Problem 3.2.1. Find all functions $f: \mathbf{Z} \to \mathbf{Z}$ that satisfy the following condition:
$$f(a+b)^3 - f(a)^3 - f(b)^3 = 3f(a)f(b)f(a+b)$$
for all $a, b \in \mathbf{Z}$.

Solution. After setting $a = b = 0$, we get that $-f(0)^3 = 3f(0)^3$ and so $f(0) = 0$. Next, after considering any $a = -b \in \mathbf{Z}$, we get that $f(a)^3 = -f(-a)^3$. Since the function $g(x) := x^3$ is a bijection, we get that $f(a) = -f(-a)$, and so the function is symmetric about the origin (point $(0,0)$). As a result, we may restrict our analysis to arguments that are natural numbers.

Suppose that $f(1) = k$ for some $k \in \mathbf{Z}$ and consider $x := f(2) \in \mathbf{Z}$ which may or may not depend on k. By considering $a = b = 1$, we get that $x^3 - 2k^3 = 3k^2 x$ and so $(x - 2k)(x+k)^2 = 0$. It follows that $x = 2k$ or $x = -k$. We will consider both cases independently.

Case 1: $x = 2k$. We will prove by induction that $f(m) = km$ for any $m \in \mathbf{N}$. The base case trivially holds: $f(1) = k$, $f(2) = 2k$. For the inductive step, suppose that $f(m) = km$ for all $1 \leq m \leq m_0$ for some natural number $m_0 \geq 2$. Our goal is to show that $f(m_0 + 1) = k(m_0 + 1)$. By considering $a = m_0$ and $b = 1$, we get that

$$f(m_0 + 1)^3 - (km_0)^3 - k^3 - 3(km_0)kf(m_0 + 1) = 0.$$

Since our goal is to show that $f(m_0 + 1) = k(m_0 + 1)$, it will be convenient to factor out the term $f(m_0 + 1) - k(m_0 + 1)$ from the left hand side of the above equality. Guided by this, we re-write the equation as follows:

$$\left(f(m_0+1) - k(m_0+1)\right)\left(f(m_0+1)^2 + k(m_0+1)f(m_0+1) + k^2(m_0^2 - m_0 + 1)\right) = 0.$$

If $k = 0$, then we see immediately that $f(m_0 + 1) = 0$. On the other hand, if $k \neq 0$, then $f(m_0 + 1)^2 + k(m_0 + 1)f(m_0 + 1) + k^2(m_0^2 - m_0 + 1)$ has no roots as the discriminant of the corresponding quadratic equation satisfies the following:

$$(k(m_0 + 1))^2 - 4k^2(m_0^2 - m_0 + 1) = -3k^2(m_0 - 1)^2 < 0.$$

(Recall that $m_0 \geq 2$.) It follows that $f(m_0 + 1) = k(m_0 + 1)$, as required, and so the proof by induction is finished.

Let us summarize our observations in this case. We obtained that one possible family of solutions is $f(m) = km$ for some fixed integer k. It is straightforward to check that indeed this family satisfies our original equation.

Case 2: $x = -k$. We may assume that $k \neq 0$, as this case was already considered above. After taking $a = 2$ and $b = 1$, we get that $f(3)^3 + k^3 - k^3 = 3(-k)kf(3)$, or equivalently that $f(3)(f(3)^2 + 3k^2) = 0$. Since the second term is positive, we get that $f(3) = 0$. Now, by considering $b = 3$, we get that $f(a + 3)^3 - f(a)^3 = 0$. Again, as $g(x) = x^3$ is a bijection, we get that

$f(a+3) = f(a)$. It follows that the only family of functions that satisfies these conditions is

$$f(a) = \begin{cases} 0 & \text{if } a \equiv 0 \pmod 3 \\ k & \text{if } a \equiv 1 \pmod 3 \\ -k & \text{if } a \equiv 2 \pmod 3, \end{cases}$$

where $k \in \mathbf{Z}$ is some fixed integer. As usual, we directly check that this family satisfies the original condition.

Problem 3.2.2. Find all pairs of functions $f \colon \mathbf{R} \to \mathbf{R}$ and $g \colon \mathbf{R} \to \mathbf{R}$ such that

$$g\big(f(x) - y\big) = f\big(g(y)\big) + x$$

for all $x, y \in \mathbf{R}$.

Solution. Let us fix $y = 0$ to get that for any $x \in \mathbf{R}$, we have that

$$g(f(x)) = f(g(0)) + x.$$

On the other hand, for an arbitrary $x \in \mathbf{R}$ and $y = f(x) \in \mathbf{R}$, we get that

$$g(0) = f(g(f(x))) + x = f(f(g(0)) + x) + x.$$

Since x is arbitrary and $f(g(0))$, $g(0)$ are constants, we get that $f(x) = a - x$ for some $a \in \mathbf{R}$. Substituting this into $g(f(x)) = f(g(0)) + x$, we get that $g(a - x) = f(g(0)) + x$ and so $g(x) = b - x$ for some $b \in \mathbf{R}$. Now, we may go back to the original equation to get that $b - (a - x - y) = a - (b - y) + x$. It follows that $b - a = a - b$ and so $a = b$. We directly check that the family of functions $f(x) = g(x) = a - x$ for some fixed $a \in \mathbf{R}$ satisfies the original condition.

Problem 3.2.3. Find all functions $f \colon \mathbf{R} \to \mathbf{R}$ such that for all $x, y \in \mathbf{R}$ we have

$$f\big(f(x) - y\big) = f(x) + f\big(f(y) - f(-x)\big) + x.$$

Solution. Let us first set $x = y = 0$ to get that $f(f(0)) = 2f(0)$. Now, set $x = 0$ and $y = f(0)$ to get that $f(0) = f(0) + f(f(f(0)) - f(0))$. Using $f(f(0)) = 2f(0)$ (twice!), we get that $0 = f(f(0)) = 2f(0)$, and so $f(0) = 0$. After fixing $x = 0$, we get that for any $y \in \mathbf{R}$, we have $f(-y) = f(f(y))$. For any $x \in \mathbf{R}$, after fixing $y = f(x)$, we get that

$$0 = f(0) = f(x) + f\big(f(f(x)) - f(-x)\big) + x.$$

Using $f(-y) = f(f(y))$, we reduce it to

$$0 = f(x) + f(f(-x) - f(-x)) + x = f(x) + f(0) + x = f(x) + x,$$

as $f(0) = 0$. It follows that $f(x) = -x$, and one can directly check that this function satisfies the original condition:

$$
\begin{aligned}
f(f(x) - y) &= -f(x) + y = x + y \\
&= -x - f(y) + f(-x) + x \\
&= f(x) + f(f(y) - f(-x)) + x.
\end{aligned}
$$

Problem 3.3.1. Prove that if a function $f : \mathbf{R} \to \mathbf{R}$ satisfies the condition $f(x) = f(2x) = f(1-x)$ for all $x \in \mathbf{R}$, then it is periodic (that is, there exists some $a \in \mathbf{R}_+$ such that $f(x+a) = f(x)$ for all $x \in \mathbf{R}$).

Solution. Let $f : \mathbf{R} \to \mathbf{R}$ be any function that satisfies the condition. We will show that f is periodic with period $a = 1/2$. Indeed, note that for any $x \in \mathbf{R}$, we have that

$$
\begin{aligned}
f(x + 1/2) &= f(1 - (x + 1/2)) = f(1/2 - x) \\
&= f(2(1/2 - x)) = f(1 - 2x) \\
&= f(1 - (1 - 2x)) = f(2x) = f(x).
\end{aligned}
$$

Problem 3.3.2. Given that the function $f(x)$ satisfies

$$
f\big(1/(1 - x)\big) = x f(x) + 1,
$$

find the value of $f(5)$.

Solution. Let us first make an observation that for any $x \neq 0$ we have

$$
f(x) = \frac{f\big(1/(1 - x)\big) - 1}{x}.
$$

Using this formula three times we get $f(5) = (f(-1/4) - 1)/5$, $f(-1/4) = -4(f(4/5) - 1)$, and $f(4/5) = 5(f(5) - 1)/4$. It follows that

$$
f(5) = (-4(5(f(5) - 1)/4 - 1) - 1)/5 = -f(5) + 8/5,
$$

and so $f(5) = 4/5$.

Problem 3.3.3. Suppose that a function $f(x, y, z)$ of three real arguments satisfies the following condition

$$
\sum_{i=1}^{5} f(x_i, x_{i+1}, x_{i+2}) = \sum_{i=1}^{5} x_i,
$$

where $x_{i+5} = x_i$. Prove that for all $n \geq 5$ we have

$$
\sum_{i=1}^{n} f(x_i, x_{i+1}, x_{i+2}) = \sum_{i=1}^{n} x_i,
$$

where $x_{i+n} = x_i$.

Solution. By considering $(x_1, x_2, x_3, x_4, x_5) = (0, 0, 0, 0, 0)$, we get that $f(0, 0, 0) = 0$. On the other hand, for $(x_1, x_2, x_3, x_4, x_5) = (a, b, c, 0, 0)$ we get

$$f(a, b, c) + f(b, c, 0) + f(c, 0, 0) + f(0, 0, a) + f(0, a, b) = a + b + c.$$

Finally, for $(x_1, x_2, x_3, x_4, x_5) = (0, b, c, 0, 0)$ we get almost the same equality, namely,

$$f(0, b, c) + f(b, c, 0) + f(c, 0, 0) + f(0, 0, 0) + f(0, 0, b) = a + b,$$

so there is hope that after subtracting the two, many values will cancel out. Indeed, after subtracting the two equalities and using the fact that $f(0, 0, 0) = 0$ we get

$$f(a, b, c) = c + f(0, 0, b) - f(0, 0, a) + f(0, b, c) - f(0, a, b).$$

It follows that

$$\sum_{i=1}^{n} f(x_i, x_{i+1}, x_{i+2}) = \sum_{i=1}^{n} x_{i+2} + \sum_{i=1}^{n} \left(f(0, 0, x_{i+1}) - f(0, 0, x_i) \right)$$

$$+ \sum_{i=1}^{n} \left(f(0, x_{i+1}, x_{i+2}) - f(0, x_i, x_{i+1}) \right).$$

Since $x_{i+n} = x_i$, all the terms in the second and the third sum cancel out and we finally get that

$$\sum_{i=1}^{n} f(x_i, x_{i+1}, x_{i+2}) = \sum_{i=1}^{n} x_i,$$

as required.

Problem 3.4.1. Let $f_1 = 0$, $f_2 = 1$, and $f_{n+2} = f_{n+1} + f_n$ for all $n \in \mathbf{N}$. Find all polynomials $P(x)$ having only integer coefficients with the property that for each $n \in \mathbf{N}$ there exists $k = k(n) \in \mathbf{Z}$ such that $P(k) = f_n$.

Solution. Suppose that $P(x)$ is a polynomial with only integer coefficients, that is, $P(x) = \sum_{i=0}^{r} c_i x^i$ for some $r \in \mathbf{N}$ and $c_i \in \mathbf{Z}$ for all $i \in [r] \cup \{0\}$. Let us start with proving the following useful property that we will use many times. Let p and q be any integers. Note that

$$P(p) - P(q) = \sum_{i=0}^{r} c_i p^i - \sum_{i=0}^{r} c_i q^i = \sum_{i=1}^{r} c_i (p^i - q^i)$$

$$= (p - q) \sum_{i=1}^{r} c_i \sum_{j=0}^{i-1} p^j q^{i-1-j}, \tag{8.8}$$

and so $(p-q)$ divides $P(p) - P(q)$.

Suppose that $P(x)$ satisfies the desired property: for each $n \in \mathbf{N}$ there exists $k = k(n) \in \mathbf{Z}$ such that $P(k) = f_n$. In particular, for $n = 1$ and $n = 2$ we get that there exist $a = k(1) \in \mathbf{Z}$ and $b = k(2) \in \mathbf{Z}$ such that $P(a) = f_1 = 0$ and $P(b) = f_2 = 1$. It follows from (8.8) that $(b - a)$ divides $P(b) - P(a) = 1$, and so $|b - a| = 1$.

Let us define the auxiliary polynomial $Q(x) := P(a + (b - a)x)$. Clearly, $Q(0) = P(a) = 0$ and $Q(1) = P(b) = 1$. We will prove by induction that $Q(f_i) = f_i$ for all $i \in \mathbf{N}$. This will finish the proof as the only polynomial that satisfies this property is $Q(x) = x$. Indeed, each polynomial $R(x)$ of degree at least 2 has the property that $|R(x)| > x$ for all $x \geq x_0$, where x_0 is a sufficiently large constant. Since $Q(x) = x$ for infinitely many natural numbers x, we get that $Q(x)$ has to be of degree at most 1. Constant polynomials are clearly ruled out and $Q(x) = x$ is the only linear function that satisfies the property. Using the fact that $|b - a| = 1$, we get that the only polynomials that satisfy the original equation are polynomials of the form $P(x) = x + c$ or $P(x) = -x + c$ for some $c \in \mathbf{Z}$. It is straightforward to check that, indeed, they satisfy the desired equation.

It remains to show that $Q(f_i) = f_i$ for all $i \in \mathbf{N}$. We already showed that this property holds for $i = 1$ and $i = 2$. For the base case, we will show that it also holds for $i \in \{3, 4, 5, 6, 7\}$. In fact, we will prove something stronger, namely, that f_i is the only integer k that satisfies $f(k) = f_i$.

Case: $i = 3$. Suppose that $Q(k) = f_3 = 2$ for some $k \in \mathbf{Z}$. Using (8.8) we get that $k - 1$ divides $Q(k) - Q(1) = f_3 - f_2 = 2 - 1 = 1$, and so $k - 1 = 1$ or $k - 1 = -1$. Since $Q(0) = 0$, $k = 0$ is ruled out and we get that $k = 2$ is the unique solution.

Case: $i = 4$. Suppose that $Q(k) = f_4 = 3$ for some $k \in \mathbf{Z}$. Using the same argument as before, we get that $k - 2$ divides $Q(k) - Q(2) = f_4 - f_3 = 3 - 2 = 1$, which implies that $k - 2 = 1$ or $k - 2 = -1$. Since $k = 1$ is ruled out ($Q(1) = 1 \neq 3$), we get that $k = 3 = f_4$ is the unique solution.

Case: $i = 5$. If $Q(k) = f_5 = 5$, then $(k - 0) \mid (5 - 0)$ and so $k \in \{-5, -1, 5\}$, as $k = 1$ is already ruled out. However, since also $(k - 3) \mid (5 - 3)$, we get that $k = 5$ is the unique solution.

Case: $i = 6$. If $Q(k) = f_6 = 8$, then $(k - 5) \mid (8 - 5)$ and so $k - 5 \in \{-3, -1, 1, 3\}$. It follows that $k \in \{4, 6, 8\}$, as $k = 2$ is already ruled out. Moreover, $(k - 1) \mid (8 - 1)$, and so $k = 8$ is the only solution.

Case: $i = 7$. If $Q(k) = f_7 = 13$, then $(k - 0) \mid (13 - 0)$ so $k \in \{-13, -1, 13\}$ as $k = 1$ is ruled out. But also $(k - 8) \mid (13 - 8)$, and so $k = 13$ is the only solution.

Let us now move to the inductive step. Suppose that for some integer $n \geq 7$, for each $i \in [n]$ we have that $k = f_i$ is the unique integer solution to

$Q(k) = f_i$. Our goal is to show that $k = f_{n+1}$ is the unique integer solution to $Q(k) = f_{n+1}$.

Let $k \in \mathbf{Z}$ be such that $Q(k) = f_{n+1}$. Since $Q(0) = 0$, we get from (8.8) that $k - 0 \mid Q(k) - Q(0)$ and so $k \mid Q(k)$. Moreover, from the same property it follows that $k - f_n$ divides $f_{n+1} - f_n = f_{n-1}$ and so $-f_{n-1} \le k - f_n \le f_{n-1}$. We conclude that $5 = f_5 \le f_{n-2} < k \le f_{n+1}$ (note that $Q(f_{n-2}) = f_{n-2} \ne f_{n+1}$ so $k = f_{n-2}$ is ruled out). Since $f_{n-2} < 2f_{n-1} < 4f_n < 8f_{n+1}$, we get that $k = f_{n+1}/x$ for some $x \in [7]$, as k divides $Q(k) = f_{n+1}$.

Recall that our goal is to show that $x = 1$. For a contradiction, suppose that $x > 1$. Applying (8.8) twice, we get that $(k-1) \mid (xk-1)$ and $(k-2) \mid (xk-2)$. In other words, there exist $a, b \in \mathbf{N}$ such that $b > a > 1$, $a(k-1) = xk-1$, and $b(k-2) = xk-2$. It follows that $a(k-1) - b(k-2) = 1$, or equivalently that $(b-a)(k-1) = b - 1$. It will be convenient to fix $c := b - a \in \mathbf{N}$. We get that $b = c(k-1) + 1$ and so $b(k-2) = (c(k-1)+1)(k-2) = xk - 2$. This means that k divides $(c(k-1)+1)(k-2) + 2 = k(ck - 3c + 1) + 2c$ and so we get that $k \mid 2c$.

Let us now summarize what we have learnt. We showed the following three things: $x \le 7$, $k \ge 6$, and $k \mid 2c$. But using these observations, we get that

$$
\begin{aligned}
c = b - a &= \frac{xk - 2}{k - 2} - \frac{xk - 1}{k - 1} = \frac{k(x-1)}{(k-1)(k-2)} \\
&\le \frac{6}{k - 3 + 2/k} \le \frac{6}{6 - 3 + 2/6} = \frac{9}{5},
\end{aligned}
$$

and so $c = 1$. But this is not possible as $k \ge 6$ and $k \mid 2c$.

Problem 3.4.2. Suppose that a polynomial $P(x)$ has all integer coefficients. Prove that if polynomials $P(P(P(x)))$ and $P(x)$ have a common real root, then $P(x)$ also has an integer root.

Solution. Let $a \in \mathbf{R}$ be a common root of $P(P(P(x)))$ and $P(x)$, that is, $P(P(P(a))) = P(a) = 0$. This implies that $P(P(0)) = 0$, that is, $P(0)$ is also a root of $P(x)$. But $P(0)$ is an integer, as $P(x)$ has all integer coefficients; in particular, the free term is an integer.

Problem 3.4.3. Consider a polynomial $P(x) = x^2 + ax + b$ with $a, b \in \mathbf{Z}$. Suppose that for every prime number p, there exists $k \in \mathbf{Z}$ such that $P(k)$ and $P(k + 1)$ are divisible by p. Prove that there exists $m \in \mathbf{Z}$ such that $P(m) = P(m + 1) = 0$.

Solution. Fix any prime number p. By our assumption, there exists $k = k(p) \in \mathbf{Z}$ such that both $P(k)$ and $P(k+1)$ are divisible by p. Our goal is to find a number which does *not* depend on k that is divisible by p. To that end, note that

$$
\begin{aligned}
P(k + 1) - P(k) &= ((k + 1)^2 + a(k + 1) + b) - (k^2 + ak + b) \\
&= 2k + (a + 1)
\end{aligned}
$$

is divisible by p and so is

$$2P(k) - k(2k + (a+1)) = 2k^2 + 2ak + 2b - (2k^2 + k(a+1))$$
$$= k(a-1) + 2b.$$

Finally, observe that

$$2(k(a-1) + 2b) - (a-1)(2k + (a+1)) = 4b - (a-1)(a+1)$$
$$= -a^2 + 1 + 4b$$

is divisible by p. Since this property holds for any p, we get that $a^2 = 4b + 1$. In particular, a is odd, that is, $a = 2s + 1$ for some $s \in \mathbf{Z}$. We get that $a^2 = 4s^2 + 4s + 1 = 4b + 1$, and so $b = s(s+1)$. Substituting this back into the quadratic polynomial, we get that

$$P(x) = x^2 + (2s+1)x + s(s+1) = (x+s)(x+s+1).$$

From this it is clear that if $m = -s - 1 \in \mathbf{Z}$, then $P(m) = P(m+1) = 0$, the desired property.

Problem 3.5.1. Find all polynomials $P(x)$ with real coefficients that satisfy the following property: if $x + y$ is rational, then $P(x) + P(y)$ is rational.

Solution. Consider any polynomial $P(x)$ with real coefficients that satisfies the desired property. Fix any rational number q and consider the polynomial $Q(x) := P(q+x) + P(q-x)$ for all $x \in \mathbf{R}$. Since for each $x \in \mathbf{R}$ we have that $q + x + (q - x) = 2q$ is rational, it follows that $Q(x)$ is rational for all $x \in \mathbf{R}$. But, since $Q(x)$ is continuous, it is only possible when $Q(x)$ is constant. In particular,

$$P(q+q) + P(q-q) = Q(q) = Q(0) = P(q+0) + P(q-0),$$

so $P(2q) + P(0) = 2P(q)$. It will be convenient to represent $P(x)$ as follows: $P(x) = x^2 R(x) + ax + b$ for some polynomial $R(x)$. It follows that

$$(2q)^2 R(2q) + a(2q) + b + b = 2(q^2 R(q) + aq + b)$$

so, assuming that $q \neq 0$, we get that $2R(2q) = R(q)$. Since this argument holds for all rational numbers q, we get that $R(2^n q) = R(q)/2^n$. It follows that $R(q)$ does *not* tend to $+\infty$ or $-\infty$ as $q \to +\infty$. This is only possible if $R(q) = 0$ for all rational numbers and so $R(x) = 0$ everywhere. As a result, we get that $P(x) = ax + b$. Now, after letting $x = 0$ we see that b must be rational, and after letting $x = 1$ we see that a must also be rational. Finally, one can directly check that if a and b are rational, then the desired condition is satisfied.

Problem 3.5.2. Let $P(x)$ be a polynomial with real coefficients. Prove that if there exists an integer k such that $P(k)$ is not an integer, then there are infinitely many such integers.

Solution. Let $P(x)$ be any polynomial such that $P(k) \notin \mathbf{Z}$ for some $k \in \mathbf{Z}$. It will be more convenient to work with the polynomial $Q(x) := P(x+k)$ instead of $P(x)$. Indeed, if there are infinitely many integers ℓ such that $Q(\ell) \notin \mathbf{Z}$, then clearly the same property holds for $P(x)$. An advantage of working with $Q(x)$ is that, by assumption, $Q(0) = P(k) \notin \mathbf{Z}$ and evaluating polynomials at $x = 0$ is easy.

For a contradiction, suppose that the set $A := \{x \in \mathbf{Z} : Q(x) \notin \mathbf{Z}\}$ is finite which implies that the set $B := \mathbf{Z} \setminus A = \{x \in \mathbf{Z} : Q(x) \in \mathbf{Z}\}$ is infinite. Suppose that the degree of $Q(x)$ is $n \in \mathbf{N} \cup \{0\}$. In fact, $n \neq 0$ as $Q(0) \notin \mathbf{Z}$ and $Q(x) \in \mathbf{Z}$ for any $x \in B$, and so $Q(x)$ is not a constant polynomial. Since B is infinite, we may consider n points (x_i, y_i), where both x_i and $y_i = Q(x_i)$ are integers. From the Lagrange interpolation formula for these points, we get that all the coefficients of $Q(x)$ are rational. It follows that $Q(x) = \sum_{i=0}^{n} \frac{n_i}{d_i} \cdot x^i$, where $n_i \in \mathbf{Z}$, $d_i \in \mathbf{N}$, and $Q(0) = n_0/d_0 \notin \mathbf{Z}$; in particular, $d_0 \geq 2$.

Let us now consider the sequence of natural numbers defined as follows: $y_t := \left(\prod_{i=0}^{n} d_i \right)^t$ for $t \in \mathbf{N}$. Since $d_0 \geq 2$, we get that $\prod_{i=0}^{n} d_i > 1$ and so the sequence $(y_t)_{t \in \mathbf{N}}$ is increasing. Moreover,

$$Q(y_t) = \sum_{i=0}^{n} \frac{n_i}{d_i} \cdot y_t^i = c_t + \frac{n_0}{d_0},$$

where c_t is some integer. It follows that $Q(y_t) \notin \mathbf{Z}$ for all $t \in \mathbf{N}$, and so we have an infinite sequence $(y_t, Q(y_t))_{t \in \mathbf{N}}$ of distinct pairs consisting of integer and non-integer which contradicts the fact that A is finite. The conclusion is that A is infinite, and so the proof is complete.

Problem 3.5.3. Let $F(x)$, $G(x)$, and $H(x)$ be some polynomials of degree at most $2n + 1$ with real coefficients. Moreover, suppose that the following properties hold:

(1) for all $x \in \mathbf{R}$, $F(x) \leq G(x) \leq H(x)$,

(2) there exist n different numbers $x_i \in \mathbf{R}$, $i \in [n]$, such that $F(x_i) = H(x_i)$ for all $i \in [n]$,

(3) there exists $x_0 \in \mathbf{R}$, different than x_i for $i \in [n]$, such that $F(x_0) + H(x_0) = 2G(x_0)$.

Prove that for all $x \in \mathbf{R}$, $F(x) + H(x) = 2G(x)$.

Solution. Consider any polynomials $F(x)$, $G(x)$, and $H(x)$ that are of degree at most $2n + 1$ and that satisfy the properties (1)–(3). It will be convenient to consider two auxiliary polynomials, $P(x) := G(x) - F(x)$ and $Q(x) := H(x) - F(x)$. It follows from property (1) that for each $x \in \mathbf{R}$, $Q(x) \geq P(x) \geq 0$. In particular, it means that both polynomials are of degree at most $2n$ as having degree $2n + 1$ would imply that either $\lim_{x \to -\infty} P(x) = -\infty$ or $\lim_{x \to \infty} P(x) = -\infty$. Moreover, from property (2) it follows that for $i \in [n]$ we have $P(x_i) = Q(x_i) = 0$. But this means that all of the x_i are roots of even multiplicity. Since we have $2n$ roots in total (including multiplicities) and $P(x)$ and $Q(x)$ have degree at most $2n$, we get that there exists $a \in [0,1]$ such as $P(x) = aQ(x)$, for all $x \in \mathbf{R}$. Now, using property (3) we get that

$$\frac{Q(x_0)}{2} = \frac{H(x_0)}{2} - \frac{F(x_0)}{2} = \frac{2G(x_0) - F(x_0)}{2} - \frac{F(x_0)}{2}$$
$$= G(x_0) - F(x_0) = P(x_0),$$

and so $a = 1/2$. It follows that for each $x \in \mathbf{R}$, $2(G(x) - F(x)) = H(x) - F(x)$, and so $2G(x) = F(x) + H(x)$, as needed.

Problem 3.6.1. Find all real numbers m for which the polynomial $f(x) = 2x^4 - 7x^3 + mx^2 + 22x - 8$ has two real roots whose product is equal to 2.

Solution. Suppose that the polynomial $f(x)$ has two real roots a and b such that $ab = 2$. It follows that $f(x) = (x-a)(x-b)(2x^2+cx+d)$ for some $a, b, c, d \in \mathbf{R}$. After comparing the corresponding coefficients, we get the following system of equations:

$$-2a - 2b + c = -7$$
$$2ab - c(a+b) + d = m$$
$$abc - (a+b)d = 22$$
$$abd = -8.$$

Since $ab = 2$, we get from the last equation that $d = -4$. Substituting it to the third equation and adding twice the first one, we get that $c = 2$. If follows that $a + b = 9/2$, and so $m = -9$ is the only possible solution. Since

$$2x^4 - 7x^3 - 9x^2 + 22x - 8 = 2(x-4)(x-1)(x-1/2)(x+2),$$

we get that, indeed, the polynomial $f(x)$ has two roots, namely 4 and $1/2$, whose product is equal to 2, as required.

Problem 3.6.2. Given the polynomial $P(x) = x^4 - 3x^3 + 5x^2 - 9x$, $x \in \mathbf{R}$, find all pairs of integers a and b such that $a \neq b$ and $P(a) = P(b)$.

Solution. Let us first note that

$$P(-x+1) - P(x) = (2x-1)(x^2 - x + 6) > 0,$$

provided $x > 1/2$. Moreover,

$$P(x+1) - P(-x+1) = 2(x-2)x(x+2) > 0,$$

provided that $x > 2$. It follows that for $x \in \mathbf{N} \setminus \{1, 2\}$, we have that

$$P(x) < P(-x+1) < P(x+1).$$

As a result, there are no $a, b \in \mathbf{Z} \setminus \{-1, 0, 1, 2\}$ such that $P(a) = P(b)$. We directly compute that $P(-1) = 18$, $P(0) = 0$, $P(1) = -6$, $P(2) = -6$, and $P(3) = 18$. Since all the values of the polynomial $P(x)$ evaluated at integers greater than 3 or smaller than -1 are greater than $P(3) = 18$, we get that there are only four solutions to the problem:

$$(a, b) \in \{(-1, 3), (3, -1), (1, 2), (2, 1)\}.$$

Problem 3.6.3. Find all polynomials $P(x)$ with real coefficients that satisfy the following property: for all $x \in \mathbf{R}$, $P(x^2) \cdot P(x^3) = (P(x))^5$.

Solution. We will independently consider two cases depending on how many terms the considered polynomial has.

Let us first assume that $P(x)$ has exactly one term (including the special case $P(x) = 0$ for $x \in \mathbf{R}$), that is, $P(x) = ax^k$ for some $a \in \mathbf{R}$ and $k \in \mathbf{N} \cup \{0\}$. Substituting this into the equation we want to hold for all $x \in \mathbf{R}$, we get that $ax^{2k} \cdot ax^{3k} = a^5 x^{5k}$. In particular, by considering $x = 1$, we get that $a^2 = a^5$ and so $a^2(a-1)(a^2+a+1) = 0$. It follows that $a = 0$ or $a = 1$, and so the only solutions in this case are: $P(x) = 0$, $P(x) = 1$, and $P(x) = x^k$ for some fixed $k \in \mathbf{N}$. It is straightforward to directly check that all of these polynomials satisfy the original equation.

Let us now assume that $P(x)$ has more than one term and has degree $k \in \mathbf{N}$. Then, it can be represented as follows: $P(x) = ax^k + bx^\ell + Q(x)$, where $a, b \in \mathbf{R} \setminus \{0\}$, $\ell \in \mathbf{Z}$ such that $0 \leq \ell < k$, and $Q(x)$ has degree less than ℓ (or $Q(x) = 0$ everywhere if $\ell = 0$). Substituting this form into the original equation we get that for all $x \in \mathbf{R}$,

$$(ax^{2k} + bx^{2\ell} + Q(x^2))(ax^{3k} + bx^{3\ell} + Q(x^3)) = (ax^k + bx^\ell + Q(x))^5.$$

Let us now compare the coefficients in front of the term $x^{4k+\ell}$ on both the left and the right hand side of the above equation. The first term on the left hand side is clearly $a^2 x^{5k}$ but, since $\ell < k$, the next non-zero term is $abx^{3k+2\ell}$. Since $3k + 2\ell < 4k + \ell$, there is no term we are looking for. Alternatively, we may say that the coefficient in front of x^{4k+l} is equal to 0. On the other hand, the right hand side after expanding is equal to $a^5 x^{5k} + 5a^4 bx^{4k+\ell} + R(x)$, where $R(x)$ has degree less than $4k + \ell$. It follows that the coefficient in front of the term x^{4k+l} is equal to $5a^4 b \neq 0$. This contradiction proves that $P(x)$ cannot have more than one term. We conclude that the only polynomials that satisfy the desired property are those that we found in the previous case.

Problem 3.7.1. Prove that there are no polynomials $P_1(x), P_2(x), P_3(x), P_4(x)$ with rational coefficients that satisfy

$$\sum_{i=1}^{4} (P_i(x))^2 = x^2 + 7 \qquad \text{for all } x \in \mathbf{R}. \tag{8.9}$$

Solution. Due to the symmetry, without loss of generality, we may assume that $n := n_1 \geq n_2 \geq n_3 \geq n_4 \geq 0$, where n_i is the degree of $P_i(x)$, $i \in [4]$. For $i \in [4]$, let $c_i \in \mathbf{R}$ be the coefficient in front of the term x^n in $P_i(x)$. Clearly, $c_1 \neq 0$. More importantly, after expanding the left hand side of (8.9), the coefficient in front of the term x^{2n} is equal to $\sum_{i=1}^{4} c_i^2 \geq c_1^2 > 0$. Since the degree of the right hand side of (8.9) is 2, we get that $n = 1$. As a result, since all the coefficients are rational, we may represented each polynomial $P_i(x)$ as follows: $P_i(x) = (a_i x + b_i)/m$, where a_i, b_i ($i \in [4]$), and m are some integers.

After comparing the coefficients in (8.9) that are in front of the term x^k for $k \in \{2, 1, 0\}$, we get the following set of equations:

$$a_1^2 + a_2^2 + a_3^2 + a_4^2 = m^2$$
$$a_1 b_1 + a_2 b_2 + a_3 b_3 + a_4 b_4 = 0$$
$$b_1^2 + b_2^2 + b_3^2 + b_4^2 = 7m^2 .$$

For $i \in [4]$, let $p_i := a_i + b_i$ and $q_i := a_i - b_i$. Adding the first, the third, and twice the second equation we get that $p_1^2 + p_2^2 + p_3^2 + p_4^2 = 8m^2$. Adding the first, the third, and subtracting the second equation twice we get that $q_1^2 + q_2^2 + q_3^2 + q_4^2 = 8m^2$. Finally, after subtracting the third equation from the first equation, we get that $p_1 q_1 + p_2 q_2 + p_3 q_3 + p_4 q_4 = -6m^2$. Summarizing, we get the following system of equations:

$$p_1^2 + p_2^2 + p_3^2 + p_4^2 = 8m^2$$
$$q_1^2 + q_2^2 + q_3^2 + q_4^2 = 8m^2$$
$$p_1 q_1 + p_2 q_2 + p_3 q_3 + p_4 q_4 = -6m^2 .$$

Let us note that, without loss of generality, we may assume that $\gcd(p_1, p_2, p_3, p_4, q_1, q_2, q_3, q_4, m) = 1$. Indeed, if all the variables involved (namely, $p_1, p_2, p_3, p_4, q_1, q_2, q_3, q_4, m$) had some common positive divisor d, one could divide the three equations by d^2 to get an equivalent system of equations. Note that for any $x \in \mathbf{Z}$, the reminder of x^2 when divided by 8 is equal to 0, 1 or 4. Hence, from the first equation we get that that all p_i are even, and from the second one it follows that all q_i are even. But this means that m is odd, since $\gcd(p_1, p_2, p_3, p_4, q_1, q_2, q_3, q_4, m) = 1$. However, if m is odd and all other variables are even, then the left hand side of the third equation is divisible by 4 wheras the right hand side is not. The conclusion is that there are no polynomials with rational coefficients that satisfy (8.9), and so the proof is complete.

Problem 3.7.2. Consider a polynomial $f(x) := x^2 + bx + c$, where $b, c \in \mathbf{Z}$. Prove that if $n \in \mathbf{N}$ divides $f(p)$, $f(q)$, and $f(r)$ for some $p, q, r \in \mathbf{Z}$, then

$$n \mid (p - q)(q - r)(r - p) .$$

Solution. Note that if $n \mid f(p)$ and $n \mid f(q)$, then n divides

$$f(p) - f(q) = p^2 + bp + c - (q^2 + bq + c) = (p - q)(p + q) + b(p - q)$$
$$= (p - q)((p + q) + b) .$$

Similarly, we get that $n \mid (q - r)((q + r) + b)$ and $n \mid (r - p)((r + p) + b)$. It follows that n also divides

$$r(p - q)((p + q) + b) + p(q - r)((q + r) + b) + q(r - p)((r + p) + b)$$
$$= r(p - q)(p + q) + p(q - r)(q + r) + q(r - p)(r + p)$$
$$= (p - q)(rp + rq) + pq^2 - pr^2 + qr^2 - qp^2$$
$$= (p - q)(rp + rq) - (p - q)(pq + r^2)$$
$$= (p - q)(rp + rq - pq - r^2)$$
$$= (p - q)(q - r)(r - p) \,,$$

and so the desired property holds.

Alternatively, one could consider the above expression as a quadratic polynomial of in terms of p. It is straightforward to check that q and r are both roots of this polynomial. Moreover, the coefficient in front of p^2 is $(r - q)$. It follows, without having to perform the laborious calculations, that the above expression is equal to $(p - q)(p - r)(r - q)$.

Problem 3.7.3. Consider a polynomial $P(x)$ with integer coefficients that satisfies the following property: if $a, b \in \mathbf{Q}$ and $a \neq b$, then $P(a) \neq P(b)$. Does it mean that $P(a) \neq P(b)$ for all $a, b \in \mathbf{R}$, $a \neq b$?

Solution. We will show that this is not true for $P(x) := x^3 - 2x$. First, let us observe that $P(x) = x(x^2 - 2) = x(x - \sqrt{2})(x + \sqrt{2})$ and so, in particular, $P(0) = P(\sqrt{2}) = 0$. Hence, it is enough to show that $P(x)$ has the desired property, namely, that there are no two different rational numbers a and b such that $P(a) = P(b)$. For a contradiction, suppose that there are $q_1, q_2 \in \mathbf{Q}$ such that $q_1 \neq q_2$ and $P(q_1) = P(q_2)$. It follows that

$$\begin{aligned} 0 = P(q_1) - P(q_2) &= q_1^3 - 2q_1 - q_2^3 + 2q_2 \\ &= (q_1 - q_2)(q_1^2 + q_1 q_2 + q_2^2 - 2) \,. \end{aligned}$$

Since $q_1 \neq q_2$, we get that

$$q_1^2 + q_1 q_2 + q_2^2 = 2 \,.$$

Since q_1 and q_2 are rational numbers, we may express these numbers as follows: $q_1 = a/c$ and $q_2 = b/c$ for some $a, b, c \in \mathbf{Z}$, and $\gcd(a, b, c) = 1$. It follows that

$$a^2 + ab + b^2 = 2c^2 \,. \tag{8.10}$$

Note that a and b cannot be both even as then the left hand side of (8.10) would be divisible by 4 whereas the right hand side would not. Similarly, if both a and b were odd, then the left hand side would be odd but the right hand side would be even. Finally, if only one of the two numbers a and b is even and the other one is odd, then the left hand side is odd, which is again impossible. We get the desired contradiction and so, indeed, the polynomial $P(x) = x^3 - 2x$ is a counter-example to our problem.

Let us make a final remark on how one can guess that $P(x) = x^3 - 2x$ is a counter-example to our problem. Let us first consider polynomials with integer coefficients that are of degree 2. Such polynomials can be expressed as follows: $P(x) := a(x - p)(x - q)$, where $a, -a(p + q), apq \in \mathbf{Z}$. Since $P(0) = P(p + q) = apq$ and both 0 and $p + q$ are rational, no polynomial of degree 2 satisfies the desired property.

Hence, we shift our attention to polynomials of degree 3 by considering polynomials of the form $P(x) := a(x - p)(x - q)(x - r)$ and with integer coefficients. No two of the three roots, say p and q, can be rational as then $P(p) = P(q) = 0$ fails the required assumption. So we have two options: all of them irrational or exactly one is rational. The second option seems easier to deal with and, without loss of generality, we may assume that $p = 0$. Indeed, if $P(x)$ is a counter-example, then so is $Q(x) := P(x - p)$. Since $P(x)$ has integer coefficients, we get that $a \in \mathbf{Z}$ and again, without loss of generality, we may assume that $a = 1$. It follows that $P(x) = x(x - q)(x - r)$ with $q, r \in \mathbf{R} \setminus \mathbf{Q}$ but $qr \in \mathbf{Z}$. A natural choice for q is an irrational square root of some natural number and $r = -q$ so that $P(x) = x(x^2 - q^2)$. Choosing $q = \sqrt{2}$ is an intuitive first guess, as it is related to a well known proof that $\sqrt{2}$ is not a rational number.

8.4 Combinatorics

Problem 4.1.1. There are $2n$ members of a chess club; each member knows at least n other members (knowing a person is a reciprocal relationship). Prove that it is possible to assign members of the club into n pairs in such a way that in each pair both members know each other.

Solution. Let us first rephrase the question in the language of graph theory. Suppose that $G = (V, E)$ is a graph on $|V| = 2n$ vertices and the minimum degree, $\delta = \delta(G) \geq n$. Our goal is to show that G has a perfect matching.

We will construct a perfect matching in n rounds, distinguishing two phases. During the first phase, we apply a trivial, greedy algorithm to construct a maximal matching, that is, a matching that cannot be extended by adding an edge. We start with an empty matching $M_0 = (\emptyset, \emptyset)$. In each round i, we consider the graph $G[V \setminus V(M_{i-1})]$ induced by unsaturated vertices. If it contains edges, then we arbitrarily pick one of them (say, edge $a_i b_i$) and add it to the current matching; that is, $V(M_i) = V(M_{i-1}) \cup \{a_i, b_i\}$ and $E(M_i) = E(M_{i-1}) \cup \{a_i b_i\}$. This phase ends if there are no more edges to pick from. If all the vertices are saturated, then we are done; otherwise, we move on to the second phase.

During the second phase, at the beginning of each round $i \leq n$, set $V \setminus V(M_{i-1})$ contains at least two vertices and it induces an independent set. We pick *any* two vertices, say, p and q from that set. We will show that there is an edge in $E(M_{i-1})$, say, rs such that $pr \in E$ and $qs \in E$. In other words, we will show that there exists a path (p, r, s, q) of length 3 (such paths are often called **augmenting paths**). The existence of such paths allows us to improve the size of our matching. Indeed, we can simply remove rs from the matching and add edges pr and qs instead. Formally, $V(M_i) = V(M_{i-1}) \cup \{p, q\}$ and $E(M_i) = (E(M_{i-1}) \setminus \{rs\}) \cup \{pr, qs\}$.

To finish the proof, let us note that p has at least n neighbors in $V(M_{i-1})$ (since $\delta \geq n$ and $V \setminus V(M_{i-1})$ induces an independent set). Let $R := N(p) \subseteq V(M_{i-1})$, and let S be the set of vertices matched with vertices from R, that is, $S = \{s \in V(M_{i-1}) : sr \in E(M_{i-1})$ for some $r \in R\}$. Clearly, $|S| = |R| \geq \delta \geq n$. Moreover, S and R can overlap (and, in fact, they do) but it causes no problem. More importantly, since q has at least n neighbors in $V(M_{i-1})$, $|S| \geq n$, and $|V(M_{i-1})| \leq 2n - 2$, it follows that q has at least one neighbor in S which finishes the argument.

Finally, let us mention that a stronger property holds. Graph G not only contains a perfect matching, but it in fact has to have a **Hamilton cycle**, that is, cycle of length $2n$ whose vertex set is precisely $V(G)$. This is a famous sufficient condition for the existence of a Hamilton cycle due to Dirac. It is indeed a stronger property since one can take every second edge of a Hamilton cycle to form a perfect matching.

Problem 4.1.2. There are 17 players in the tournament in which each pair of two players compete against each other. Every game can last 1, 2, or 3 rounds. Prove that there exist three players who have played exactly the same number of rounds with one another.

Solution. As before, let us rephrase this problem in the language of graph theory. The tournament can be represented as coloring of the edges of K_{17}, the complete graph on 17 vertices, with three colors (say, red, blue, and green). Coloring edge vw red indicates that the game between players corresponding to vertices v and w lasted one round. Similarly, blue and green indicate that the corresponding game lasted two and, respectively, three rounds. Our goal is to show that, regardless how the graph is colored, it must contain a monochromatic triangle (that is, the edges of some K_3 are all the same color).

In order to warm up, let us prove something slightly easier. Suppose that only two colors are available (it does not matter which ones; without loss of generality, we may assume that we use red and blue). We claim that, regardless how the edges of K_6 are colored with these two selected colors, there is a monochromatic triangle. Indeed, pick *any* vertex v and consider the 5 edges incident to v. Clearly, at least three of them (say va, vb, and vc) must be of the same color, say red. If any one of ab, ac, bc is red, then we have a red triangle. If none of these edges is red, then we have a blue triangle. This proves the claim about two colors.

Now, let us come back to the original problem with three colors and K_{17}. Pick *any* vertex v and consider the 16 edges incident to v. Since, $3 \cdot 5 < 16$, at least 6 of them must be of the same color, say, green. Let N be the set of neighbors of v that are adjacent to v by a green edge. If any edge of $G[N]$, the graph induced by N, is colored green, then we have a green triangle. If none of these edges is green, then all the edges of $G[N]$ are colored red and blue. By the previous claim, this also generates a monochromatic triangle and so we are done.

Let us mention that this result is sharp, namely, one can color the edges of K_{16} with three colors and avoid monochromatic triangle. Finally, let us mention that this is a specific case of the classic and famous problem of Ramsey numbers (for three colors and triangles). Indeed, this observation can be generalized to any number of colors and any order of a monochromatic complete graph.

Problem 4.1.3. Consider a group of people with the following property. Some of them know each other, in which case the corresponding pair of people mutually like each other or dislike each other. Moreover, there is a person who knows at least six other people. Interestingly, for each person the number of people he or she likes is equal to the number of people he or she dislikes. Prove that it is possible to remove some, but not all, like/dislike links such that it is still the case that each person has the same number of liked and disliked acquaintances.

Solution. As usual, let us rephrase this problem in the language of graph theory. Note first that acquaintances can be modelled by a graph G: if v and w know each other, then we put an edge between v and w; otherwise, v and w are not adjacent. Then, likes and dislikes can be represented by coloring edges red and, respectively, blue. We assume that the maximum degree is at least 6. More importantly, we assume that the following property holds: for each vertex $v \in V(G)$, the number of red edges incident to v is equal to the number of blue edges incident to V (in particular, it implies that each vertex has even degree). Our goal is to show that it is possible to remove some edges (but not all of them) such that this property is preserved.

It will be convenient to use a notion of a walk on graphs. A **walk** $W = (v_0, \ldots, v_k)$ of length k is a sequence of vertices such that $v_{i-1}v_i \in E$ for any $i \in [k]$. Note that walks are allowed to revisit some vertices and edges but they do not have to. As a result, a path is a walk but not every walk is a path.

In order to show the result, let us select any vertex v_0 that has degree at least 6 and start walking from there, first using a red edge and then alternate colors. Note that, because of the property of our coloring, each time we enter some vertex $v \neq v_0$ we may continue using some other edge of the other color. Hence, at some point, we need to get back to v_0; let W_1 be the walk we did so far. If W_1 has even length, then we get the desired property after removing edges of this walk (note that each vertex on the walk is incident to the same number of red and blue edges used in this walk). On the other hand, if the length of W_1 is odd, then the two edges used by W_1 that are incident to v_0 are red. We continue walking from v_0 starting from blue edge and oscillating colors. However, this time, we are not allowed to use any edges of W_1. As before, we are guaranteed that we will not get stuck and we need to get back to v_0; let W_2 be the second walk. If W_2 is even, then removing this walk gives us the desired property. If it is odd, then removing both W_1 and W_2 does the trick. (Recall that W_1 and W_2 are edge disjoint.) Finally, let us mention that the condition about the maximum degree is at least 6 is needed. One can easily construct a counter-example when $\Delta(G) = 4$.

Problem 4.2.1. Consider a square grid of size 25×25 that has a smaller square grid of size 5×5 cut out from its bottom left corner. Can you cover the remaining cells with 100 blocks of size 1×6 or 2×3?

Solution. Label the grid so that the bottom left cell has label $(1, 1)$ and the top right one has label $(25, 25)$. Put 1 in a cell with label (i, j) if $i+j$ is divisible by 3; otherwise, put 0—see Figure 8.4. Observe that each block (regardless whether it is of size 1×6 or 2×3) covers precisely two 1's. There are 100 such blocks but the number of 1's to cover is 199. To see this note that in each row we have either 8 or 9 ones (before removing 5×5 square). We have exactly 17 rows with 8 ones and 8 rows with 9 ones. It follows that the 25×25 grid contains $17 \cdot 8 + 8 \cdot 9 = 208$ ones. After removing 9 of them from the bottom left 5×5 square grid we are left with 199 ones. The conclusion is that no matter how hard we try, we will not be able to cover the remaining cells with 100 blocks.

Problem 4.2.2. Prove that it is impossible to cover a square grid of size 9×9 with tiles of size 1×5 or 1×6.

Solution. Let us first observe that any potential covering would have to use at least 14 blocks (as $13 \cdot 6 = 78 < 81 = 9 \cdot 9$). Hence, there must be at least 7 blocks that are vertical or at least 7 that are horizontal. Without loss of generality, we may assume that there are at least 7 horizontal blocks. However, this means that there must be exactly 9 horizontal blocks because the middle column (column 5) is covered by each of such blocks and so no vertical block can intersect it. Consider now the middle row (row 5). At least 5 cells are covered by the horizontal block so there are at most 4 vertical blocks. But $4 + 9 = 13 < 14$.

Problem 4.2.3. Can you cover a square grid of size 10×10 with 25 "T-shaped" blocks consisting of 4 small squares?

Solution. As usual, label the grid so that the bottom left cell has label $(1, 1)$ and the top right one has label $(10, 10)$. Put 1 in a cell with label (i, j) if $i+j$ is even, and 0 otherwise. The number of 1's covered by each block is 1 or 3. Since there are 25 blocks they are going to cover an odd number of 1's. On the other hand, the 10×10 grid contains an even number of 1's (precisely 50). Hence, our task is not possible.

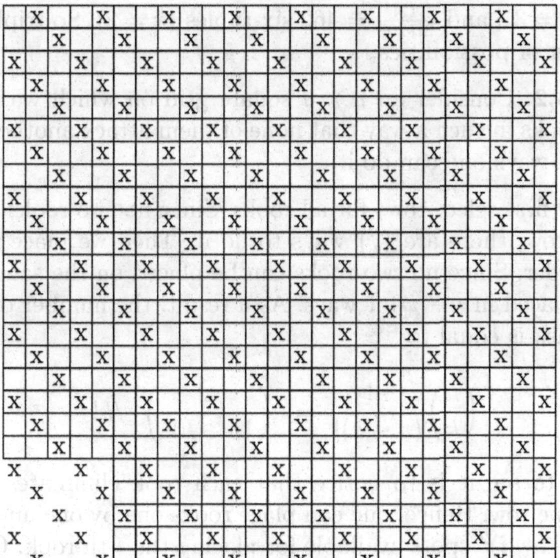

FIGURE 8.4: Illustration for Problem 4.2.1. We put 'x' in places where 1 should be placed. We also shown 'x' in the 5×5 grid that was removed.

Problem 4.3.1. The class consists of 12 people. Count in how many ways one can divide them into 6 pairs, 4 triples, 3 quadruples, and 2 six-tuples. Which option yields the largest number of possibilities?

Solution. Suppose that we have n people and we want to divide them into k groups. Assume that $k \mid n$ so that there will be $s = n/k$ people in each group. In order to generate a division, we can first assign unique numbers from the set $[n]$ to all the people. This can be done in $n!$ ways. Now, people with numbers from 1 to s form the first group, people with numbers from $s + 1$ to $2s$ form the second group, and so on.

The problem is that a given group is generated multiple times. First of all, we do not care if $\{1, 2, \ldots, s\}$ form the first or the second group. That means that we can rearrange the k groups in any way we want. There are $k!$ ways to do it. Moreover, in any particular group such as $\{1, 2, \ldots, s\}$, it does not matter who has 1 assigned and who has 2. This gives us another factor of $s!$ per group. Combining these observations together we get that there are $\frac{n!}{k!(s!)^k}$ such divisions. (To be slightly more formal, one can construct a bijection between the family of unique divisions and the partition of the set of permutations into sets of size $k!(s!)^k$.) Alternatively, one can count it as $\prod_{i=0}^{k-1} \binom{n-is}{s}/k!$, because we can iteratively select s element sets and then observe that each division is counted $k!$ times.

In our particular situation, we have $n = 12$ people so we can divide them into $\frac{12!}{6!2^6} = 10{,}395$ pairs ($k = 6$), $\frac{12!}{4!6^4} = 15{,}400$ triples ($k = 4$), $\frac{12!}{3!24^3} = 5{,}775$

quadruples ($k = 3$), and $\frac{12!}{2!720^2} = 462$ six-tuples ($k = 2$). So triples give us the largest number of possibilities.

Problem 4.3.2. Consider an $n \times n$ square grid on which we want to place $k \le n$ chess rooks in such a way that none of them attack another rook. Count the number of ways one can do it.

Solution. We first select rows for all rooks. Since no two rooks can be placed on the same row, there are $\binom{n}{k}$ ways to do it. Then we place rooks, one by one, in any order. Since no two rooks can be placed on the same column, ith rook can be placed in $n - i + 1$ ways. As a result, the number of ways we can achieve our task is equal to

$$\binom{n}{k} \frac{n!}{(n-k)!} = \left(\frac{n!}{(n-k)!}\right)^2 / k! .$$

Another way to see it is to notice that each rook eliminates precisely one column and one row. Hence, one can place rooks one by one and observe that there are $(n + 1 - i)^2$ spots available for placing the i-th rook. Once we finish the process, there are $k!$ duplicates because of $k!$ possible permutations of placing rooks.

Problem 4.3.3. Create all possible 4-digit numbers using digits from set $[9] = \{1, 2, 3, 4, 5, 6, 7, 8, 9\}$. Find the sum of those numbers.

Solution. Let us first notice that there are 9^4 numbers that satisfy our requirement. Then, notice that number $c_1 c_2 c_3 c_4$ can be associated with number $d_1 d_2 d_3 d_4$, where $d_i = 10 - c_i$. As a result, we get a bijection from the set of possible numbers to itself. Additionally, the sum of $c_1 c_2 c_3 c_4$ and $d_1 d_2 d_3 d_4$ is equal to 11,110, independently of the number used. Therefore, the sum of all numbers is equal to $9^4 \cdot 11{,}110/2 = 36{,}446{,}355$. (We had to divide the value by 2 because each number was counted twice.)

It is easy to verify that our result is correct using the following one line Julia code: `sum(x for x in 1000:9999 if !(0 in digits(x)))`.

Problem 4.3.4. Alice has 20 balls, all different. She first splits them into two piles and then she picks one of the piles with at least two balls, and splits it into two. She repeats this until each pile has only one ball. Find the number of ways in which she can carry out this procedure.

Solution. The number of ways this splitting procedure can be carried out is the same as the number of ways to do it backward; that is, Alice can start with 20 piles, each of them containing only one ball, and then keep merging piles together. Indeed, to see this let us note that any sequence of splits of one set of 20 balls that results in 20 sets can be uniquely reversed to get a sequence of merges from 20 sets to one set. In other words, there is a bijection between the two sets corresponding to these two operations and so it does not matter which one we concentrate on. Working backward is slightly easier. In the i-th move, Alice has $21 - i$ sets to choose from so she can do $\binom{21-i}{2} = (21 - i)(20 - i)/2$

different merges. As she does 19 moves in total, we get that the number of ways is equal to

$$\prod_{i=1}^{19} \frac{(21-i)(20-i)}{2} = 20! \cdot 19!/2^{19}.$$

Problem 4.4.1. There is a club with 100 members where there are 1,000 pairs of friends. We want to pick a three person team from the club with one team member selected as a team leader. The procedure is that one club member first becomes a leader. The leader then chooses two followers from his/her friends and the team is formed. Show that it is possible to pick a team from the club in at least 19,000 ways.

Solution. Suppose that the ith club member has d_i friends. Since there are 1,000 friends in the club, we get that $\sum_{i=1}^{100} d_i = 2,000$. Therefore, if we choose the ith member as a leader, he/she can form $\binom{d_i}{2} = d_i(d_i - 1)/2$ unique teams. After taking the sum over all club members we get the number of possible teams is equal to

$$\sum_{i=1}^{100} \binom{d_i}{2} = \frac{1}{2} \sum_{i=1}^{100} (d_i^2 - d_i) = \frac{1}{2} \sum_{i=1}^{100} d_i^2 - \frac{1}{2} \sum_{i=1}^{100} d_i = \frac{1}{2} \sum_{i=1}^{100} d_i^2 - 1,000.$$

Now we see from Jensen's inequality applied to $f(x) = x^2$ (see Section 1.1) that

$$\sum_{i=1}^{100} d_i^2 = 100 \sum_{i=1}^{100} \frac{d_i^2}{100} \geq 100 \left(\sum_{i=1}^{100} d_i/100 \right)^2 = 40,000.$$

Let us note that this problem can be reformulated in the language of graph theory. One can consider a "friendship graph" consisting of 100 vertices corresponding to the club members and edges that represent friendship relationships. Our goal is to show that any graph on 100 vertices with the average degree 20 has at least 19,000 paths of length 2. Indeed, each path abc of length 2 corresponds to a team with b being the leader of the team.

Let us also note that the lower bound we just proved is best possible as it is achieved when every member of the club has precisely 20 friends. The corresponding arrangement exists and an underlying graph is called a 20-regular graph. In order to see one possible example, imagine all members of the club sitting in a circle. Each member is a friend with 10 people to the left and with 10 people to the right.

Problem 4.4.2. Consider the following combinatorial game between two players, Builder and Painter. The game starts with the empty graph on 400 vertices. In each round, Builder presents an edge uv between two non-adjacent vertices u and v which has to be immediately colored red or blue by Painter. Show that Builder can force Painter to create a monochromatic (that is, either red or blue) path on 100 vertices in 400 rounds.

Solution. Let r_t and b_t be the number of vertices in a longest red and, respectively, blue path after t rounds of the game. Clearly, both r_t and b_t are nondecreasing functions of t. We will show that Builder has a strategy that in two rounds increases the sum of r_t and b_t by 1; that is, for each $t \in \mathbf{N}$, $r_{2t} + b_{2t} \geq t$. In particular, it will follow that $r_{400} + b_{400} \geq 200$ and so $\max\{r_{400}, b_{400}\} \geq 100$, as required.

In order to see this, we will prove slightly stronger claim and insist that the two paths (red and blue) are vertex disjoint, that is, have no common vertices. Moreover, we will require that there are two endpoints, one in each path, that are not adjacent. At time 0, we initiate the process by picking two different vertices. The desired property as well as the desired lower bound trivially holds: $r_0 + t_0 = 2 \geq 0$.

Suppose now that at time $2t$ we have two disjoint paths, a red path $R_t = (v_1, \ldots, v_{r_t})$ and a blue path $B_t = (w_1, \ldots, w_{b_t})$. Moreover, the desired property (there is no edge between v_{r_t} and w_{b_t}) and the desired condition ($r_{2t} + b_{2t} \geq t$) are met. Bulder presents edge $v_{r_t} w_{b_t}$. Without loss of generality, we may assume that Painter paints it red. Builder now presents and edge from w_{b_t} to a new vertex v. If Painter paints it blue then we discard the edge $v_{r_t} w_{b_t}$ but keep $w_{b_t} v$ to extend blue path. We get

$$r_{2t+2} + b_{2t+2} \;=\; r_{2t} + (b_{2t} + 1) \;=\; r_{2t} + b_{2t} + 1,$$

the desired lower bound holds, the two paths are disjoint, and the corresponding endpoints are not adjacent. Suppose then that Painter paints it red. This time, we discard vertex w_{b_t} from the blue path, making it shorter. If, as a result, the blue path becomes empty, we choose any unused vertex as initialization of the blue path. We get that

$$r_{2t+2} + b_{2t+2} \;=\; (r_{2t} + 2) + (b_{2t} - 1) \;=\; r_{2t} + b_{2t} + 1,$$

and the desired bounds holds too.

Problem 4.4.3. Consider a chess club consisting of 4^t members for some $t \in \mathbf{N}$; some of the members know each other. Show that there exist t members that all know each other, or there exist t members such that no two of them know each other.

Solution. Let us first reformulate the problem in the language of graph theory. Friendships between members of the club can be represented by coloring the edges of the complete graph on $4^t = 2^{2t}$ vertices using two colors, say, red and blue. Edge uv is colored red if the members corresponding to u and v know each other; otherwise, uv is colored blue. Our goal is to show that no matter how the edges of the complete graph on 2^{2t} vertices are colored, there exists a set of t vertices that induces a monochromatic graph; that is, all edges of this induced graph are red or all of them are blue.

Start the process with selecting an arbitrary vertex v. Note that v has an odd number of neighbors, namely $2^{2t} - 1$. As a result, either at least $2^{2t}/2 =$

2^{2t-1} of them are adjacent to v via red edge or at least 2^{2t-1} of them are adjacent to v via blue edge. If v is adjacent to more red edges than blue ones, then we assign label R to v, remove v and all of its neighbors that are adjacent to v via blue edges. For simplicity, if needed, we additionally and arbitrarily remove some vertices to keep the number of them to be exactly 2^{2t-1}. On the other hand, if majority of neighbors of v are blue, then v gets label B assigned. This time, we remove v and its neighbors that are adjacent to v via red edges, and remove some additional vertices so that the number of vertices left is 2^{2t-1}.

We repeat the process on the remaining subset of vertices until we exhaust all of them. Since the number of vertices decreases by a factor of 2 each time, the process lasts $2t$ rounds. The last round, round $2t$, is slightly different as there is only one vertex left. The last vertex can get any label assigned, say, B. It follows that there are at least t vertices with label R assigned or at least t vertices with label B. Due to symmetry, we may assume without loss of generality that at least t vertices have label R assigned. It is straightforward to see that all edges in the complete graph induced by these vertices are red. The desired property is satisfied.

Finally, let us mention that this problem is a famous and an extremely difficult problem. The corresponding numbers that we tried to bound in this problem are called the Ramsey numbers. In fact, with slightly more work, one can replace 4^t by $\binom{2t}{t} \le 4^t/\sqrt{2t}$. However, perhaps surprisingly, it is not known if it can be replaced by $(4 - \epsilon)^t$ for some $\epsilon > 0$.

Problem 4.5.1. Let $k \in \mathbf{N}$ and fix $N = N(k) := \lfloor 2^{k/2} \rfloor$. Show that it is possible to partition set $X := [N] = \{1, 2, \ldots, N\}$ into two subsets A and B such that neither A nor B contains an arithmetic progression of length k.

Solution. Let us first make an obvious observation. If $a_1, a_2, \ldots a_k$ is an arithmetic progression, then so is $a_k, a_{k-1}, \ldots, a_1$. Hence, without loss of generality, we may restrict ourselves to increasing progressions.

Since the first two terms of an increasing arithmetic progression uniquely defines it, the number of increasing arithmetic progressions of length k in X is at most $\binom{N}{2} = \frac{N(N-1)}{2} < N^2/2 \le 2^{k-1}$. Consider then a random partition of X into two subsets A and B; that is, each element of X is independently put into A with probability $1/2$. Clearly, the probability that a given k-element sequence is in A is equal to $(1/2)^k$ and the same holds for B. It follows that the expected number of arithmetic sequences of length k entirely contained in one of the two sets is less than $2^{k-1} \cdot 2 \cdot (1/2)^k = 1$. By the probabilistic method, we get that the desired partition exists.

Problem 4.5.2. Show that for any $n \in \mathbf{N}$, there is a tournament for which n basketball teams are participating in, and for which there are at least $k = n!/2^{n-1}$ orderings $t_1, \ldots, t_n$ such that team t_i won against team t_{i+1}, for all $i \in [n-1]$.

Solution. We will compute the expected number of desired orderings in a random tournament where for each pair A, B of teams, team A wins against team B with probability $1/2$, independently from all other games. Let us fix any of the $n!$ permutations of teams: $t_1, \ldots, t_n$. The probability that it satisfies the desired property is equal to $(1/2)^{n-1}$. Hence, the expected number of desired orderings is equal to $n!/2^{n-1}$ and, by the probabilistic method, there must exist a tournament for which there are at least $n!/2^{n-1}$ such orderings. (Surprisingly, this trivial argument gives the results that is almost as best as possible. It is known that in *any* tournament, the number of such orderings is $O(n^{3/2} n!/2^{n-1})$.)

Problem 4.5.3. Consider a graph with T triangles. Show that it is possible to color the edges of this graph with two colors so that the number of monochromatic triangles is at most $T/4$.

Solution. Let us color the edges of this graph at random, uniformly and independently. The probability that a given triangle is monochromatic is equal to $2 \cdot (1/2)^3 = 1/4$. Therefore, the expected number of monochromatic triangles is $T/4$. It follows that there must exist a coloring for which the number of monochromatic triangles is less than or equal to $T/4$. (Moreover, if T is not divisible by 4, then a strict inequality holds.)

Problem 4.5.4. There are 100 people invited to the party; 450 pairs of people know each other. Show that it is possible to select 10 people so that no two of them know each other.

Solution. Since the acquaintances between people invited to the party can be represented as a graph, our problem can be reformulated in the language of graph theory. Our goal is to show that any graph G on $n = 100$ vertices and $m = 450$ edges has an independent set of size 10.

For a given permutation π of the vertices, we put vertex v into set $S = S(\pi)$ if no neighbor of v follows it in the permutation. Clearly, S forms an independent set. Let π now be a random permutation of the vertices of G taken with uniform distribution; that is, each permutation occurs with probability $1/n!$. For a given vertex $v \in V$, let $d^+(v)$ be the number of neighbors of v that follow it in the permutation. The random variable $d^+(v)$ attains each of the values $0, 1, \ldots, \deg(v)$ with probability $1/(\deg(v)+1)$. Indeed, this follows from the fact that the random permutation π induces a uniform random permutation on the set of $\deg(v) + 1$ vertices consisting of v and its neighbors (to see this one can fix the positions of non-neighbors of v first, and then fixing one of $\deg(v) + 1$ free positions for the vertex v will yield desired values of $d^+(v)$ with uniform distribution). Therefore the expected number of vertices with $d^+(v) = 0$ is equal to $C := \sum_{v \in V} \frac{1}{\deg(v)+1}$. This implies that there exists a specific permutation with at least C vertices of this type, which form an independent set.

The last part is an optimization problem. Notice that the average degree is equal to $d := \sum \deg(v)/n = 2m/n = 2 \cdot 450/100 = 9$. It follows from Jensen's inequality (see Section 1.1) applied to function $1/t$ that

$$C = \sum_{v \in V} \frac{1}{\deg(v) + 1} = n \frac{\sum_{v \in V} \frac{1}{\deg(v)+1}}{n} \geq n \frac{1}{\frac{\sum_{v \in V}(\deg(v)+1)}{n}}$$

$$= \frac{n}{\frac{nd+n}{n}} = \frac{n}{d+1} = \frac{100}{9+1} = 10.$$

Problem 4.6.1. Consider an urn that initially contains one white and one black ball. We repeatedly perform the following process. In a given round, one ball is drawn randomly from the urn and its color is observed. The ball is then returned to the urn, and an additional ball of the same color is added to the urn. We repeat this selection process for 50 rounds so that the urn contains 52 balls. What number of white balls is the most probable?

Solution. Let $p_{k,n}$ be the probability of seeing exactly k white balls in an urn having n balls in total. One could write down the recursion for $p_{k,n}$ and then solve it but it seems that it is easier to perform calculations for the few first rounds to make a natural conjecture that can be then proved by induction. During the first round, we select a white ball with probability $1/2$ so we end up with 2 white balls with probability $1/2$ and, otherwise, we stay with 1 white ball. It follows that $p_{1,3} = p_{2,3} = 1/2$. In order to see 1 white ball after two rounds, we have to chose black balls during the two rounds and so $p_{1,4} = (1/2) \cdot (2/3) = 1/3$. Similarly, to see 1 black ball, we have to select white balls twice and so $p_{3,4} = 1/3$ as well. It follows that $p_{2,4} = 1 - p_{1,4} - p_{3,4} = 1/3$. The pattern occurs naturally and we conjecture that for any $n \geq 2$ and any $1 \leq k \leq n - 1$, $p_{k,n} = 1/(n-1)$. We also see that $p_{0,n} = p_{n,n} = 0$, as there is always at least one black and one white ball in the urn.

We prove the claim by induction. The base case ($n = 2$) is trivial: $p_{1,2} = 1$ (in fact, we already checked it for $n = 3$ and $n = 4$). For the inductive step, suppose that $p_{k,n} = 1/(n-1)$ for some $n \geq 2$ and all $1 \leq k \leq n - 1$. Fix k such that $1 \leq k \leq n$. Our goal is to show that $p_{k,n+1} = 1/n$. Note that in order to see k white balls at the end of some round we need to have k white balls in the previous round and select a black ball, or have $k - 1$ white balls and select a white one. We get that if $k > 1$, then

$$p_{k,n+1} = \frac{n-k}{n} p_{k,n} + \frac{k-1}{n} p_{k-1,n}$$

$$= \frac{n-k}{n} \cdot \frac{1}{n-1} + \frac{k-1}{n} \cdot \frac{1}{n-1} = \frac{n-1}{n} \cdot \frac{1}{n-1} = \frac{1}{n},$$

while for $k = 1$ we have

$$p_{1,n+1} = \frac{n-1}{n} p_{1,n} + \frac{0}{n} p_{0,n} = \frac{n-1}{n} \cdot \frac{1}{n-1} = \frac{1}{n}.$$

The inductive hypothesis holds and the proof is finished.

Finally, let us mention that this problem is an easy, specific case of the famous Pólya urn model. This endows the urn with a self-reinforcing property sometimes expressed as the *rich get richer*. In some sense, the Pólya urn model is the "opposite" of the model of sampling without replacement, where every time a particular value is observed, it is less likely to be observed again, whereas in the Pólya urn model, an observed value is more likely to be observed again.

Problem 4.6.2. There are 65 participants competing in a ski jumping tournament. They take turns and perform their jumps in a given sequence. We assume that no two jumpers obtain the same result and that each final resulting order of participants is equally probable. At each given round of the tournament, the person that has obtained the best result thus far is called a leader. Prove that the probability that the leader changed exactly once during the whole tournament is greater than $1/16$.

Solution. Let $\pi : [n] \to [n]$ be the final order/permutation of jumpers. In particular, $\pi(1)$ is the winner of the tournament. Our assumption is that π is a random permutation; that is, for a given permutation π_0 of $[n]$, we have that $\pi = \pi_0$ with probability $1/n!$. Let $p_{i,n}$ be the probability that the leader changed exactly i times during the tournament of n participants. Our goal is to show that $p_{1,65} > 1/16$.

Let $q_n(k)$ be the probability that the kth participant won the tournament of n ski jumpers and that the leader changed exactly once during the whole event. If the first participant wins, then he is the leader from the very beginning and no change in the leadership occurs. It follows that $q_n(1) = 0$ and so

$$p_{1,n} = \sum_{k=1}^{n} q_n(k) = \sum_{k=2}^{n} q_n(k).$$

There are many ways to generate random permutations of $[n]$. The following one will be very convenient to compute $q_n(k)$. We start with 1 and then place 2 before 1 with probability $1/2$; otherwise, it will be placed after 1. After that we place 3 in a random place and move on to 4. Formally, given a partial (random) permutation of elements $1, \ldots, k-1$ (for some integer $k \geq 2$), we place k uniformly at random in one of the k possible places. This point of view has an important implication for our problem. We immediately get that the kth participant becomes a leader (at least till the next participant jumps) with probability $1/k$. It follows that

$$q_n(k) = \frac{1}{2} \cdot \frac{2}{3} \cdots \frac{k-2}{k-1} \cdot \frac{1}{k} \cdot \frac{k}{k+1} \cdot \frac{k+1}{k+2} \cdots \frac{n-1}{n} = \frac{1}{n(k-1)}.$$

As a result,

$$p_{1,n} = \sum_{k=2}^{n} \frac{1}{n(k-1)} = \frac{1}{n} \sum_{k=2}^{n} \frac{1}{k-1} = \frac{1}{n} H_{n-1},$$

where H_{n-1} is the $(n-1)$-st harmonic number.

One way to prove the desired lower bound for $p_{1,n}$ is to compare the harmonic series with another divergent series where each denominator is replaced with the next largest power of two:

$$
\begin{aligned}
H_{2^i} &= \frac{1}{1} + \frac{1}{2} + \frac{1}{3} + \frac{1}{4} + \frac{1}{5} + \frac{1}{6} + \frac{1}{7} + \frac{1}{8} + \frac{1}{9} + \ldots + \frac{1}{2^i} \\
&> \frac{1}{1} + \frac{1}{2} + \frac{1}{4} + \frac{1}{4} + \frac{1}{8} + \frac{1}{8} + \frac{1}{8} + \frac{1}{8} + \frac{1}{16} + \ldots + \frac{1}{2^i} \quad (8.11) \\
&= 1 + \frac{1}{2} + 2 \cdot \frac{1}{4} + 4 \cdot \frac{1}{8} + \ldots + 2^{i-1} \cdot \frac{1}{2^i} = 1 + \frac{i}{2}.
\end{aligned}
$$

It follows that

$$
p_{1,65} = \frac{1}{65} H_{2^6} > \frac{1}{65}\left(1 + \frac{6}{2}\right) = \frac{4}{65},
$$

which is very close to the desired bound of $1/4$ but, unfortunately, slightly smaller than that. Fortunately, it is easy to improve the bound (8.11) for H_{2^i} to, for example, $H_{2^i} > 1 + i/2 + 1/3 - 1/4$. This time we get the desired bound:

$$
p_{1,65} = \frac{1}{65} H_{2^6} > \frac{1}{65}\left(1 + \frac{6}{2} + \frac{1}{3} - \frac{1}{4}\right) = \frac{49}{780} > 0.0628 > \frac{1}{16}.
$$

We are done with this problem but let us make two additional comments. Let us first note that another way to lower bound H_n is to do the following. We know from Section 1.5 that for any $i \in \mathbf{N}$,

$$
\left(\frac{i+1}{i}\right)^i = \left(1 + \frac{1}{i}\right)^i < e.
$$

After taking the natural logarithm of both sides of this inequality, we get

$$
\ln(i+1) - \ln(i) < \frac{1}{i}
$$

and so

$$
H_n = \sum_{i=1}^{n} \frac{1}{i} > \sum_{i=1}^{n}\left(\ln(i+1) - \ln(i)\right) = \ln(n+1) - \ln(1) = \ln(n+1).
$$

This bound is quite good as one can show that $H_n < \ln(n) + 1$ (see solution to Problem 2.6.1) and asymptotically $H_n = \ln(n) + \gamma + o(1)$, where $\gamma \approx 0.577216$ is the Euler-Mascheroni constant. It follows that

$$
p_{1,65} = \frac{1}{65} H_{64} > \frac{\ln(65)}{65} > 0.0642.
$$

Let us also mention about $p_{i,n}$ for some values of $i \neq 1$. Clearly, $p_{0,n} =$

$(n-1)!/n! = 1/n$ as there are $(n-1)!$ permutations that correspond to situations when the first participant is the winner. On the other hand, for $i \in \mathbf{N}$ such that $2 \le i \le n-1$, we can get a recursive formula by independently considering cases when the last change of the leader occurred at round $k+1$. We get

$$p_{i,n} = \sum_{k=i}^{n-1} p_{i-1,k} \cdot \frac{1}{k+1} \cdot \frac{k+1}{k+2} \cdot \frac{k+2}{k+2} \cdots \frac{n-1}{n} = \frac{1}{n} \sum_{k=i}^{n-1} p_{i-1,k}.$$

Using this recursion, with computer support, we can easily find $p_{i,n}$ for some given parameters. Here is a simple program written in Julia that does this.

```
function probs65()
    probs = Dict{Tuple{Int, Int}, Float64}()

    function prob(j,k)
        if !haskey(probs, (j,k))
            if j == 0
                probs[(j,k)] = 1/k
            else
                probs[(j,k)] = 1/k*sum(prob(j-1,i) for i in j:(k-1))
            end
        end
        return probs[(j,k)]
    end
    [prob(j, 65) for j in 0:64]
end
```

You can run it by writing `probs65()` in the Julia session.

The first few probabilities are $p_{0,65} \approx 0.0154$, $p_{1,65} \approx 0.073$, $p_{2,65} \approx 0.1606$, $p_{3,65} \approx 0.2204$, $p_{4,65} \approx 0.2138$, $p_{5,65} \approx 0.157$, and $p_{6,65} \approx 0.0913$. It follows that the most probable case is to see 3 changes of the leader in the tournament, and 4 changes are only slightly less probable.

If one is unsure about our computations, it is relatively easy to check them using a computer simulation. Here is another Julia code that simulates the tournament.

```
function sim_tournament()
    # the jump length of first jumper drawn uniformly from [0,1) interval
    best_length = rand()
    best_changes = 0
    # simulate jumps of consecutive jumpers
    # and count the number of leader changes
    for i in 2:65
        jump_length = rand()
        if jump_length > best_length
            best_length = jump_length
            best_changes += 1
```

```
            end
        end
        return best_changes
    end

    function run_simulation()
        # simprobs65 will hold the counts of observed tournament results
        simprobs65 = zeros(65)
        sim_runs = 10_000_000
        for i in 1:sim_runs
            # we have to add 1 to sim_tournament() result as
            # in Julia arrays are 1-based and
            # a minimal number of leader changes in the tournament is 0
            simprobs65[sim_tournament() + 1] += 1
        end
        simprobs65 / sim_runs
    end

    simprobs65 = run_simulation()
```

The first few probabilities estimated by the simulation are $p_{0,65} \approx 0.0155$, $p_{1,65} \approx 0.0729$, $p_{2,65} \approx 0.1604$, $p_{3,65} \approx 0.2207$, $p_{4,65} \approx 0.2138$, $p_{5,65} \approx 0.1569$, and $p_{6,65} \approx 0.0914$ (you might get a slightly different results because this time we use a rand simulation to generate the outputs). They are close to the exact values calculated earlier and so we are quite confident that no mistake was made.

Problem 4.6.3. Three random events satisfy the following three conditions: (a) their probabilities are all equal to p for some $p \in [0, 1]$, (b) they are pairwise independent, and (c) all of them cannot happen at the same time. What is the maximum value that p may take?

Solution. Denote the the events by A_i (for $i \in [3]$). It follows from condition (a) that there exists $p \in [0, 1]$ such that $p = \mathbf{P}(A_i)$ for all i. By condition (b) we get that $\mathbf{P}(A_i \cap A_j) = p^2$ for $i \neq j$. Finally, condition (c) implies that $\mathbf{P}(A_1 \cap A_2 \cap A_3) = 0$. It follows immediately from the inclusion–exclusion principle (4.9) that

$$
\begin{aligned}
\mathbf{P}(A_1 \cup A_2 \cup A_3) &= \mathbf{P}(A_1) + \mathbf{P}(A_2) + \mathbf{P}(A_3) \\
&\quad - \mathbf{P}(A_1 \cap A_2) - \mathbf{P}(A_1 \cap A_3) - \mathbf{P}(A_2 \cap A_3) \\
&\quad + \mathbf{P}(A_1 \cap A_2 \cap A_3) \\
&= 3p - 3p^2 + 0.
\end{aligned}
$$

On the other hand, clearly $(A_1 \cap A_2) \cup (A_1 \cap A_3) \cup (A_2 \cap A_3) \subseteq A_1 \cup A_2 \cup A_3$. Using the inclusion–exclusion principle the same way as before, we get that $\mathbf{P}((A_1 \cap A_2) \cup (A_1 \cap A_3) \cup (A_2 \cap A_3)) = 3p^2$. It follows that

$$
3p^2 = \mathbf{P}((A_1 \cap A_2) \cup (A_1 \cap A_3) \cup (A_2 \cap A_3)) \leq \mathbf{P}(A_1 \cup A_2 \cup A_3) = 3p - 3p^2,
$$

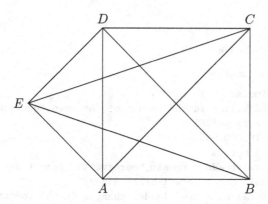

FIGURE 8.5: Configuration of five points forming two obtuse triangles. We take $|AB| = |BC| = |CD|$ and the following angles are right $\sphericalangle DEA$, $\sphericalangle ABC$, $\sphericalangle BCD$, $\sphericalangle CDA$, $\sphericalangle DAB$, $\sphericalangle BDE$ and $\sphericalangle EAC$. Out of the 10 triangles created by points A, B, C, D, and E only triangles EDC and EAB are obtuse.

and so $3p(1 - 2p) \geq 0$. This implies that $p \leq 1/2$.

It remains to show that $p = 1/2$ is achievable, that is, that one can design an experiment and the three events A_1, A_2, and A_3 such that $\mathbf{P}(A_i) = 1/2$ for all i, and conditions (b) and (c) are met. Suppose that we roll an 8-sided fair die. A_1 represents the event that an even number is rolled, A_2 represents the event that the number rolled is less than or equal to 4, and A_3 represents the event than a number from the set $\{1, 3, 6, 8\}$ is rolled. It is straightforward to check that all conditions are met.

Problem 4.7.1. Let P be a set of five points on a plane with the property that no three of them lie on the same line. Denote by $a(P)$ the number of obtuse triangles whose vertices lie in P. Find the minimum and the maximum value that $a(P)$ can attain over all possible sets P.

Solution. First note that 5 points, A, B, C, D, and E, create 10 triangles as one can select 3 points from the set of 5 points in $\binom{5}{3} = 10$ ways and each choice yields a unique triangle. We will first prove that $a(P) \geq 2$ for any configuration P of 5 points. This bound is best possible as shown in Figure 8.5.

We will independently consider two cases. Let us first assume that some point, say point A, lies inside a convex hull of the remaining four points. As the points are not colinear, it must lie inside a triangle formed by some other three points, say, B, C, and D. Note that the sum of the three angles $\sphericalangle BAC$, $\sphericalangle BAD$, and $\sphericalangle CAD$ is equal to 2π and all of them are less than π. As a result, at least two of them are obtuse. These two angles form the two obtuse triangles, as required. Suppose now that the five points form a convex pentagon. Note that the sum of the interior angles of the pentagon is equal to

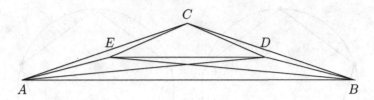

FIGURE 8.6: Configuration of five points forming ten obtuse triangles.

3π. As each individual angle is less than π, we get that at least two of them are obtuse and, as before, at least two obtuse triangles are present.

On the other hand, the maximum number of obtuse triangles formed by five points is 10, that is, it is possible that all triangles are obtuse. Such configuration is shown in Figure 8.6.

Problem 4.7.2. Every point on a circle is painted with one of three colors. Prove that there are three points on the circle that have the same color and form an isosceles triangle.

Solution. In order to warm up, let us consider a much simpler version of this problem when only two colors are available. Our goal is the same, we want to show that there are three points on the circle that have the same color and form an isosceles triangle. In order to see this one can take any 5 points that form a regular pentagon inscribed in the circle. It is enough to concentrate on these 5 points as no matter how they are colored, the desired triangle has to be created. Indeed, observe that at least three of these vertices must be painted with the same color. It follows that two of them, say A and B, must be adjacent. If the third vertex, say C, is adjacent to either A or B, then they form an isosceles triangle (note that they form the two sides of the pentagon). On the other hand, if C is not adjacent neither to A nor to B, then $|AC| = |BC|$ as they are diagonals of the pentagram. See Figure 8.7 for an illustration of both cases.

Let us now come back to the original problem with three colors. Our proof technique is the same as before. However, instead of regular pentagon we will use 13-sided regular polygon inscribed in the circle. Let us label the vertices of this polygon with numbers from 1 to 13, anticlockwise. Clearly, at least 5 of the vertices must be painted with the same color. We will concentrate on them and disregard the remaining vertices of the polygon (and an infinite number of other points from the circle). Let us denote their unique labels as $a_1, a_2, \ldots, a_5 \in [13]$. We will say that two vertices are at distance k if the number of vertices (from our polygon) that separate them is equal to k. Note that the smallest distance is 0 (corresponding to the situation when the two vertices are adjacent) and the largest distance is 5.

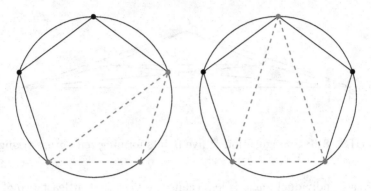

FIGURE 8.7: Two possible scenarios of two-colorings with at least three gray points.

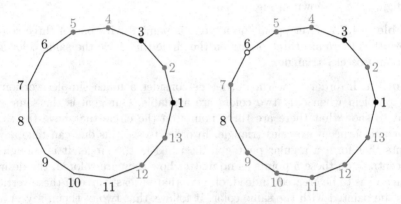

FIGURE 8.8: coloring 13-gon. Case 1.

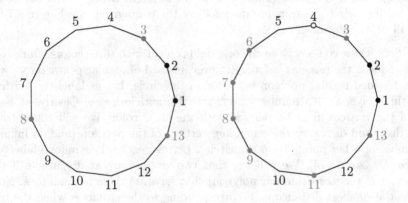

FIGURE 8.9: coloring 13-gon. Case 2a.

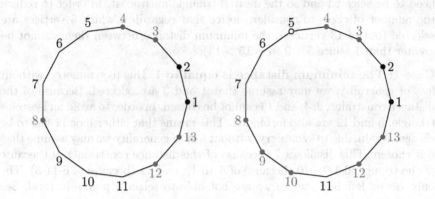

FIGURE 8.10: coloring 13-gon. Case 2b.

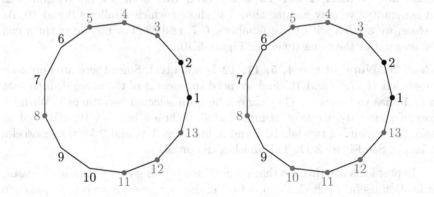

FIGURE 8.11: coloring 13-gon. Case 2c.

It is easy to see that three vertices form an isosceles triangle if and only if the distance from one of them to the remaining two is the same. We will do case analysis to show that such situation cannot be avoided when 5 vertices need to be selected and so the desired triangle must exist. In order to reduce the number of cases to consider, notice that regardless which 5 vertices are selected (out of 13 vertices), the minimum distance between them cannot be greater than 1 (since $5 + 2 \cdot 5 = 15 > 13$).

Case 1: The minimum distance is equal to 1. Due to symmetry, without loss of generality, we may assume that 1 and 3 are selected. Because of the distance constraint, 2, 4, and 13 cannot be chosen. In order to avoid an isosceles triangle, 5 and 12 are also forbidden. This means that either 6 or 11 has to be chosen. Again, due to symmetry, without loss of generality we may assume that 6 is chosen. This disallows 7 (because of the distance constraint), 9 (because of the triangle 6-1-9), 10 (because of 6-10-1), and 11 (because of 6-11-3). The only vertex left is 8 and so we are not able to select 5 points in total. See Figure 8.8.

Case 2: The minimum distance is equal to 0. Without loss of generality, we may assume that 1 and 2 are chosen which eliminates 3, 8 and 13. We consider the following sub-cases.

Case 2a: 4 or 12 is selected. Without loss of generality, we may assume that 4 is chosen which disallows 6, 7, 9, and 11. We are left with three numbers, 5, 10, and 12, but no two of them can be selected at the same time. Indeed, 5 and 10 cannot be together because of 1, 5 and 12 cannot be together because of 2, and finally 10 and 12 cannot be together because of 1. See Figure 8.9.

Case 2b: Neither 4 nor 12 is selected but 5 or 11 is. Without loss of generality, we may assume that 5 is chosen which disallows 9 and 10. As before, we are left with three numbers, 6, 7, and 11, but no two of them can be selected at the same time. See Figure 8.10.

Case 2c. None of the 4, 5, 11, 12 is selected. Since there are only four labels left (6, 7, 9, and 10) and we need to select 3 of them, we deduce that 6 or 10 has to be taken. They cannot be both selected because of 1. Without loss of generality, we may assume that 6 is chosen but not 10. We need to take the remaining two labels, 7 and 9, but then 1, 6, and 9 form an isosceles triangle. See Figure 8.11. This finishes the proof.

In problems like this one that require investigating a large number of cases, it is often useful to check the proof using the computer. Here is a simple code written in Julia language that verifies that for any selection of 5 vertices from the 13 sided regular polygon, there always exists an isosceles triangle. Running this code by calling test13gon() returns true, ans so we have a computational confirmation of our claim.

```
using Combinatorics

function isisosceles(points)
```

```
    # make sure points are in ascending order
    sort(points)
    # calculate their distances
    d1 = points[2] - points[1]
    d2 = points[3] - points[2]
    d3 = 13 - d1 - d2
    # check if any of their distances is equal
    return d1 == d2 || d2 == d3 || d3 == d1
end


function test13gon()
    # pick all 5 element subsets from the set 1:13
    for p5 in combinations(1:13, 5)
        # check if any 3 element subset of the picked
        # 5 element subset forms an isosceles triangle
        if !any(isisosceles(p3) for p3 in combinations(p5, 3))
            return false
        end
    end
    return true
end
```

Additionally, we might search for a coloring of the 13-gon that yields a minimum number of monochromatic isosceles triangles. Below is an additional function, also written in Julia, that calculates it.

```
function isosceles_count(i)
    # convert number to its representation in base 3
    c = string(i, base=3, pad=13)
    # count monochromatic triangles that are isosceles
    count(t -> c[t[1]] == c[t[2]] == c[t[3]] && isisosceles(t),
          combinations(1:13, 3)), c
end


function best13gon()
    # initialize the sequence with monochromatic colorings
    # then traverse all non monochromatic colorings
    # we may then assume (without loss of generality)
    # that they start with digits 0 and 2
    mapreduce(isosceles_count, min, 2*3^11:3^12-1)
end
```

Running this code by calling best13gon() returns (2, "0200011022112"). We also check that init=isosceles_count(0) produces a larger number (a monochromatic coloring). It implies that it is almost possible to avoid isosceles triangles when coloring 13-gon with three colors. The returned coloring creates only two monochromatic isosceles triangles—see Figure 8.12. Let us note that it is only one example of such coloring; in other words, this example is not unique.

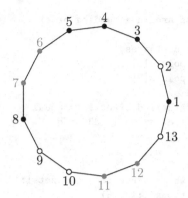

FIGURE 8.12: Optimal coloring of the 13-gon. Only two monochromatic isosceles triangles are created: 3-4-5 and 1-3-5.

Let us also notice that 13 sided regular polygon is the smallest polygon that can be used in our method. Indeed, one can color vertices of the 12 sided regular polygon so that it contains no isosceles triangle. For example, consider the following coloring: vertices $\{1, 2, 4, 5\}$ are colored red, $\{8, 9, 11, 12\}$ are colored green, and $\{3, 6, 7, 10\}$ are colored blue. Similar patterns can be found for smaller regular polygons.

Finally, let us mention about natural generalization of this problem to more than 3 colors. It was easy to show that with 2 colors it is impossible to avoid three monochromatic points on the circle that form an isosceles triangle. Generalizing this observation to 3 colors was more delicate and tedious. However, it feels that with, say, 10^{10} colors one might be able to actually avoid it. It turns out that this is impossible, regardless how many colors we have available!

We already made the key observation that we need to deal with an arbitrary number of colors: three vertices of a regular n-gon form an isosceles triangle if and only if the distance from one of them to the remaining two is the same. In other words, if the vertices are labelled with consecutive numbers from $[n]$, then it is enough that the corresponding labels a, b, and c form an arithmetic progression. Clearly, this is a sufficient condition but *not* a necessary one. As a result, we get a natural connection between our problem and the famous Van der Waerden numbers.

Let us start with a striking observation made by Van der Waerden. For any given natural numbers r and k, there is some number $n = n(r, k)$ such that if the integers from $[n]$ are colored, each with one of r available colors, then there are at least k integers, all of the same color, which form an arithmetic progression. The least such n is the **van der Waerden number** $W(r, k)$.

In our problem, we do not actually need to know $W(r, k)$, all we need is to make sure it exists. It is guaranteed by the original observation of Van der Waerden and can be proved by induction. Indeed, despite the fact that we have so powerful computers these days, only 6 nontrivial numbers and known: $W(2,3) = 9$ (easy exercise), $W(2,4) = 35$ (Chvátal (1970)), $W(2,5) = 178$ (Stevens and Shantaram (1978)), $W(2,6) = 1,132$ (Kouril and Paul (2008)), $W(3,3) = 27$ (Chvátal (1970)), and $W(4,3) = 76$ (Beeler and O'Neil (1979)).

Let us come back to the original problem. We start with a regular n-gon with $n = W(3,3) = 27$, vertices of which are labelled with numbers from $[n]$. Regardless how we color the vertices, a monochromatic arithmetic sequence of length 3 must be created and the corresponding vertices form an isosceles triangle. This argument is not optimal (27-gon is used instead of 13-gon) but it trivially generalizes to *any* number of colors. For an arbitrary number of r colors, one needs to start with $n = W(r, 3)$ and the same argument follows.

Problem 4.7.3. Take a set of $n \geq 2$ points with the property that no three of them lie on the same line. We paint all line segments formed by those points in such a way that no two line segments that have a common vertex have the same color. Find the minimum number of colors for which such coloring exists.

Solution. Since no three points lie on the same line, the number of lines is equal to $f(n)$, the number of two element subsets of an n-element set; $f(n) = \binom{n}{2} = n(n - 1)/2$. Let $g(n)$ be the maximum number of disjoint two element subsets of such a set; $g(n) = \lfloor n/2 \rfloor$ ($g(n) = n/2$ if n is even and $g(n) = (n - 1)/2$ if n is odd). Clearly, if the two line segments created by points a, b and, respectively, points c, d are colored with the same color, then all of these points are different. It follows that the maximum number of line segments that are in the same color is at most $g(n)$. Combining the two observations we get that the minimum number of colors for which the desired coloring exists is then at least

$$f(n)/g(n) = \begin{cases} n - 1 & \text{if } n \text{ is even} \\ n & \text{otherwise}. \end{cases}$$

In order to see that this bound is achievable, let us consider the following simple construction. Suppose first that n is odd. We start with n points on the circle, equally spaced (that is, these points are vertices of a regular n-gon). Clearly, there are n directions defined by the line segments formed by those points, and in each direction there are exactly $(n - 1)/2$ line segments. All line segments associated with the same direction receive the same color. See Figure 8.13 (left) for an example with $n = 5$. Since $g(n) = (n - 1)/2$ line segments have the same direction, only $f(n)/g(n) = n$ colors are used.

For an even value of n, we simply use the previous construction with $n - 1$ points that can be delt with $n - 1$ colors. Note that there are $n - 1$ colors (or, equivalently, directions) but each vertex is part of $n - 2$ line segments. Observe that, as a result, each vertex is missing one unique color. We add the

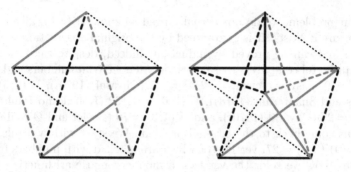

FIGURE 8.13: coloring the line segments for $n = 5$ (left) and $n = 6$ (right).

n-th point in the center of the circle and connect it to the $n - 1$ points using the missing color. See Figure 8.13 (right) for an example with $n = 5$.

Problem 4.8.1. Twenty five boys and 25 girls sit around a table. Prove that it is always possible to find a person both of whose neighbors are girls.

Solution. Let us label seats with numbers from the set [50]. Consider two subsets of people, those sitting in even and odd positions at the table. In one of those sets there must be at least 13 girls. However, this implies that there are two girls that are separated by one person. That person is the one that we are looking for.

Problem 4.8.2. A person takes at least one aspirin a day for 30 days. Show that if the person takes 45 aspirin altogether, then in some sequence of consecutive days that person takes exactly 14 aspirin.

Solution. For $i \in [30]$, let a_i be the cumulative number of aspirins taken up to and including day i. We know that $a_1 > 0$ and for all i, $a_{i+1} > a_i$ (since a person takes at least one aspirin a day). Moreover, $a_{30} = 45$ (the person takes 45 aspirin altogether). Now, for $i \in [30]$, let $b_i = a_i + 14$. The properties that we determined for a_i's imply that $b_1 > 14$, $b_{30} = 59$ and for all i, $b_{i+1} > b_i$. Putting these two sequences together, we get 60 numbers in total, all of them are positive and smaller than 60. By the pigeonhole principle we get that there are two numbers, k and ℓ, for which $a_k = b_\ell = a_\ell + 14$. It follows that $a_k - a_\ell = 14$, so the person takes exactly 14 aspirin between day $\ell + 1$ and day k.

Problem 4.8.3. Prove that, if we take $n + 1$ numbers from the set from 1 to $2n$, then in this subset there exist two numbers such that one divides the other.

Solution. Each number from the set $[2n]$ can be uniquely represented in the form $2^p q$, where $p \in \mathbf{N} \cup \{0\}$ and $q \in \mathbf{N}$ is an odd number. We say that number represented as $2^p q$ is of type q. Clearly, two numbers of the same type have the desired property, that is, one divides the other. So our goal is to show that regardless which $n + 1$ numbers are selected from $[2n]$, there will be two numbers of the same type. Since the number of types is equal to n (there are n odd numbers in $[2n]$), this follows immediately from the pigeonhole principle. Finally, let us mention that this result is sharp in the sense that one can select n numbers from $[2n]$ (namely, all odd numbers) and avoid this property.

Problem 4.9.1. Consider the Sicherman dice problem in which the restriction that each side is labelled with a positive integer is relaxed to any integer, not necessarily positive. Can you design more pairs of dice?

Solution. Notice that by adding 1 to all sides on one die and subtracting 1 from all sides on the other die does not affect the distribution for their sum. So there are infinitely many solutions, for example, $((0, 1, 1, 2, 2, 3), (2, 4, 5, 6, 7, 9))$ or $((2, 3, 4, 5, 6, 7), (0, 1, 2, 3, 4, 5))$.

Problem 4.9.2. Solve the recurrence $x_{n+1} = x_n + 2x_{n-1}$ for $n \in \mathbf{N}$, with $x_0 = 0$ and $x_1 = 1$. Verify your solution using induction.

Solution. In order to find the corresponding generating function we follow the same strategy as for the Fibonacci sequence (see the example above). We get that

$$\frac{G(x) - x}{x} = G(x) + 2xG(x).$$

Therefore

$$G(x) = \frac{1}{3}\left(\frac{1}{-1 - x} + \frac{1}{1 - 2x}\right) = \frac{1}{3}\sum_{i=0}^{\infty}\left(-(-1)^n x^n + 2^n x^n\right),$$

and so $x_n = (2^n - (-1)^n)/3$.

Verifying the solution by induction is straightforward. Since

$$x_0 = (1 - 1)/3 = 0 \quad \text{and} \quad x_1 = (2 - (-1))/3 = 1,$$

the initial conditions hold. Assuming that $x_i = (2^i - (-1)^i)/3$ for $i \geq n$, we get that

$$\begin{aligned}
x_{n+1} &= x_n + 2x_{n-1} \\
&= (2^n - (-1)^n)/3 + 2(2^{n-1} - (-1)^{n-1})/3 \\
&= (2^{n+1} - (-1)^{n+1})/3.
\end{aligned}$$

Problem 4.9.3. Your friend wants to play the following game with you. You toss three 6-sided fair dies and calculate the sum of outcomes. For every game you have to pay \$1. If the sum is 10 or 11 you get \$4, otherwise you get nothing. Is this game fair?

Solution. We will compute the probability of getting 10 and 11 by investigating $f(x)^3$, where $f(x)$ is the generating function for a fair die we have introduced in the solution. We note that

$$f(x)^3 = \left(\sum_{i=1}^{6} x^i\right)^3 = \sum_{i=3}^{10} a_i(x^i + x^{21-i}),$$

where $(a_3, a_4, a_5, a_6, a_7, a_8, a_9, a_{10}) = (1, 3, 6, 10, 15, 21, 25, 27)$. It follows that the probability of winning (that is, earning $\$4 - \$1 = \$3$) is equal to $p :=
2a_{10}/6^3$. On the other hand, the probability of losing \$1 is equal to $q :=
2\sum_{i=3}^{9} a_i/6^3$. Let us note that $f(1)^3 = 6^3 = 216 = 4 \cdot 2a_{10}$. Therefore, $4a_{10} =
\sum_{i=3}^{10} a_i = \sum_{i=3}^{9} a_i + a_{10}$, and so $q + p = 4p$ or $q = 3p$. The expected number of dollars earned in each game is then equal to

$$3p - q = 0,$$

so the game is fair.

Here is a simple Julia program that allows us to verify this claim:

```
count(x->10<=x<=11, i+j+k for i in 1:6, j in 1:6, k in 1:6)/6^3
```

The output it produces is 0.25, as expected.

8.5   Number Theory

Problem 5.1.1. A positive fraction a/b is said to be in lowest terms if $\gcd(a, b) = 1$. Prove that, if a positive fraction a/b is in lowest terms, then fraction

$$(a + b)/(a^2 + ab + b^2)$$

is also in lowest terms.

Solution. Suppose that a positive fraction a/b is in lowest terms. Since $a^2 +
ab + b^2 = (a + b)^2 - ab$, our goal is prove that

$$\gcd(a + b, a^2 + ab + b^2) = \gcd(a + b, (a + b)^2 - ab) = 1.$$

Since $a + b \mid (a + b)^2$, it follows that

$$\gcd(a + b, (a + b)^2 - ab) = \gcd(a + b, ab).$$

Consider any prime p that divides ab. Since a/b is in lowest terms, p cannot divide both a and b. By symmetry, without loss of generality, we may assume that it divides a but not b. It follows that p does not divide $a + b$ and so $\gcd(a + b, ab) = 1$, and the proof is finished.

Problem 5.1.2. You are given two natural numbers a and b. Prove that if $a + b \mid a^2$, then $a + b \mid b^2$.

Solution. Let a and b be any two natural numbers such that $a + b \mid a^2$. Let $c = \gcd(a, b)$ and set $a' := a/c$, $b' := b/c$ so that $\gcd(a', b') = 1$. Note that our assumption $a + b \mid a^2$ can be rewritten as $c(a' + b') \mid c^2 a'^2$, or equivalently as $a' + b' \mid ca'^2$. However, since $\gcd(a', b') = 1$, we get that $\gcd(a' + b', a'^2) = 1$. It follows that $a' + b' \mid c$, and so $a + b \mid c^2$, but this implies that $a + b$ also divides $c^2 b'^2 = (cb')^2 = b^2$, as required.

Problem 5.1.3. Consider a set A of four digit numbers whose decimal representation uses precisely two digits; moreover, both of them are non-zero. Let $f : A \to A$ be the function such that $f(a)$ flips the digits of $a \in A$ (for example, $f(1333) = 3111$). Find $n > f(n)$ for which $\gcd(n, f(n))$ is as large as possible.

Solution. Let us first note that $\gcd(8484, 4848) = 1212$. We will show that this is the maximum possible value of $\gcd(n, f(n))$ and so 8484 is the value of n we are looking for. In fact, we will prove that it is the unique value of n such that $n > f(n)$ that maximizes $\gcd(n, f(n))$.

Suppose that $n > f(n)$ is such that $k = \gcd(n, f(n)) \geq 1212$. Note that $k = \gcd(n, f(n)) = \gcd(n, n + f(n))$ and so, in particular, k divides $n + f(n)$. Suppose that the representation of n uses digits a and b, $1 \leq a, b \leq 9$ and $a \neq b$. It is easy to see that the property of the function f implies that

$$n + f(n) = (a + b)1111 = (a + b)11 \cdot 101.$$

Since $k \geq 1212$ divides $(a+b)11 \cdot 101$, 101 is a prime number, and $11(a+b) \leq 11(9 + 8) = 187 < 1212$, we conclude that 101 divides k and so it also divides n. Therefore, n must have the form $abab$, as numbers of the form $baaa$, $abaa$, $aaba$, $aaab$ and $aabb$ are *not* divisible by 101. In order to see this, note that

- $baaa$: $1000b + 111a = 10(a - b) + 101(10b + a)$ and $0 < 10 \leq |10(a - b)| \leq 80 < 101$,

- $abaa$: $1011a + 100b = (a - b) + 101(10a + b)$ and $0 < 1 \leq |a - b| \leq 8 < 101$,

- $aaba$: $1101a + 10b = 10(b - a) + 101 \cdot 11a$ and $0 < 10 \leq |10(b - a)| \leq 80 < 101$,

- $aaab$: $1110a + b = (b - a) + 101 \cdot 11a$ and $0 < 1 \leq |b - a| \leq 8 < 101$,

- $aabb$: $1100a + 11b = 11(b - a) + 101 \cdot 11a$ and $0 < 11 \leq |11(b - a)| \leq 88 < 101$.

The remaining case that is left to deal with is when n is of the form $abab$ which can be written as $101(10a + b)$. The corresponding value of $f(n)$ (that is of the form $baba$) can be written as $101(10b + a)$ and so

$$k = \gcd(n, n + f(n)) = 101 \gcd(10a + b, 11(a + b)).$$

Since $a \neq b$, $10a + b$ is *not* divisible by 11 and we get that

$$k = 101 \gcd(10a + b, a + b) = 101 \gcd(9a, a + b).$$

We will independently consider two cases. Suppose first that $9 \mid a + b$. Since $0 < 1 + 2 \leq a + b \leq 9 + 8 < 9 \cdot 2$, we get that $a + b = 9$ and so $101 \cdot 9 < 1212$. Suppose then that $a + b$ is not divisible by 9. Let $c := \gcd(a, b)$, $a' = a/c$, and $b' = b/c$. We get that $\gcd(9a, a + b) = c \gcd(9a', a' + b') = c \gcd(9, a' + b')$. It is straightforward to see that $c \leq 4$ and $\gcd(9, a' + b') \leq 3$. As a result, the maximum value $\gcd(n, f(n))$ can attain is less than or equal to $101 \cdot 4 \cdot 3 = 1212$. Our initial example shows that it is achievable by $n = 8484$.

Actually, one can show that that this is the unique value of n such that $n > f(n)$ that maximizes $\gcd(n, f(n))$. Indeed, in order to achieve the maximum value of 1212, we must have that $c = 4$ and $3 \mid a' + b'$. Since $a + b = c(a' + b') = 4(a' + b') \leq 9 + 8 = 17$, we get that $a' + b' \leq 4$. It follows that $a' + b' = 3$, or equivalently, that $a + b = 12$. Since $1 \leq b < a \leq 9$ (as $n > f(n)$), $4 \mid a$, and $4 \mid b$, we get that $a = 8$ and $b = 4$. The proof of uniqueness is finished.

Note that this problem is relatively easy to solve using the computer. Here is a simple program in the Julia language that finds all values of n that satisfy the required condition and maximize $\gcd(n, f(n))$:

```julia
julia> function check(n)
           s = unique(digits(n))
           if length(s) != 2 || minimum(s) == 0
               # return a small value as n is invalid
               return 0
           end
           return gcd(1111*sum(s) - n, n)
       end
check (generic function with 1 method)

julia> gcds = [check(n) for n in 1000:9999];

julia> max_gcd = maximum(gcds)
1212

julia> [n for n in 1000:9999 if check(n) == max_gcd]
2-element Array{Int64,1}:
 4848
 8484
```

And we have a confirmation that the pair $(4848, 8282)$ yields the maximum greatest common divisor that is equal to 1212.

Problem 5.2.1. Find all primes p for which $p^2 + 2$ is also prime.

Solution. We will independently consider the following three cases.

Case 1: Suppose first that $p \equiv 0 \pmod 3$. There is only one prime number that is divisible by 3, namely, 3 itself. If $p = 3$, then $p^2 + 2 = 11$ is a prime. So $p = 3$ is a solution to our problem.

Case 2: Suppose now that $p \equiv 1 \pmod 3$. Since $p^2 + 2 > 3$ and $p^2 + 2 \equiv 1^2 + 2 = 3 \pmod 3$, $p^2 + 2$ is not a prime.

Case 3: Finally, suppose that $p \equiv 2 \pmod 3$. Arguing as before, since $p^2 + 2 > 3$ and $p^2 + 2 \equiv 2^2 + 2 = 6 \equiv 3 \pmod 3$, $p^2 + 2$ is not a prime.

Combining all three cases together we conclude that the only solution is $p = 3$.

Problem 5.2.2. You are given three consecutive natural numbers (say, a, $a + 1$, and $a + 2$) such that the middle one is a cube (that is, $a + 1 = \ell^3$ for some $\ell \in \mathbf{N}$). Prove that their product is divisible by 504.

Solution. As $504 = 8 \cdot 9 \cdot 7$, it is enough to show that $k := (\ell^3 - 1)\ell^3(\ell^3 + 1)$ is divisible by 8, 9, and 7. We will independently deal with each case.

Case: $8 \mid k$. We will consider two sub-cases. If ℓ is even, then $8 \mid \ell^3$ and we immediately get that $8 \mid k$. Suppose then that ℓ is odd. It follows that ℓ^3 is also odd and so $\ell^3 - 1$ and $\ell^3 + 1$ are two consecutive even numbers. One of them must be divisible by 4 and so $8 \mid (\ell^3 - 1)(\ell^3 + 1)$. We conclude that $8 \mid k$ in this sub-case too.

Case: $9 \mid k$. It is easy to check that ℓ^3 is congruent to 0, 1, or 8 modulo 9. So one of the three numbers, ℓ^3, $\ell^3 - 1$, or $\ell^3 + 1$, is divisible by 9. We get that their product is also divisible by 9.

Case: $7 \mid k$. As before, it is easy to check that ℓ^3 is congruent to 0, 1, or 6 modulo 7. One of the three numbers, ℓ^3, $\ell^3 - 1$, or $\ell^3 + 1$, is divisible by 7 and so is k.

Problem 5.2.3. Prove that for any natural $n \in \mathbf{N}$ that is not divisible by 10 there exists $k \in \mathbf{N}$ such that n^k has in its decimal representation the same digit at the first and the last position.

Solution. The property trivially holds for $n < 10$ as $n^1 = n$ has only one digit. In order to deal with $n > 10$, let us first show the following useful property:

$$n \equiv n^{4k+1} \pmod{10} \qquad \text{for all } k \in \mathbf{N}.$$

As a result, we will be able to restrict ourselves to the subsequence $(n^{4k+1})_{k \in \mathbf{N}}$, that has the property that all terms have the same last digit, and concentrate exclusively on the first digit.

Let

$$\ell := n^{4k+1} - n = n(n^k - 1)(n^k + 1)((n^k)^2 + 1).$$

Our goal is to show that $10 \mid \ell$. We will show independently that $2 \mid \ell$ and that $5 \mid \ell$. The first task is easy: it is clear that $2 \mid n(n^k - 1)$ and so $2 \mid \ell$. Divisibility by 5 requires considering a few cases. For each case, we will show that 5 divides some term in the above representation of ℓ. If $n^k \equiv 0 \pmod 5$, then $5 \mid n$. If $n^k \equiv 1 \pmod 5$, then $5 \mid (n^k - 1)$. If $n^k \equiv 4 \pmod 5$ then $5 \mid (n^k + 1)$. If $n^k \equiv 2 \pmod 5$, then $(n^k)^2 + 1 \equiv 4 + 1 = 5 \equiv 0 \pmod 5$. Finally, if $n^k \equiv 3 \pmod 5$, then $(n^k)^2 + 1 \equiv 9 + 1 = 10 \equiv 0 \pmod 5$. This finishes the proof that in the sequence $(n^{4k+1})_{k \in \mathbf{N}}$ all numbers have the same last digit.

Let us now consider numbers of the form n^{4i}, for $i \in [91]$, and concentrate on their first two digits. Clearly, there are 90 possibilities for the first two digits, from 10 to 99. Hence, by the pigeonhole principle, there exist $1 \le i_1 < i_2 \le 91$ such that n^{4i_1} and n^{4i_2} have the same two first digits; that is, $n^{4i_1} = (d + r_1)10^{p_1}$ and $n^{4i_2} = (d + r_2)10^{p_2}$, for some integer d such that $10 \le d \le 99$, some real numbers r_1, r_2 such that $0 \le r_1, r_2 < 1$, and some integers $0 \le p_1 < p_2$. In fact $r_1, r_2 > 0$ as n is *not* divisible by 10 (which will be important soon). Recall that $i_1 < i_2$, $p_2 < p_3$, and note that

$$n^{4(i_2-i_1)} = \frac{(d+r_2)10^{p_2}}{(d+r_1)10^{p_1}} = \left(1 + \frac{r_2 - r_1}{d + r_1}\right)10^{p_2 - p_1}.$$

Since $-1/10 < (r_2 - r_1)/(d + r_1) < 1/10$,

$$n^{4t} = x \cdot 10^s \quad \text{for a given} \quad x \in (0.9, 1.1) \setminus \{1\}, \tag{8.12}$$

where $t = i_2 - i_1 \in \mathbf{N}$ and $s = p_2 - p_1 \in \mathbf{N}$. Let us stress it again that the assumption that n is *not* divisible by 10 is important and allows us to exclude the case $x = 1$.

We will now restrict ourselves even further, and concentrate on the subsequence $(n^{4tk+1})_{k \in \mathbf{N}}$ of the subsequence $(n^{4k+1})_{k \in \mathbf{N}}$. It will be convenient to represent each term in its (normalized) **scientific notation** which is a standard way of expressing numbers that are too large or too small to be conveniently written in decimal form. All terms can be written in the form $m \cdot 10^n$, where the exponent n (called the **order of magnitude**) is an integer, and the coefficient m (called the **significand** or **mantissa**) is a real number with absolute value at least one but less than ten. The first digit of the term is equal to the floor of the corresponding mantissa. We start with the original number n (the term corresponding to $k = 0$) that has the last digit $c \ne 0$. To get the next term, we multiply the current term by $n^{4t} = x \cdot 10^s$. Because of the property (8.12), the mantissa does not change much after that (unless, of course, it "switches" from a value from the interval $[1, 2)$ to a value from the interval $[9, 10)$, or vice versa). As a result, the first digit never "skips" any digit. Indeed, if $1 < x < 1.1$, then the mantissa keeps geometrically increasing (until it eventually "switches"). More importantly, since $9 \cdot x < 9.9 < 10$, it never "skips" any digit (the extreme case is when digit 8 changes to 9). Similarly, if $0.9 < x < 1$, then the mantissa keeps geometrically decreasing (again, until it eventually "switches"). Since $10 \cdot x > 9$, it never "skips" any digit (this time the extreme case is when digit 1 changes to 9). Hence, for some $k \in \mathbf{N}$, the floor of the mantissa is equal to c and so the first and the last digits of the term n^{4tk+1} are the same. This finishes the proof.

Problem 5.3.1. You are given two integers, a and b, and a prime $p > 2$. Prove that if $p \mid a + b$ and $p \mid a^2 + b^2$, then $p^2 \mid a^2 + b^2$.

Solution. Suppose that $p \mid a + b$ and $p \mid a^2 + b^2$. Since $p \mid a + b$, we get that $p \mid (a+b)^2$. From this and the fact that $p \mid a^2 + b^2$, we get that p also divides $(a+b)^2 - (a^2 + b^2) = 2ab$. As $p > 2$, it follows that $p \mid a$ or $p \mid b$. By symmetry, without loss of generality, we may assume that $p \mid a$. Since $p \mid a + b$ and $p \mid a$, we get that p also divides $(a + b) - a = b$. Hence, $p^2 \mid a^2$, $p^2 \mid b^2$ and, as a consequence, $p^2 \mid a^2 + b^2$.

Problem 5.3.2. You are given four integers a, b, c, and d. Prove that if $a - c \mid ab + cd$, then $a - c \mid ad + bc$.

Solution. Suppose that $a - c \mid ab + cd$ for some integers a, b, c, and d. Since

$$(ab + cd) - (ad + bc) = a(b - d) + c(d - b) = (a - c)(b - d)$$

we get that $a - c$ divides $(ab + cd) - (ad + bc)$. As $a - c \mid ab + cd$, we get that $a - c$ divides $(ab + cd) - (ad + bc) - (ab + cd) = -(ad + bc)$. We conclude that $a - c \mid ad + bc$.

Problem 5.3.3. Consider any natural number $n \geq 2$. Prove that $n^{12} + 64$ has at least four different non-trivial natural factors; that is, $n^{12} + 64 = a \cdot b \cdot c \cdot d$ for some $a, b, c, d \in \mathbf{N}$ such that $1 < a < b < c < d < n^{12} + 64$.

Solution. We will deal with the case $n = 2$ independently. We have that

$$2^{12} + 64 = 64 \cdot (64 + 1) = 64 \cdot 13 \cdot 5 = 2 \cdot 32 \cdot 13 \cdot 5,$$

and so the desired property holds.

From now on, we will assume $n \geq 3$. Using the formula

$$a^4 + 4b^4 = (a^2 + 2b^2 - 2ab)(a^2 + 2b^2 + 2ab),$$

we can factor our expression as follows (by setting $a = n^3$ and $b = 2$):

$$n^{12} + 64 = (n^6 - 4n^3 + 8)(n^6 + 4n^3 + 8).$$

Then, we observe that

$$n^6 - 4n^3 + 8 = (n^2 + 2n + 2)(n^4 - 2n^3 + 2n^2 - 4n + 4)$$

and that

$$n^6 + 4n^3 + 8 = (n^2 - 2n + 2)(n^4 + 2n^3 + 2n^2 + 4n + 4).$$

It is obvious that

$$n^2 - 2n + 2 < n^2 + 2n + 2$$

and that

$$n^4 - 2n^3 + 2n^2 - 4n + 4 < n^4 + 2n^3 + 2n^2 + 4n + 4.$$

So, in order to finish the proof, it is enough to show that for any $n \geq 3$

$$n^2 + 2n + 2 \; < \; n^4 - 2n^3 + 2n^2 - 4n + 4,$$

or equivalently that

$$n^4 - 2n^3 + n^2 - 6n + 2 \; > \; 0.$$

The desired inequality thus holds, as for any $n \geq 3$, we have that

$$
\begin{aligned}
n^4 - 2n^3 + n^2 - 6n + 2 \; &= \; n^3(n-3) + n(n^2 + n - 6) + 2 \\
&\geq \; 0 + n(9 + 3 - 6) + 2 \geq 6n + 2 \geq 20 > 0.
\end{aligned}
$$

This finishes the proof.

Let us make some final remark. The reader might wonder how one can find a factorization like the one we used in this problem:

$$n^6 - 4n^3 + 8 \; = \; (n^2 + 2n + 2)(n^4 - 2n^3 + 2n^2 - 4n + 4).$$

One first needs to notice that $n^6 - 4n^3 + 8$ has no integer roots (see Section 3.4 for a discussion on the **rational root theorem**). Therefore, the next step is to try to find a factorization of the form

$$(n^2 + a_1 n + a_0)(n^4 + b_3 n^3 + b_2 n^2 + b_1 n + b_0).$$

By comparing the coefficients associated with a given power of n, we get the following system of equations:

$$
\begin{aligned}
b_3 + a_1 &= 0 \\
a_1 b_3 + b_2 + a_0 &= 0 \\
a_0 b_3 + a_1 b_2 + b_1 &= -4 \\
a_0 b_2 + a_1 b_1 + b_0 &= 0 \\
a_0 b_1 + a_1 b_0 &= 0 \\
a_0 b_0 &= 8.
\end{aligned}
$$

One can consecutively remove b_i's from this systems to get a system of only two equations in a_1 and a_2. Then, as we know that $a_0 \mid 8$, it is enough to check 8 possible values of a_0 ($\pm 1, \pm 2, \pm 4, \pm 8$) to find out that the only solution is $a_1 = a_0 = 2$.

Problem 5.4.1. Prove that for all $a \in \mathbf{N}$, we have $35 \mid a^{64} - a^4$.

Solution. Let us first note that

$$a^{64} - a^4 \; = \; a^4((a^{15})^4 - 1) \; = \; a^4((a^{10})^6 - 1).$$

In order to see that $a^{64} - a^4$ is divisible by 5, let us consider two cases. If $5 \mid a$, then we are immediately done. Suppose then that a is *not* divisible by 5. It follows that $n = a^{15}$ is also *not* divisible by 5 and so, since 5 is a prime, we get from Fermat's little theorem that $p = 5$ divides $n^{p-1} - 1 = (a^{15})^4 - 1$. Exactly the same argument shows that either $7 \mid a$ or $7 \mid (a^{10})^6 - 1$.

Problem 5.4.2. Prove that for any odd integer n, we have that $n \mid \prod_{i=1}^{n} \sum_{j=0}^{i} 2^j$.

Solution. Let us first note that

$$\prod_{i=1}^{n} \sum_{j=0}^{i} 2^j = \prod_{i=1}^{n} (2^{i+1} - 1).$$

If n is prime, then it follows immediately from Fermat's little theorem that $n \mid 2^{n-1} - 1$ and clearly $2 \leq n - 1 \leq n + 1$. The desired property holds. Let us then assume that n is composite: $n = \prod_{k=1}^{s} p_k^{w_k}$, where p_k's are unique prime numbers and w_k's are natural numbers. Since n is odd, then all p_k's are at least 3.

Fix $k \in [s]$. To get the desired property, it is enough to show that at least w_k terms in the product $\prod_{i=1}^{n} (2^{i+1} - 1)$ are divisible by p_k. If $w_k = 1$, then we immediately get that $p_k \mid 2^{p_k - 1} - 1$ (by Fermat's little theorem) and clearly $2 \leq p_k - 1 \leq n + 1$. So we need to concentrate on $w_k > 1$. In order to deal with this case, we will use the fact that for any natural number x, $2^{x(p_k - 1)} - 1$ is divisible by p_k. Indeed, using Fermat's little theorem one more time we get that

$$2^{x(p_k-1)} - 1 \equiv 1^x - 1 = 0 \pmod{p_k}.$$

Clearly, for any $x \in \mathbf{N}$, $x(p_k - 1) \geq 2$. Hence, in order to see that the number of terms in the product that are of the form $2^{x(p_k-1)} - 1$ is at least w_k, it is enough to check that $w_k(p_k - 1) \leq n + 1$. But this inequality holds as

$$w_k(p_k - 1) \leq (p_k - 1)^{w_k} \leq p_k^{w_k} \leq n \leq n + 1.$$

This finishes the proof.

Problem 5.4.3. Find the last two digits in the decimal representation of 7^{123}.

Solution. Using Property 4 of the Euler's Totient Function, we get that

$$\phi(100) = 100 \left(1 - \frac{1}{2}\right) \left(1 - \frac{1}{5}\right) = 40.$$

Hence, since 7 and 100 are co-prime, it follows from Euler's theorem that $100 \mid 7^{40} - 1$ or, equivalently, that $7^{40} \equiv 1 \pmod{100}$. Finally, since

$$7^{123} = (7^{40})^3 \cdot 7^3 \equiv 1^3 \cdot 343 = 343 \equiv 43 \pmod{100},$$

the last two digits of 7^{123} are 43.

Problem 5.5.1. Decide if there exists $k \in \mathbf{N}$ with the property that in the decimal representation of 2^k each of the 10 digits $(0, 1, 2, \ldots, 9)$ is present the same number of times.

Solution. We will show that no such k exists. For a contradiction, suppose that for some $k \in \mathbf{N}$, the number of times each digit appears in the decimal representation of 2^k is s for some $s \in \mathbf{N}$. Note that the sum of these digits is equal to $s \sum_{i=0}^{9} i = 45s$. Since $45s$ is divisible by 3, we get that 2^k is divisible by 3. But this is clearly impossible and we get the desired contradiction.

Problem 5.5.2. Find the minimum of $|20^m - 9^n|$ over all natural numbers m and n.

Solution. Let us first note that $|20^1 - 9^1| = 11$. We will show that it is impossible to achieve smaller values. Clearly,

$$20^m - 9^n \equiv 0^m - (-1)^n = \pm 1 \pmod{10},$$

so the last digit in the decimal representation of $|20^m - 9^n|$ is either 1 or 9. Hence, the only potentially possible values of $|20^m - 9^n|$ that are less than 11 are 9 or 1. We will independently rule them out.

Clearly $|20^m - 9^n| = 9$ is not possible as 20 is not divisible by 9. In order to rule out the case $|20^m - 9^n| = 1$, we need to consider two sub-cases. Since $19 \mid (20^m - 1)$, it is impossible that $20^m - 1 = 9^n$. It remains to deal with the sub-case $20^m + 1 = 9^n$. Since $20^m + 1 \equiv (-1)^m + 1 \pmod 3$, in order for $20^m + 1$ to be divisible by 3, m would have to be odd. But if m is odd then $21 \mid 20^m + 1$ (since $20^m + 1 \equiv (-1)^m + 1 \equiv -1 + 1 = 0 \pmod{21}$) and so also $7 \mid 20^m + 1$. It follows that it is impossible that $20^m - 1$ is equal to 9^n as 9^n is *not* divisible by 7.

Problem 5.5.3. Given $m, n, d \in \mathbf{N}$, prove that if $m^2 n + 1$ and $mn^2 + 1$ are divisible by d, then $m^3 + 1$ and $n^3 + 1$ are also divisible by d.

Solution. If $d = 1$, then the desired property trivially holds. Hence, we may assume that $d \geq 2$. Suppose that $m^2 n + 1$ and $mn^2 + 1$ are divisible by d. Let us note that, due to the symmetry, it is enough to show that $m^3 + 1$ is divisible by d. Since $d \mid m^2 n + 1$ and $d \mid mn^2 + 1$, d also divides $(m^2 n + 1) - (mn^2 + 1) = mn(m - n)$. Note that it is *not* the case that $d \mid n$ or $d \mid m$, as otherwise d would divide $m^2 n$ and so it would *not* divide $m^2 n + 1$. As such, we must have that d divides $m - n$ and so $m^2(m - n)$ as well. It follows that d divides

$$m^2(m - n) + (m^2 n + 1) = m^3 + 1,$$

as desired.

Problem 5.6.1. Find all $x, y \in \mathbf{N}$ such that $2^x + 5^y$ is a square.

Solution. Suppose that $2^x + 5^y = z^2$ for some natural numbers x, y, and z. Let us first note that z is *not* divisible by 5. We will split the proof into two independent cases depending on the parity of x.

The case when x is odd is easy to deal with. If $x = 2k + 1$ for some non-negative integer k, then $2^x = 2^{2k+1}$ gives the reminder of 2 or 3 when divided by 5 ($2^1 = 2$ (mod 5), $2^3 = 8 \equiv 3$ (mod 5), $2^5 = 32 \equiv 2$ (mod 5), etc.). On the other hand, z^2 gives the reminder of 1 or 4 when divided by 5 as z is not divisible by 5 ($1^2 = 1$ (mod 5), $2^2 = 4$ (mod 5), $3^2 = 9 \equiv 4$ (mod 5), $4^2 = 16 \equiv 1$ (mod 5), $6^2 \equiv 1^2 = 1$ (mod 5), etc.). It follows that there is no solution in this case.

The case when x is even requires more work. If $x = 2k$ for some $k \in \mathbf{N}$, then

$$5^y = z^2 - 2^{2k} = (z + 2^k)(z - 2^k).$$

It is straightforward to see that it is impossible that both $(z + 2^k)$ and $(z - 2^k)$ are divisible by 5. It follows that the second term, namely, $z - 2^k$ is equal to 1 and so $5^y = 2 \cdot 2^k + 1$. Note that there exists one solution corresponding to $k = 1$: $x = 2$ and $y = 1$. We will show that this is the only solution.

For a contradiction, suppose that for some $y, k \in \mathbf{N} \setminus \{1\}$, we have that

$$5^y - 2 \cdot 2^k = 1 = 5 - 4,$$

or equivalently that

$$5(5^{y-1} - 1) = 4(2^{k-1} - 1).$$

In order for the right hand side to be divisible by 5, we must have that $k = 4t + 1$ for some non-negative integer t; that is, 2^{k-1} has to be of the form 16^t. But it means that both sides are also divisible by 3, since $16^t - 1 \equiv 1^t - 1 = 0$ (mod 3). Now, in order for the left hand side to be divisible by 3, we must have that $y = 2s + 1$ for some $s \in \mathbf{N}$; that is, 5^{y-1} has to be of the form 25^s. Recall that the case $y = 1$ ($s = 0$) corresponds to a feasible solution and is excluded now. But this implies that the left hand side is divisible by 8, as $25^s - 1 \equiv 1^s - 1 = 0$ (mod 8), wheres the right hand side is clearly not. We get the desired contradiction and the proof is finished.

Problem 5.6.2. Prove that for any two sequences, $(x_i)_{i=1}^{2011}$ and $(y_i)_{i=1}^{2011}$, of natural numbers, $\prod_{i=1}^{2011}(2x_i^2 + 3y_i^2)$ is *not* a square.

Solution. Let us first note that, without loss of generality, we may assume that $\gcd(x_i, y_i) = 1$ for all $i \in [2011]$. Indeed, it is easy to see that one could factor out $\gcd(x_i, y_i)^2$ (that is clearly a square) from the ith term and move it in front of the product. As a result, two sequences $(x_i)_{i=1}^{2011}$ and $(y_i)_{i=1}^{2011}$ satisfy the desired property if and only if $(x_i / \gcd(x_i, y_i))_{i=1}^{2011}$ and $(y_i / \gcd(x_i, y_i))_{i=1}^{2011}$ do.

Now, assuming that $\gcd(x_i, y_i) = 1$, we will analyze the reminder of $2x_i^2 + 3y_i^2$ when divided by 3. If $3 \mid x_i$ then, by our assumption, 3 does *not* divide y_i and so $y_i^2 \equiv 1$ (mod 3). Indeed, if $y_i \equiv 1$ (mod 3), then $y_i^2 \equiv 1^1 = 1$ (mod 3) whereas if $y_i \equiv 2$ (mod 3), then $y_i^2 \equiv 2^2 = 4 \equiv 1$ (mod 3). It follows that $2x_i^2 + 3y_i^2 = 3(3t_i + 1)$ for some $t_i \in \mathbf{N}$. On the other hand, if 3 does not divide x_i then $x_i^2 \equiv 1$ (mod 3) and so $2x_i^2 + 3y_i^2 = 3t_i + 2$ for some $t_i \in \mathbf{N}$. These two cases naturally define a partition of the 2011 terms of the product. Let A be the subset of $[2011]$ that consists of those indices i for which $3 \mid x_i$, and let $B = [2011] \setminus A$.

For a contradiction, suppose that the product $\prod_{i=1}^{2011}(2x_i^2 + 3y_i^2)$ is a square. Since 3 does not divide the term $2x_i^2 + 3y_i^2$ when $i \in B$ and each term corresponding to $i \in A$ has precisely one 3 in its unique factorization, we get that $|A|$ is even, that is, $|A| = 2s$ for some non-negative integer s. Hence, $|B| = 2011 - |A|$ is odd. But this means that

$$\prod_{i \in B}(2x_i^2 + 3y_i^2) \equiv \prod_{i \in B}(3t_i + 2) \equiv 2^{|B|} \equiv (-1)^{|B|} \equiv -1 \equiv 2 \pmod 3$$

gives the reminder of 2 when divided by 3. It follows that

$$\prod_{i=1}^{2011}(2x_i^2 + 3y_i^2) = 3^{2s}\prod_{i \in A}(3t_i + 1)\prod_{i \in B}(3t_i + 2) = 3^{2s}(3p + 2)$$

for some $p \in \mathbf{N}$. But it means that $3p + 2$ is a square but this is impossible as no square gives a reminder of 2 when divided by 3. This finishes the proof.

Problem 5.6.3. Consider any integer $n \geq 2$ and any subset S of the set $N := \{0, 1, 2, \ldots, n - 1\}$ that has more than $\frac{3}{4}n$ elements. Prove that there exist integers a, b, c such that the remainders when numbers a, b, c, $a + b$, $a + c$, $b + c$, $a + b + c$ are divided by n are all in S.

Solution. Consider any subset S of the set $N = \{0, 1, 2, \ldots, n - 1\}$ of size $s = |S| > \frac{3}{4}n = \frac{3}{4}|N|$. Let us start with the following, simple but useful, observation that we will use a few times. Let $x, y, z \in N$ be such that $x < y$. Then, $z + x$ and $z + y$ yield two different reminders when divided by n. Indeed, if the remainders of $z + x$ and $z + y$ are equal, then $(z + y) - (z + x) = y - x$ would be divisible by n which is impossible as $0 < y - x < n$.

We will select the three numbers a, b, c from S that satisfy the desired properties in a greedy way. Let us start with selecting any number $a \in S$, arbitrarily chosen. Because of our property (used with $z = a$), the s numbers of the form $a + x$ for $x \in S$ all give unique reminders when divided by n. At most $n - s < \frac{1}{4}n$ of them are *not* in S. Since $s > \frac{3}{4}n > \frac{1}{4}n$, we can select $b \in S$ such that the reminder of $a + b$ when divided by n is in S.

Similarly, having a and b fixed, we observe that there are less than $\frac{1}{4}n$ values of $x \in S$ for which the reminder of $a + x$ when dividing by n is not in S (property used with $z = a$), less than $\frac{1}{4}n$ values that create a problem for $b + x$ (property used with $z = b$), and less than $\frac{1}{4}n$ values not satisfying the condition for $(a + b) + x$ (property used with $z = a + b$). It follows that there are less than $\frac{3}{4}n$ values that do *not* satisfy *some* condition but we have more than $\frac{3}{4}n$ values to choose from. Hence, we are guaranteed that there exists $c \in S$ that, together with a and b, satisfy the desired conditions.

Problem 5.7.1. Prove that if the sum of positive divisors of some natural number n is odd, then either n is a square or $n/2$ is a square.

Solution. Let us consider the unique factorization of n. That is, we write $n = \prod_{i=1}^{k} p_i^{\ell_i}$ for some sequence of prime numbers $2 \leq p_1 < p_2 < \ldots < p_k$ and $\ell_i \in \mathbf{N}$ for $i \in [k]$. Note that each positive divisor of n has unique representation $\prod_{i=1}^{k} p_i^{j_i}$, where $j_i \in \{0, 1, \ldots, \ell_i\}$, and two different divisors have different representations. It follows that the sum of all positive divisors of n is equal to

$$S := \sum_{j_1=0}^{\ell_1} \cdots \sum_{j_k=0}^{\ell_k} \prod_{i=1}^{k} p_i^{j_i} = \prod_{i=1}^{k} \sum_{j=0}^{\ell_i} p_i^{j}.$$

Since our assumption is that S is odd, we get that $\sum_{j=0}^{\ell_i} p_i^{j}$ is odd for each $i \in [k]$.

Consider any $i \in [k]$. Suppose first that $p_i > 2$ and so it is odd. Since $\sum_{j=0}^{\ell_i} p_i^{j}$ is odd, the sum has $\ell_i + 1$ terms, and each term is odd, it follows that the number of terms is odd, that is, ℓ_i is even. In this case we get that $p_i^{\ell_i}$ is a square. On the other hand, if $p_1 = 2$, then the first term in the corresponding sum is odd ($2^0 = 1$) and the remaining terms are even. As a result, the sum is always odd and ℓ_1 could be even or odd. If ℓ_1 is even, then $p_1^{\ell_1}$ is a square; otherwise, $p_1^{\ell_1 - 1}$ is a square. Putting these observations together we conclude that if n is odd (that is, $p_1 \neq 2$), then n is a square. If n is even ($p_1 = 2$), then either n is a square or $n/2$ is.

Problem 5.7.2. Find all natural numbers n for which there exist $2n$ pairwise different numbers $a_1, a_2, \ldots, a_n, b_1, b_2, \ldots, b_n$ such that $\sum_{i=1}^{n} a_i = \sum_{i=1}^{n} b_i$ and $\prod_{i=1}^{n} a_i = \prod_{i=1}^{n} b_i$.

Solution. It is clear that there is no solution for $n = 1$. For $n = 2$, we have the following two conditions: $a_1 + a_2 = b_1 + b_2$ and $a_1 a_2 = b_1 b_2$. From the first equation we have that $a_2 = b_1 + b_2 - a_1$. After substituting this to the second equation we get that $a_1(b_1 + b_2 - a_1) - b_1 b_2 = 0$ or, equivalently, that $(a_1 - b_2)(b_1 - a_1) = 0$. It follows that $a_1 = b_2$ or $a_1 = b_1$ and so it is impossible to have a solution with pairwise different numbers.

For $n = 3$, 4, and 5 we have the following solutions:

$$(a_1, a_2, a_3) = (2, 8, 9) \quad \text{and} \quad (b_1, b_2, b_3) = (3, 4, 12)$$
$$(a_1, a_2, a_3, a_4) = (1, 6, 7, 10) \quad \text{and} \quad (b_1, b_2, b_3, b_4) = (2, 3, 5, 14)$$
$$(a_1, a_2, a_3, a_4, a_5) = (1, 5, 7, 8, 12) \quad \text{and} \quad (b_1, b_2, b_3, b_4, b_5) = (2, 3, 4, 10, 14).$$

We will now show that if there is a solution for $n = n_0$, then there is one for $n = n_0 + 3$ (inductive step). Since we already showed that there is a solution for $n \in \{3, 4, 5\}$ (base case), by mathematical induction we will get that there is a solution for any natural number $n \geq 3$.

In order to prove the claim, let us make two simple observations. First of all, let us note that the solution that we have for $n = 3$ (($(a_1, a_2, a_3) = (2, 8, 9)$ and $(b_1, b_2, b_3) = (3, 4, 12)$) can be easily generalized to get an infinite family of solutions. Indeed, it is obvious that for any $x \in \mathbf{N}$, $(a_1, a_2, a_3) = (2x, 8x, 9x)$ and $(b_1, b_2, b_3) = (3x, 4x, 12x)$ is also a solution to our problem. The second ingredient that we need is the fact that any two solutions that consist of non-overlapping values can be merged together to get another solution. Formally, suppose that the pair $(a_1, \ldots, a_n)$ and $(b_1, \ldots, b_n)$ is the solution for some $n \geq 3$. Then, one can take x large enough such that $2x > m := \max\{a_1, \ldots, a_n, b_1, \ldots, b_n\}$ (for example, $x = m$). It follows that $(a_1, \ldots, a_n, 2x, 8x, 9x)$ and $(b_1, \ldots, b_n, 3x, 4x, 12x)$ is a solution for $n + 3$.

Let us make some final remarks. Finding solutions for $n \in \{3, 4, 5\}$ by hand can be tedious. However, with access to a computer, one can easily do it. Below is a short Julia script that was used to find the solutions given above.

```
using Combinatorics

function f(n, k)
    for x in combinations(1:k, n),
        y in combinations(setdiff(1:k, x), n)
        if x < y # avoid printing duplicates
            if sum(x) == sum(y) && prod(x) == prod(y)
                println((x, y))
            end
        end
    end
end
```

and now we can run it to get the desired solutions:

```
julia> f(3, 12)
([2, 8, 9], [3, 4, 12])
([3, 8, 10], [4, 5, 12])
([4, 9, 10], [5, 6, 12])

julia> f(4, 14)
([1, 6, 7, 10], [2, 3, 5, 14])
([1, 7, 8, 9], [2, 3, 6, 14])
([2, 7, 9, 10], [3, 5, 6, 14])
([3, 7, 10, 12], [4, 5, 9, 14])
([4, 7, 10, 12], [5, 6, 8, 14])

julia> f(5, 14)
([1, 5, 7, 8, 12], [2, 3, 4, 10, 14])
```

Finally, note that the printed solutions show that actually $x = 12$, $x = 14$ and $x = 14$ are minimal ranges of the sets of the form $[x]$ that produce the solutions for n equal to, respectively, 3, 4 and 5.

Problem 5.7.3. Call a natural number **white** if it is equal to 1 or is a product of an even number of prime numbers; otherwise, call it **black**. Is there an integer for which the sum of its white divisors is equal to the sum of its black divisors?

Solution. Let x be any natural number. Let $W(x)$ be the set of white divisors of x, and let $B(x)$ be the set of black divisors of x. Finally, let

$$D(x) := \sum_{w_x \in W(x)} w_x - \sum_{b_x \in B(x)} b_x .$$

We will show that $D(x) \neq 0$ which gives a negative answer to the question; that is, there is no integer for which the sum of its white divisors is equal to the sum of its black divisors.

Let us start with proving the following useful property. For any p and q that are co-prime, we have that

$$D(p \cdot q) = D(p) \cdot D(q) . \tag{8.13}$$

Indeed, note that

$$
D(p) \cdot D(q) = \left(\sum_{w_p \in W(p)} w_p - \sum_{b_p \in B(p)} b_p \right) \left(\sum_{w_q \in W(q)} w_q - \sum_{b_q \in B(q)} b_q \right)
$$

$$
= \left(\sum_{w_p \in W(p)} w_p \sum_{w_q \in W(q)} w_q + \sum_{b_p \in B(p)} b_p \sum_{b_q \in B(q)} b_q \right)
$$

$$
- \left(\sum_{w_p \in W(p)} w_p \sum_{b_q \in B(q)} b_q + \sum_{b_p \in B(p)} b_p \sum_{w_q \in W(q)} w_q \right) .
$$

Note also that $w_p w_q$ and $b_p b_q$ are white divisors of $p \cdot q$ (as both the sum of two even numbers and the sum of two odd numbers is even), and $w_p b_q$ and $b_p w_q$ are black divisors of $p \cdot q$ (as the sum of an even and an odd number is odd). Also, in the expression above, all divisors of $p \cdot q$ are present exactly once since p and q are co-prime. This shows that, indeed, (8.13) holds.

Let us now come back to our task of showing that $D(x) \neq 0$. Let $x = \prod_{i=1}^{t} p_i^{s_i}$ be the unique prime factorization of x: $2 \leq p_1 < \ldots < p_t$, $s_i \in \mathbf{N}$ for $i \in [t]$. Using (8.13), we get that

$$D(x) = D\left(\prod_{i=1}^{t} p_i^{s_i} \right) = \prod_{i=1}^{t} D(p_i^{s_i}) .$$

Finally, note that all positive divisors of $p_i^{s_i}$ have the form p_i^k for $0 \leq k \leq s_i$. Moreover, the corresponding divisor is white if and only if k is even. It follows that

$$D(p_i^{s_i}) = \sum_{j=0}^{\lfloor s_i/2 \rfloor} p_i^{2j} - \sum_{j=0}^{\lfloor (s_i-1)/2 \rfloor} p_i^{2j+1} = \sum_{k=0}^{s_i} (-p_i)^k = \frac{1 - (-p_i)^{s_i+1}}{1 + p_i} \neq 0 .$$

As a result, $D(x) \neq 0$ and the proof is complete.

Problem 5.8.1. Find all natural solutions of the following equation: $x^4 + y = x^3 + y^2$.

Solution. We can re-write the equation as $x^4 - x^3 = y^2 - y$. In order to aggregate similar factors, we multiply both sides by 4 and then add 1 to get that

$$(2y - 1)^2 = (2x^2 - x)^2 - (x^2 - 1).$$

We will show that $x = 1$. For a contradiction, suppose that $x \geq 2$. It follows that

$$(2y - 1)^2 \leq (2x^2 - x)^2 - (2^2 - 1) < (2x^2 - x)^2.$$

On the other hand, note that

$$
\begin{aligned}
((2x^2 - x) - 1)^2 &= (2x^2 - x)^2 - 2(2x^2 - x) + 1 \\
&= (2x^2 - x)^2 - (x^2 - 1) - x(3x - 2) \\
&< (2x^2 - x)^2 - (x^2 - 1) = (2y - 1)^2.
\end{aligned}
$$

Combining the two observations together, we get that

$$((2x^2 - x) - 1)^2 < (2y - 1)^2 < (2x^2 - x)^2.$$

But there is no natural number such that its square is between squares of two consecutive natural numbers and so we get the desired contradiction. It follows that $x = 1$.

If $x = 1$, then we get that $(2y - 1)^2 = 1^2$ which implies that $y = 1$. Therefore, the only solution of our equation in natural numbers is $(x, y) = (1, 1)$.

Problem 5.8.2. Find all pairs of natural numbers that satisfy $(x - y)^n = xy$.

Solution. We will show that there is no solution when $n = 1$ or $n = 2$. For $n = 1$, for any $x, y \in \mathbf{N}$, we have $x - y < x \leq xy$ and so there are no natural numbers that satisfy $x - y = xy$.

For $n = 2$, for a contradiction, suppose that $x^2 + y^2 = 3xy$ for some $x, y \in \mathbf{N}$. Moreover, let us assume that this is a smallest example (in terms of variable x), that is, there is no other pair $x', y' \in \mathbf{N}$ that satisfy the desired equality and $x' < x$. Suppose first that both x and y are divisible by 3. It is easy to see that $x' = x/3 \in \mathbf{N}$ and $y' = y/3 \in \mathbf{N}$ also form a solution, which contradicts our assumption. Similarly, it is not possible that one of the numbers is divisible by 3 and other is not, as then the left hand side is not divisible by 3 while the right hand side is. Finally, if both x and y are not divisible by 3, then $x^2 + y^2$ gives the reminder of 2 when divided by 3 whereas the right hand side is clearly divisible by 3. All cases lead to a contradiction and so there is no solution when $n = 2$.

For $n \geq 3$, let $z := x - y$. Note that $(x - y)^n = xy \geq 1$ and so $x > y$ if

n is odd. If n is even, due to the symmetry, we may assume that $x > y$ and potential solutions will come in pairs; that is, if $(x, y) = (x_0, y_0)$ is a solution, then so is $(x, y) = (y_0, x_0)$. Either way, we may assume that $z \geq 1$. After substitution, our equation becomes $z^n = y^2 + zy$. Now, multiply both sides by 4 and add z^2 to both sides to get that

$$z^2(4z^{n-2} + 1) = (2y + z)^2.$$

Note that the right hand side of this equation is a square. Since z^2 is a square, it follows that $4z^{n-2}+1$ is also a square. As $4z^{n-2}+1$ is clearly an odd number, it must be a square of an odd natural number, that is, $4z^{n-2} + 1 = (2t + 1)^2$ for some $t \in \mathbf{N}$. It follows that $z^{n-2} = t(t+1)$. Since t and $t+1$ are co-prime, $t = a^{n-2}$ and $t + 1 = b^{n-2}$ for some $a, b \in \mathbf{N}$. We get that $b^{n-2} - a^{n-2} = (t+1) - t = 1$. If $n \geq 4$, we get that $b^{n-2} - a^{n-2} > b - a \geq 1$ and so it must be the case that $n = 3$ and then also $x > y$. Since $z = t(t + 1)$ and $z^3 = y^2 + zy$, after substitution we get that

$$\left(t^2(t+1)\right)\left(t(t+1)^2\right) = t^3(t-1)^3 = y^2 + t(t+1)y = y^2 + (t(t+1)^2 - t^2(t+1))y$$

or, alternatively, that

$$(y - t^2(t + 1))(y + t(t + 1)^2) = 0.$$

Since both y and t are at least 1 (both are natural numbers), we get that $y = t^2(t+1)$. Then $x = z + y = t(t+1) + t^2(t+1) = (t+1)^2 t$ for some $t \in \mathbf{N}$.

Finally, we have to check that $y = t^2(t + 1)$ and $x = (t + 1)^2 t$ do, in fact, yield a solution. We get $xy = t^3(t+1)^3$ and $(x - y)^3 = (t(t+1))^3$ and so both sides are equal.

Problem 5.8.3. Find all natural numbers satisfying the following system of equations:

$$a + b + c = xyz,$$
$$x + y + z = abc,$$

and such that $a \geq b \geq c \geq 1$ and $x \geq y \geq z \geq 1$.

Solution. Adding the two equations together we get

$$a + b + c + x + y + z = abc + xyz.$$

Observe that

$$abc - (a + b + c) = c(ab - 1) - a - b$$
$$= c(ab - 1) - a - b + ab + 1 - (ab - 1) - 2$$
$$= (c - 1)(ab - 1) + (a - 1)(b - 1) - 2.$$

Similarly,

$$xyz - (x + y + z) = (z - 1)(xy - 1) + (x - 1)(y - 1) - 2.$$

It follows that our equation can be rewritten as follows:

$$(c-1)(ab-1) + (a-1)(b-1) + (z-1)(xy-1) + (x-1)(y-1) = 4.$$

Let us note that all the 4 terms at the right hand side are non-negative.
Now, observe that if $c \geq 2$ (and so a and b are also at least 2), then

$$(c-1)(ab-1) + (a-1)(b-1) \geq 4.$$

This implies that $(z-1)(xy-1) + (x-1)(y-1) = 0$ and so $x = y = z = 1$. However, such potential solutions would not be able to satisfy the second equation of the original system. We conclude that $c = 1$. Similarly, by symmetry, we get that $z = 1$.

If $c = z = 1$, then our equation reduces to

$$(a-1)(b-1) + (x-1)(y-1) = 4.$$

Now, if $b = 1$, then we have $(x-1)(y-1) = 4$ and so $(x,y) = (3,3)$ or $(x,y) = (5,2)$. If $(x,y) = (3,3)$, then $xyz = 9$ and $x + y + z = 7$. But this would mean that $a+2 = 9$ and $2a = 7$, which is not possible. If $(x,y) = (5,2)$, then $xyz = 10$ and $x + y + z = 8$. But this would mean that $a + 2 = 10$ and $2a = 8$, which is also not possible. Therefore, we conclude that $b \geq 2$ and, by symmetry, that also $y \geq 2$.

Now if $b \geq 3$ and $y \geq 2$, then

$$(a-1)(b-1) + (x-1)(y-1) \geq (3-1)(3-1) + (2-1)(2-1) = 5,$$

which is impossible. By symmetry, it is also not possible that $y \geq 3$ and $c \geq 2$.

We are left with only one possibility: $b = y = 2$. Then, we have $a + x = 6$. Additionally, from the $a + b + c = xyz$ condition we also get that $a + 3 = 2x$, and so $3x = 9$. As a result, we get that $a = x = 3$. We conclude that the only possible solution is $(a,b,c,x,y,z) = (3,2,1,3,2,1)$. We directly check that, indeed, it satisfies our system of equations.

8.6   Geometry

Problem 6.1.1. We are given an acute triangle ABC with $\sphericalangle ACB = \pi/3$. Let A' be the orthogonal projection of A on BC, let B' be the orthogonal projection of B on AC, and let M be the middle point of line segment AB. Prove that $|A'B'| = |A'M| = |B'M|$.

Solution. Since $A'AC$ is a right triangle, $\sphericalangle B'AA' = \sphericalangle CAA' = \pi/2 - \pi/3 = \pi/6$. Since $AA'B$ and $AB'B$ are right triangles, points A, B, A', and B' lie on a circle whose center is M. But this means that $\sphericalangle B'MA' = 2\sphericalangle B'AA' =$

$2(\pi/6) = \pi/3$. Since $|B'M| = |A'M|$, we get that the triangle $A'B'M$ is equilateral, which finishes the proof.

Problem 6.1.2. Consider a square $ABCD$. Choose point P outside of this square such that $\sphericalangle CPB$ is the right angle. Denote by Q the intersection of AC and BD. Prove that $\sphericalangle QPC = \sphericalangle QPB$.

Solution. Since $\sphericalangle BQC$ and $\sphericalangle BPC$ are both right angles, points Q, B, C, and P lie on a circle. Since $|QB| = |QC|$, we get that $\sphericalangle QPB = \sphericalangle QPC$, as required.

Problem 6.1.3. Point O is the center of a circumcircle of a triangle ABC. Point C' is the orthogonal projection of C on AB. Prove that $\sphericalangle ACC' = \sphericalangle OCB$.

Solution. Let us first note that $\sphericalangle OCB = \pi/2 - \sphericalangle COB/2 = \pi/2 - \sphericalangle CAB$. On the other hand, since triangle ACC' is a right triangle, $\sphericalangle ACC' = \pi/2 - \sphericalangle CAC' = \pi/2 - \sphericalangle CAB$. It follows that $\sphericalangle ACC' = \sphericalangle OCB$.

Problem 6.2.1. Suppose that points P and Q lie on sides BC and CD of a square $ABCD$ such that $\sphericalangle PAQ = \pi/4$. Prove that $|BP| + |DQ| = |PQ|$.

Solution. Consider point R inside the square such that $|AR| = |AB| = |AD|$ and $\sphericalangle BAP = \sphericalangle PAR$. Note that R lies inside the angle $\sphericalangle PAQ$. After considering congruent triangles BAP and PAR, we get that $|BP| = |PR|$. Now, notice that $\sphericalangle BAP + \sphericalangle DAQ = \pi/2 - \pi/4 = \pi/4$. Using this we have $\sphericalangle QAR = \pi/4 - \sphericalangle PAR = \pi/4 - \sphericalangle BAP = \pi/4 - (\pi/4 - \sphericalangle DAQ) = \sphericalangle DAQ$. It follows that $|DQ| = |QR|$.

It is left to show that R lies on the line segment PQ, as then we will conclude that $|BP| + |DQ| = |PR| + |QR| = |PQ|$. But $\sphericalangle QRA = \sphericalangle QDA = \pi/2$. Similarly, $\sphericalangle PRA = \sphericalangle PBA = \pi/2$, and so $\sphericalangle PRQ = \pi$, as required.

Problem 6.2.2. Point P lies on a diagonal AC of a square $ABCD$. Points Q and R are the orthogonal projections of P on lines CD and DA, respectively. Prove that $|BP| = |RQ|$.

Solution. Since $RPQD$ is a rectangle, $|RQ| = |PD|$. Since $\sphericalangle DCP = \sphericalangle BCP(= (\pi/2)/2 = \pi/4)$ and $|DC| = |BC|$, triangles PDC and PBC are congruent. It follows that $|PB| = |PD| = |RQ|$, and the proof is finished.

Problem 6.2.3. Consider an acute triangle ABC where $\sphericalangle ACB = \pi/4$. Point B' is the orthogonal projection of B on AC and point A' is the orthogonal projection of A on BC. Let H be the intersection point of AA' and BB'. Prove that $|CH| = |AB|$.

Solution. Since triangle $BB'C$ is a right triangle and $\sphericalangle B'CB = \pi/4$, we get that $|BB'| = |CB'|$. Since $\sphericalangle CB'H = \sphericalangle CA'H = \pi/2$, points H, B', C, and A' lie on a circle. It follows that $\sphericalangle B'CH = \sphericalangle B'A'H$. Similarly, since $\sphericalangle AB'B = \sphericalangle AA'B = \pi/2$, points A, B', A', and B lie on a circle. It follows that $\sphericalangle B'A'H = \sphericalangle B'BA$. As a result, we get that triangles $BB'A$ and $HB'C$ are congruent, and so $|AB| = |HC|$, as desired.

Problem 6.3.1. You are given a rectangle that can be covered with n disks of radius r. Prove that it can be also covered by $4n$ disks of radius $r/2$.

Solution. Clearly, if we scale the rectangle down by a factor of 2, then it can be covered by n disks of radius $r/2$. Since we can put 4 such rectangles together to recreate the original rectangle, it is enough to use $4n$ disks to do the desired covering.

Problem 6.3.2. You are given an acute triangle ABC. Let B' be the projection of B on AC and C' be the projection of C on AB. Show that ABC and $AB'C'$ are similar.

Solution. Since $\sphericalangle BB'C = \sphericalangle BC'C = \pi/2$, points B, C, B', and C lie on a circle. Thus, $AB'C' = \pi/2 - \sphericalangle C'B'B = \pi/2 - \sphericalangle C'CB = \sphericalangle C'BC$. This means that also $\sphericalangle AC'B' = \sphericalangle BCB'$. It follows that triangles ABC and $AB'C'$ are similar.

Problem 6.3.3. Consider two circles, o_1 and o_2, that intersect at two points, A and B. Let P be a point on o_1 such that AP goes through the center of o_1 and Q be a point on o_2 such that AQ goes through the center of o_2. Prove that if $\sphericalangle PAQ = \pi/2$, then $|PB|/|BQ| = [o_1]/[o_2]$, where $[x]$ denotes the area of figure x.

Solution. Let us first note that the centers of o_1 and o_2 cannot lie inside the other circle as then $\sphericalangle PAQ$ could not be equal to $\pi/2$. Note then that $\sphericalangle ABQ = \sphericalangle ABP = \pi/2$, and so Q, B and P are colinear. It follows that AQP is a right triangle and B is an orthogonal projection of A on PQ. So $|QB|/|AB| = |QA|/|AP|$. Similarly $|PB|/|AB| = |PA|/|AQ|$. We conclude that $|PB|/|BQ| = |PA|^2/|AQ|^2 = [o_1]/[o_2]$, and the proof is finished.

   Alternatively, for the last step one could use the power of the point property we introduce in Section 6.6. Using it one gets that $|AQ|^2 = |QB||QP|$ and that $|AP|^2 = |PB||PQ|$, and so $[o_1]/[o_2] = |AP|^2/|AQ|^2 = |PB|/|QB|$.

Problem 6.4.1. Points D, E, and F lie on sides BC, CA, and AB of a triangle ABC in such a way that lines AD, BE, and CF intersect in a single point P. Prove that $|AF|/|FB| + |AE|/|EC| = |AP|/|PD|$.

Solution. After applying Menelaus's theorem to triangle ABD and line FP, we get that

$$\frac{|BC|}{|DC|} \cdot \frac{|DP|}{|PA|} \cdot \frac{|AF|}{|FB|} = 1,$$

and so

$$\frac{|AF|}{|FB|} = \frac{|DC|}{|BC|} \cdot \frac{|PA|}{|DP|}.$$

Similarly, applying it to triangle ACD and line EP we get

$$\frac{|CB|}{|DB|} \cdot \frac{|DP|}{|PA|} \cdot \frac{|AE|}{|EC|} = 1,$$

and so

$$\frac{|AE|}{|EC|} = \frac{|DB|}{|CB|} \cdot \frac{|PA|}{|DP|}.$$

It follows that

$$\frac{|AF|}{|FB|} + \frac{|AE|}{|EC|} = \frac{|DC|}{|BC|} \cdot \frac{|PA|}{|DP|} + \frac{|DB|}{|CB|} \cdot \frac{|PA|}{|DP|}$$

$$= \frac{|DC| + |DB|}{|BC|} \cdot \frac{|PA|}{|DP|} = \frac{|PA|}{|DP|},$$

as required.

Problem 6.4.2. You are given a triangle ABC where $\sphericalangle ACB = \pi/2$. On side AC build a square $ACGH$, externally to the triangle. Similarly, on side BC build a square $CBEF$, externally to the triangle. Show that the point of intersection of AE and BH lies on the line orthogonal to AB that goes through point C.

Solution. Let A' be the intersection point of AE and BC, B' be the intersection point of BH and AC, and C' be the orthogonal projection of C on AB. Since AC is parallel to BE, $|CA'|/|A'B| = |AC|/|BE| = |AC|/|BC|$. Similarly, we argue that $|AB'|/|B'C| = |AC|/|BC|$. Let us now observe that triangles $AC'C$ and $BC'C$ are similar, and so $|AC'|/|CC'| = |CC'|/|BC'|$. It follows that $|BC'|/|C'A| = (|BC'|/|CC'|)^2$. But $BC'C$ and ACB are similar, and so $|BC'|/|C'A| = (|CB|/|AC|)^2$. We get that $|CA'|/|A'B| \cdot |BC'|/|C'A| \cdot |AB'|/|B'C| = 1$. Using Ceva's theorem, we conclude that lines AA', BB', and CC' intersect in one point, which finishes the proof.

Problem 6.4.3. You are given a convex quadrilateral $ABCD$ and a line that intersects lines DA, AB, BC, and CD in points K, L, M, and N, respectively. Prove that $|DK| \cdot |AL| \cdot |BM| \cdot |CN| = |AK| \cdot |BL| \cdot |CM| \cdot |DN|$.

Solution. Let us add an auxiliary line BD to the plot. Let X be the intersection point of line BD with the new line going through K, L, M, and N. Applying Menelaus's theorem twice, the first time to triangle ABD, and the second time to triangle BDC, we get that

$$\frac{|AL|}{|LB|} \cdot \frac{|BX|}{|XD|} \cdot \frac{|DK|}{|AK|} = 1,$$

and, respectively, that

$$\frac{|BM|}{|MC|} \cdot \frac{|CN|}{|ND|} \cdot \frac{|DX|}{|XB|} = 1.$$

We get the desired result after multiplying these two equations together.

Problem 6.5.1. Consider a quadrilateral $ABCD$. Prove that the sum of distances from any point P inside this quadrilateral to the lines AB, BC,

CD, and DA is constant (that is, does not depend on the choice of P) if and only if $ABCD$ is a parallelogram.

Solution. Let us consider any two half-lines, ℓ_1 and ℓ_2, that are not parallel and have a common origin, point A. We will show that for any two points $B \in \ell_1$ and $C \in \ell_2$ with $|AB| = |AC|$ the following property holds: all points lying on the line segment BC have the same total distance to lines ℓ_1 and ℓ_2. To see this, let us consider any point P on the line segment BC. Clearly, $[ABC] = [ABP] + [ACP]$, where $[x]$ denotes the area of figure x. Since $|AB| = |AC|$, we immediately get that the sum of heights of the two triangles ABP and ACP, projected from P to AB and, respectively, from P to AC is constant (namely, equal to $2[ABC]/|AB| = 2[ABC]/|AC|$). From this argument, we immediately get the following important observation. Any point P lies on the unique line segment BC defined as above; in particular $|AB| = |AC|$. More importantly, for any two points P_i ($i \in \{1, 2\}$) and the associated line segments B_iC_i, the total distances from P_i to ℓ_1 and ℓ_2 are equal if and only if the two corresponding line segments B_1C_1 and B_2C_2 are identical. Moreover, if we extend half-lines ℓ_1 and ℓ_2 to lines, then the set of all points having the same distance from these two lines forms a rectangle with point A being the intersection of its diagonals. All points *inside* this rectangle have sums of the distances from these two lines strictly *smaller* than the points on this rectangle.

Let us now go back to our problem. Clearly, if $ABCD$ is a parallelogram, then the desired property holds. Suppose then that $ABCD$ is *not* a parallelogram. Our goal is to show that there are two points inside of $ABCD$ with different sums of distances.

Let us first deal with convex quadrilaterals. Without loss of generality, we may assume that AB and CD are *not* parallel. Select any point P inside of $ABCD$. From the observation above it follows that the set of points that are at the same distance as P from the two lines yielded by line segments AB and CD lie on some rectangle $\mathcal{R}$. We will independently consider the following two cases.

Case 1: AD and BC are parallel. Note that the sum of distances from AD and BC for all points inside of $ABCD$ is the same. Thus, we may select any point P' not lying on rectangle $\mathcal{R}$ but lying inside of $ABCD$ (note that such point always exists) to conclude that its total distance from the sides of $ABCD$ is different than the total distance of P.

Case 2: AD and BC are not parallel. Let S be the rectangle formed by all points that have the same total distance to AD and BC as point P. Clearly, $P \in \mathcal{R} \cap S$. Note that both $\mathcal{R}$ and S are non-degenerate as $ABCD$ is convex. (Recall that a single point can be viewed as a degenerate rectangle.) It is easy to see that the intersection of interiors of $\mathcal{R}$ and S is non-empty. Let us select any point P' from this intersection that also lies inside of $ABCD$ (since P lies inside of $ABCD$ we may always select P' close enough to P). The sum of distances of P' from the sides of $ABCD$ is smaller than of the corresponding sum of P, and we are done.

Finally, let us handle the case when $ABCD$ is *not* convex; that is, one of the vertices lies in the triangle formed by the remaining ones. Without loss of generality, we may assume that A lies inside a triangle BCD. Let B' be the intersection of line AB and line CD. Note that B' lies on the line segment DC. Similarly, let D' be the intersection of lines DA and BC. Again, note that D' lies on the line segment CB. Observe now that $AB'CD'$ is defined by the same lines as $ABCD$, but is convex; in particular, $AB'CD'$ is not a parallelogram as $ABCD$ is not. By the previous argument, we get that there are points inside of $AB'CD'$ (and so also inside of $ABCD$) with different sums of the distances from the sides. This finishes the proof.

Problem 6.5.2. Consider a triangle ABC such that $|AB| = |AC|$ (that is, an isosceles triangle), AD is the height of this triangle, and E is in the middle of AD. Let F be the orthogonal projection of D on BE. Prove that $\sphericalangle AFC = \pi/2$.

Solution. Let us introduce an auxiliary point X such that that $ADCX$ forms a rectangle. Note that $2|DE| = |CX|$ and $2|BD| = |BC|$ so B, F, E, and X lie on the same line; in particular, $\sphericalangle DFX = \sphericalangle DFE = \pi/2$. It follows that points D, F, A, and X lie on some circle. But C lies on the circle on which A, D, and X lie. It follows that they all lie on the same cycle and so $\sphericalangle CFA = \sphericalangle CDA = \pi/2$.

Problem 6.5.3. Consider a triangle ABC. Outside of the triangle, on sides AB and AC, we built squares $ABDE$ and, respectively, $ACFG$. Let M and N be the middle points of DG and, respectively, EF. What are the possible values of the rato $|MN|/|BC|$?

Solution. Let us add an auxiliary point P such that $EAGP$ is a parallelogram; in particular, PE and GA are parallel and have equal length. On the other hand, since $ACFG$ is a square, FC and GA are also parallel and have equal length. It follows that $CFPE$ is a parallelogram. But this means that N lies in the middle of the line segment PC as the diagonals of a parallelogram intersect in their middles. Using the same argument we get that $GPDB$ is a parallelogram and M lies in the middle of the line segment BP. It follows that $|PN|/|PC| = |PM|/|PB| = 1/2$ which means that a triangles PBC and PMN are similar and thus $|NM|/|CB| = 1/2$. Hence, this is the only possible ratio.

Problem 6.6.1. Two circles intersect in points A and B. Point P is selected on line AB outside of the circles. Points C and D are locations where tangent lines going through point P touch both circles. Prove that $\sphericalangle PCD = \sphericalangle PDC$.

Solution. Let us first note that points C and D are *not* uniquely defined (there are two possible locations). However, regardless of their location, $|PC|^2 = |PA| \cdot |PB| = |PD|^2$. It follows that PCD is an isosceles triangle; in particular, $\sphericalangle PCD = \sphericalangle PDC$.

Problem 6.6.2. Consider a convex hexagon $ABCDEF$ such that $|AB| = |BC|$, $|CD| = |DE|$, and $|EF| = |FA|$. Prove that lines containing altitudes of triangles BCD, DEF, and FAB from vertices C, E, and A, respectively, intersect in one point.

Solution. Consider a circle k_1 with center in D and radius $|DE| = |DC|$ and circle k_2 with center in F and radius $|FE| = |FA|$. These circles intersect in point E and some other point E' (such point must exist as $\sphericalangle FED \neq \pi$ and so $|FD| < |FE| + |ED|$). Let us now note that line EE' coincides with the altitude of a triangle FED as $DEFE'$ is a kite ($|FE| = |FE'|$ and $|DE| = |DE'|$). The crucial observation now is the fact that all points lying on a line going through E and E' have the same power with respect to circles k_1 and k_2, as can be seen by calculating this power along EE' line.

Let us now consider a circle k_1 and a circle k_3 with center in B and radius $|BA| = |BC|$. As before, we define point C' that is a second intersection (the first one is C) of k_1 and k_3. We note that CC' contains the altitude of BCD, and all points on the line CC' have the same power with respect to circles k_1 and k_3. Let us now observe that it is *not* possible that EE' and CC' are parallel as then $\sphericalangle FDB$ would have to be 0, which is not the case. Therefore, lines EE' and CC' have an unique intersection point Z.

It follows that the power of point Z with respect to circles k_2 and k_3 is the same. Let us draw a line going through Z and A. Because of the above fact, it must also go through point A' that is the other intersection point of circles k_2 and k_3 (the first one is A). We conclude that AA' contains the altitude of ABF, which finishes the proof.

Problem 6.6.3. Consider two points A and B. Take two circles o_1 and o_2 such that o_1 is tangent to AB in point A, o_2 is tangent to AB in point B, and o_1 and o_2 are externally tangent in point X. If we allow o_1 and o_2 to vary, then what is the set of points that contains all possible locations of X?

Solution. Draw a line tangent to o_1 (and so also to o_2) in point X. Denote by Y the point it intersects line AB. Note that $|YA| = |YX| = |YB|$. Therefore, Y is in the middle of line segment AB and $|YX|$ is constant. It follows that the only possible locations of point X are on the semi-circle with a center in Y and radius $|AY| = |BY|$ (excluding points A and B). It remains to show that for each such point, it is possible to generate the two cycles that satisfy the desired properties. (Let us mention that, indeed, points A and B are excluded, as for them one of the circles would be degenerated to a point.)

Select any point X on such a semi-circle (again, excluding points A and B). It is easy to see that it is possible to select then a point P such that $|PA| = |PX|$ and $\sphericalangle YAP = \sphericalangle YXP = \pi/2$. Similarly, we select Q such that $|QB| = |QX|$ and $\sphericalangle YBQ = \sphericalangle YXQ = \pi/2$. It remains to show that the two circles with centers in P and Q and radiuses $|PA|$ and, respectively, $|QB|$ are tangent. In order to prove it is is enough to show that X lies on a line segment PQ. But this is indeed true as $\sphericalangle PXY + \sphericalangle QXY = \pi/2 + \pi/2 = \pi$.

Problem 6.7.1. Let P be an interior point of a triangle ABC. Let lines AP, BP, and CP intersect sides BC, CA, and AB in points A', B' and, respectively, C'. Prove that $|PA|/|AA'| + |PB|/|BB'| + |PC|/|CC'| = 2$.

Solution. Let us consider triangles ABC and BPC. We get that $|PA'|/|AA'|$ is proportional to the ratio between heights of these triangles projected on BC, and so

$$\frac{|PA'|}{|AA'|} = \frac{[BPC]}{[ABC]}.$$

Using the argument for triangles APC and ABP, we get that

$$\frac{|PA'|}{|AA'|} + \frac{|PB'|}{|BB'|} + \frac{|PC'|}{|CC'|} = \frac{[BPC]}{[ABC]} + \frac{[APC]}{[ABC]} + \frac{[APB]}{[ABC]} = 1.$$

It follows that

$$\frac{|PA|}{|AA'|} + \frac{|PB|}{|BB'|} + \frac{|PC|}{|CC'|} = \left(1 - \frac{|PA'|}{|AA'|}\right) + \left(1 - \frac{|PB'|}{|BB'|}\right) + \left(1 - \frac{|PC'|}{|CC'|}\right)$$

$$= 3 - 1 = 2.$$

Problem 6.7.2. Points E and F lie on sides BC and, respectively, DA of a parallelogram $ABCD$ such that $|BE| = |DF|$. Select any point K on side CD. Let P and Q be intersection points of line FE with lines AK and, respectively, BK. Prove that $[APF] + [BQE] = [KPQ]$.

Solution. Since $|BE| = |FD|$, $|BC| = |AD|$, and BC and AD are parallel, we get that $ABEF$ and $CDFE$ are congruent trapezoids. It follows that $[ABEF] = [CDFE] = [ABCD]/2$. On the other hand, since triangle AKB has the same base (namely, AB) and the height projected on this base as the parallelogram $ABCD$, $[AKB] = [ABCD]/2$. It follows that

$$[APF] + [BEQ] = [ABEF] - [ABQP] = [AKB] - [ABQP] = [KPQ],$$

and the proof is finished.

Problem 6.7.3. Consider a convex quadrilateral $ABCD$. Select points K and L on side AB such that $|AK| = |KL| = |LB| = |AB|/3$. Similarly, select points N and M on side DC such that $|DN| = |NM| = |MC| = |CD|/3$. Show that $[KLMN] = [ABCD]/3$.

Solution. Note that $[DAK] = [DAB]/3$ as the corresponding triangles have the same height projected on the bases that have proportion 1 to 3. Similarly, note that $[BMC] = [BDC]/3$. Clearly, $[ABCD] = [DAB] + [BDC]$. Combining all of these observations together, we get that

$$[KBMD] = [ABCD] - [AKD] - [BMC]$$
$$= [ABCD] - [DAB]/3 - [BDC]/3$$
$$= [ABCD] - [ABCD]/3 = 2[ABCD]/3.$$

Let us now note that $[KLM] = [LBM]$ as these triangles have the same height projected on bases of equal length. Similarly, $[MKN] = [KND]$. But this implies that

$$
\begin{aligned}
[KLMN] &= [KLM] + [KMN] \\
&= \frac{1}{2} \Big([KLM] + [LBM] + [MKN] + [KND] \Big) \\
&= \frac{1}{2} [KBMD] = \frac{1}{2} \cdot \frac{2}{3} [ABCD] = [ABCD]/3 .
\end{aligned}
$$

Problem 6.8.1. Given a parallelogram $ABCD$, consider points M and N that are in the middle of sides BC and CD, respectively. Section BD intersects with AN in point Q, and with AM in point P. Prove that $3|QP| = |BD|$.

Solution. Since AB is parallel to DN, triangles ABQ and DQN are similar. It follows from Thales' theorem that $|QD|/|QB| = |DN|/|AB| = 1/2$, and so $2|QD| = |QB|$. As a result, $|QD| = |BD|/3$. Similarly, by considering triangles APD and BMP we conclude that $|PB| = |BD|/3$. Combining these two things together, we get that $|QP| = |BD| - |QD| - |PB| = |BD|/3$.

Problem 6.8.2. Points K, L, M, and N are the middle points of sides AB, BC, CD and, respectively, DA of a parallelogram $ABCD$ whose area is equal to 1. Let P be the intersection point of KC and NB, Q be the intersection point of LD with KC, R be the intersection point of MA with LD, and, finally, S be the intersection point of NB with MA. Calculate the area of $PQRS$.

Solution. Denote by C' the point of intersection of line CK with line AD. Since $2|AK| = |CD|$, we get from Thales's theorem that $|AC'| = |DA|$. But this means, again by Thales' theorem, that triangles $C'PN$ and BCP are similar with ratio of $3/2$. Consider now heights of these triangles projected from P onto $C'N$ and BC. Denote their lengths as h_1 and, respectively, as h_2. Since $h_1/h_2 = 3/2$ (by similarity of the corresponding triangles), we get that $|BC|(h_1 + h_2) = 1$. It follows that $|BC|h_2/2 = 1/5$, and so $[BPC] = 1/5$. Similarly, we conclude that $[DQC] = [ARD] = [BSA] = 1/5$, and so $[PQRS] = 1 - 4/5 = 1/5$.

Problem 6.8.3. Points E and F are on sides AB and, respectively, AD of rhombus $ABCD$. Lines CE and CF intersect line BD in points K and L, respectively. Line EL intersects side CD in point P. Line FK intersects side BC in point Q. Prove that $|CP| = |CQ|$.

Solution. Consider triangles FLD and BCL. By Thales' theorem we get that $|FD|/|LD| = |BC|/|LB|$. Consider now triangles LPD and LEB. Using Thales' theorem one more time we get that $|DP|/|LD| = |BE|/|LB|$. Combining those two facts together, we conclude that $|DP| = |BE| \cdot |LD|/|LB| = |BE| \cdot |FD|/|BC|$. Analogously, by analyzing triangles FBK and DKC, and then triangles BKQ and FKD, we get that $|BQ| = |BE| \cdot |FD|/|DC|$. We conclude that $|BQ| = |DP|$ as $|BC| = |DC|$, and so also $|CP| = |CQ|$.

Further Reading

We do hope that our book increased appetite for more problems to solve and the readers will search for more books to expand her or his knowledge and skills. Here is a list of books that we have on our shelves and like to read but, of course, this list is not complete. There are many more books that are worth reading. Moreover, the mathematical level of these books varies a lot. In any case, we hope that the readers will enjoy reading some of them and keep solving interesting problems.

- *102 Combinatorial Problems* by Titu Andreescu, Birkhäuser, 2003.

- *Are You Smart Enough to Work at Google?* by William Poundstone, Little Brown, 2012.

- *Asymptopia* by Joel Spencer with Laura Florescu, Orient Blackswan, 2017.

- *Concrete Mathematics: A Foundation for Computer Science* by Ronald L. Graham, Donald E. Knuth and Oren Patashnik, Addison-Wesley Professional, 1994.

- *Discrete Mathematics—Elementary and Beyond* by L. Lovasz, J. Pelikan, K. Vesztergombi, Springer, 2006.

- *How To Prove It—A Structured Approach* by Daniel J. Velleman, Cambridge University Press, 2006.

- *How to Read and Do Proofs* by Daniel Solow, Wiley, 2014.

- *Lessons in Play—An Introduction to Combinatorial Game Theory* by Michael H. Albert, Richard J. Nowakowski, David Wolfe, CRC Press, 2007.

- *Magical Mathematics—The Mathematical Ideas That Animate Great Magic Tricks* by Persi Diaconis and Ron Graham, Princeton University Press, 2011.

- *Mathematical Mind-Benders* by Peter Winkler, Routledge, 2007.

- *Mathematical Puzzles: A Connoisseur's Collection* by Peter Winkler, Routledge, 2003.

- *Moscow Mathematical Olympiads, 1993-1999* by Roman Fedorov, Alexei Belov, Alexander Kovaldzhi, Ivan Yashchenko, American Mathematical Society, 2011.

- *Moscow Mathematical Olympiads, 2000-2005* by Roman Fedorov, Alexei Belov, Alexander Kovaldzhi, Ivan Yashchenko, American Mathematical Society, 2011.

- *Pearls of Discrete Mathematics* by Martin Erickson, CRC Press, 2009.

- *Professor Stewart's Casebook of Mathematical Mysteries* by Ian Stewart, Basic Books, 2014.

- *Proofs from the book* by Martin Aigner, Gunter M. Ziegler, Springer, 2013.

- *The Art of Mathematics—Coffee Time in Memphis* by Bela Bollobas, Cambridge University Press, 2006.

- *The Art of Proof—Basic Training for Deeper Mathematics* by Matthias Beck, Ross Geoghegan, Springer, 2010.

- *The Math Book—From Pythagoras to the 57th Dimension, 250 Milestones in the History of Mathematics* by Clifford A. Pickover, Sterling, 2012.

- *The Nikola Tesla Puzzle Collection—An Electrifying Series of Challenges, Enigmas and Puzzles* by Richard Galland, Carlton Books, 2001.

- *Thirty-three Miniatures—Mathematical and Algorithmic Applications of Linear Algebra* by Jiri Matousek, American Mathematical Society, 2010.

One might also consider visiting the website of International Mathematical Olympiad at `https://www.imo-official.org/problems.aspx` where one can find a collection of challenging problems to solve. Another website worth mentioning is an on-line resource maintained by Evan Chen that contains problems and solutions to several USA contests `https://web.evanchen.cc/problems.html`.

Good luck!

Index

Printed in the United States
by Baker & Taylor Publisher Services